Recent Advances in Scar Biology

Recent Advances in Scar Biology

Special Issue Editor

Rei Ogawa

MDPI • Basel • Beijing • Wuhan • Barcelona • Belgrade

MDPI

Special Issue Editor
Rei Ogawa
Nippon Medical School
Japan

Editorial Office
MDPI
St. Alban-Anlage 66
4052 Basel, Switzerland

This is a reprint of articles from the Special Issue published online in the open access journal *International Journal of Molecular Sciences* (ISSN 1422-0067) from 2017 to 2018 (available at: http://www.mdpi.com/journal/ijms/special_issues/scar_biology)

For citation purposes, cite each article independently as indicated on the article page online and as indicated below:

LastName, A.A.; LastName, B.B.; LastName, C.C. Article Title. *Journal Name* **Year**, *Article Number*, Page Range.

ISBN 978-3-03897-398-0 (Pbk)
ISBN 978-3-03897-399-7 (PDF)

Contents

About the Special Issue Editor

Rei Ogawa, M.D., Ph.D. is currently a faculty member at the Nippon Medical School in Tokyo, Japan, in the position of Professor and Chief of the Department of Plastic, Reconstructive, and Aesthetic Surgery. He is a fellow of the American College of Surgeons. In addition, he now directs the Mechanobiology and Mechanotherapy Laboratory at his medical school. He joined the Tissue Engineering and Wound Healing Laboratory, Brigham and Women's Hospital, Harvard Medical School, Boston, USA, where he worked between 2007 and 2009 as a Research Fellow. He has focused his recent studies on mechanobiology and its application to tissue engineering, wound healing, and anti-aging medicine. Moreover, his clinical specialty is reconstructive surgery and scar management, including abnormal scar (keloid and hypertrophic scars) prevention and treatment. In relation to this, he studied the mechanobiology of scarring, and is a world leader in this area. He has been the recipient of several awards, including the Award of Japanese Society of Plastic Surgery, and many research grants (e.g., grant-in-aid for scientific research in Japan). Moreover, he holds several national patents in the field of tissue engineering. He is an Editorial Board Member of many international/local scientific journals (e.g., *Plastic and Reconstructive Surgery*) and is a Board Member of international/local medical societies (e.g., Japanese Society of Plastic and Reconstructive Surgery (JSPRS)). In addition, he is an active member of many international medical societies (e.g., American College of Surgeons (ACS), American Association of Plastic Surgeons (AAPS), American Society of Plastic Surgeons (ASPS), Plastic Surgery Research Council (PSRC)). He has coauthored over 50 chapters in international/national books, coauthored over 500 papers in international/national scientific journals, and presented over 1500 coauthored papers at international/national conferences, including over 250 invited lectures.

Preface to "Recent Advances in Scar Biology"

Scars develop in the final stage of wound healing. The biological pathways that underlie wound healing and scarring are complex. In particular, the exact mechanisms that initiate and regulate them and lead to their progression remain to be fully elucidated. A major goal of medical science is scarless wound healing. To achieve this goal, it is necessary to elucidate the relevant clinical, histopathological, and molecular manifestations of scars, and to understand how these manifestations relate to each other. The purpose of the Special Issue "Recent Advances in Scar Biology" that was published in the *International Journal of Molecular Sciences* was to illustrate the biological mechanisms that underpin scarring and effective clinical treatments. This Special Issue included a selection of recent research topics and current review articles in the field of scar research for all kinds of tissues and organs.

Rei Ogawa
Special Issue Editor

International Journal of
Molecular Sciences

MDPI

Editorial

Recent Advances in Scar Biology

Rei Ogawa [ID]

Department of Plastic, Reconstructive and Aesthetic Surgery, Nippon Medical School, 1-1-5 Sendagi, Bunkyo-ku, Tokyo 113-0022, Japan; r.ogawa@nms.ac.jp; Tel.: +81-3-5814-6208; Fax: +81-5685-3076

Received: 30 May 2018; Accepted: 12 June 2018; Published: 13 June 2018

Scars develop in the final stage of wound healing. The biological pathways that underlie wound healing and scarring are complex. In particular, the exact mechanisms that initiate and regulate them and lead to their progression remain to be fully elucidated. A major goal of medical science is scar-less wound healing. To achieve this goal, it is necessary to elucidate the relevant clinical, histopathological, and molecular manifestations of scars, and to understand how these manifestations relate to each other. The purpose of the Special Issue "Recent Advances in Scar Biology" that was published recently in the *International Journal of Molecular Sciences* was to illustrate the biological mechanisms that underpin scarring and effective clinical treatments. This Special Issue included a selection of recent research topics and current review articles in the field of scar research for all kinds of tissues and organs.

Normally, the cutaneous wound healing process closes skin gaps by inducing the formation of granulation tissue and epithelialization, which re-establishes an effective epidermal barrier. The complex biochemical events that underlie wound closure can be categorized into four overlapping processes: coagulation, inflammation, proliferation and remodeling. Coagulation and the inflammatory process begin immediately after injury, while the proliferative phases start within a few days. The remodeling phase commences within a week of injury and continues for months. If the inflammatory and proliferative phases are feeble, wound healing may be delayed and chronic wounds may develop. In relation to this, Horng et al. [1] showed in the Special Issue that estrogen deficiency, such as that in postmenopausal women, has detrimental effects on wound-healing processes, particularly inflammation and re-granulation, and that exogenous estrogen treatment may reverse these effects [1]. Conversely, if the inflammatory and proliferative phases are excessively vigorous and prolonged due, for example, to infection or burn, heavy scars can develop. Clinical interventions that target these phases can, therefore, improve wound healing. For example, Jeong et al. [2] showed in a rat incisional wound-healing model that injections with polydeoxyribonucleotide (a mixture of nucleotides from trout sperm) have anti-inflammatory effects and, therefore, reduce the size of the scar [2].

After full-thickness burning, necrotized tissues (eschars) develop. These eschars delay wound healing, thereby promoting the formation of hypertrophic scars. Monsuur et al. [3] showed in the Special Issue that, while acellular extracts of burn eschars stimulate the proliferation and migration of adipose mesenchymal stromal cells and fibroblasts, they also inhibit the basic fibroblast growth factor-induced proliferation and sprouting of endothelial cells. This inhibitory effect may explain why the presence of an eschar blocks the formation of excessive granulation tissue by full-thickness burn wounds [3]. Akita et al. also showed that proper epithelialization plays an important role in the healing of burn wounds: when patients with extensive burns received cultured epithelial autografts (CEA) along with either highly expanded (over 1:6 ratio) or less expanded (gap 1:6) mesh, the former combination was associated with accelerated wound healing. Moreover, scoring by experts using the Vancouver and Manchester Scar Scales showed that CEA with the highly expanded mesh led to better scar formation [4]. The exhaustive review of Mostaço-Guidolin et al. also showed that proper formation of the extracellular matrix plays a key role in the epithelialization and other wound-healing events that lead to a smooth wound-healing course: studies that used second harmonic generation microscopy to image the fibrillar collagens in wounded and repaired skin, lung, cardiovascular, tendon and ligaments, and eye tissue indicate that the balance between extracellular matrix synthesis and degradation determines the degree of scarring after wounding [5].

Multiple wound-healing processes, including granulation tissue formation and wound contraction and epithelialization, are influenced by mechanical forces [6,7]. The mechanisms by which these forces shape wound healing remain to be fully elucidated, but the review of Januszyk et al. [8] in the Special Issue suggests that focal adhesion kinase (FAK), which is a mediator of mechanotransduction pathways, plays a central role in both the inflammation and fibrosis that characterizes aberrant wound healing [8]. Moreover, multiple lines of evidence suggest that the formation of granulation tissue and numerous functions of fibroblasts, myofibroblasts, endothelial cells, and epithelial cells are affected by intrinsic and extrinsic mechanical stimuli. In recent years, many mechanosensors in these cells and tissues and the mechanosignaling pathways that they trigger have been elucidated [9–11]. These mechanosensors include mechanosensitive ion channels, cell-adhesion molecules (including integrins), and actin filaments in the cytoskeleton. When these structures and molecules sense mechanical stimuli, signaling pathways are activated and gene expression is altered. An important family of mechanosensitive ion channels is the transient receptor potential cation channel (TRP channel) family. Its members include TRP vanilloid (TRPV) 4, which is a mechanosensor in the skin [9], and TRPV3, which is a temperature sensor and vasoregulator. Park et al. reported that TRPV3 may contribute to the pruritus in burn scars by increasing the expression of thymic stromal lymphopoietin by epidermal keratinocytes. Thymic stromal lymphopoietin is a cytokine that has been linked to allergic and fibrotic diseases. Thus, thymic stromal lymphopoietin may be a potential therapeutic target for post-burn pruritus [12]. In relation to the signaling pathways that are triggered by mechanosensors, one may be the transforming growth factor (TGF)-β/SMAD pathway. This pathway plays a very well known role in collagen synthesis and fibrosis, but several lines of evidence suggest that it is also a mechanosignaling pathway [10,11]. This is supported by the study of Maeda et al., who subjected canine eyes to, first, glaucoma filtration surgery and, then, subconjunctival implantation of gelatin hydrogel with and without an anti-TGF-β antibody. They found that the controlled release of the anti-TGF-β antibody was associated with better intraocular pressure and less bleb formation and conjunctival scarring [13]. Other important mechanosignaling pathways are the mitogen-activated protein kinase (MAPK) and NF-κB interaction signaling pathways [10].

In relation to cutaneous scarring specifically, keloids and hypertrophic scars develop when the inflammation process is prolonged. Common initiators of these scars are cutaneous injury (including trauma) and irritation, insect bites, burn, surgery, vaccination, skin piercing, acne, folliculitis, chicken pox, and herpes zoster infection. These injuries and infections appear to result in chronic inflammation of the reticular layer of the dermis, which then drives the aberrant growth of keloids and hypertrophic scars [14]. The involvement of the reticular dermal layer is crucial: superficial injuries that do not reach the reticular dermis never cause these heavy scars. Reticular dermal inflammation may be promoted by a number of external and internal post-wounding stimuli, including mechanical tension on the wound edge, systemic factors such as sex hormones, and genetic factors [14]. Several molecules that play important roles in excessive cutaneous scarring have been identified. Kim et al. reported in this Special Issue that one of these may be high-mobility group box 1 (HMGB1): when normal and keloid fibroblasts were treated with HMGB1 or its inhibitor, their migration was accelerated and inhibited, respectively [15]. Moreover, Yamawaki et al. reported that the serine protease HtrA1 not only participates in the development of diseases such as osteoarthritis and age-related macular degeneration, it may also play an important role in keloid pathogenesis: they showed that keloid tissue fibroblasts express higher levels of this protein than surrounding normal skin and that silencing HtrA1 expression inhibits keloid fibroblast proliferation [16].

Numerous preventive and treatment strategies for keloids and hypertrophic scars have been reported. They include corticosteroid injection/tape/ointment, radiotherapy, cryotherapy, compression therapy, stabilization therapy, 5-fluorouracil (5-FU) therapy, and surgical methods [17–19]. In their review in this Special Issue, Lee et al. [20] summarized these methods after describing the wound-healing phases, the proteins and cytokines that play important roles in each phase, and some recently discovered anti- and pro-fibrotic pathways (e.g., hypoxia) [20]. Moreover, Park et al. reported that −79 °C spray-type cryotherapy effectively treats keloids [21]. Similarly, Cui et al. [22] reported that extracorporeal shock-wave therapy markedly improves the appearance and symptoms of post-burn hypertrophic scars, apparently by inhibiting the epithelial–mesenchymal transition [22].

Thus, the Special Issue "Recent Advances in Scar Biology" that was published in the *International Journal of Molecular Sciences* provides intriguing glimpses into the current wound healing/scarring field. It will be of interest for researchers and physicians who wish to understand the mechanisms that underlie wound healing and scarring and how these mechanisms can be manipulated to yield effective treatments of wounds and scars.

Conflicts of Interest: The author declares no conflicts of interest.

References

1. Horng, H.C.; Chang, W.H.; Yeh, C.C.; Huang, B.S.; Chang, C.P.; Chen, Y.J.; Tsui, K.H.; Wang, P.H. Estrogen Effects on Wound Healing. *Int. J. Mol. Sci.* **2017**, *18*, 2325. [CrossRef] [PubMed]
2. Jeong, W.; Yang, C.E.; Roh, T.S.; Kim, J.H.; Lee, J.H.; Lee, W.J. Scar Prevention and Enhanced Wound Healing Induced by Polydeoxyribonucleotide in a Rat Incisional Wound-Healing Model. *Int. J. Mol. Sci.* **2017**, *18*, 1689. [CrossRef] [PubMed]
3. Monsuur, H.N.; van den Broek, L.J.; Jhingoerie, R.L.; Vloemans, A.F.P.M.; Gibbs, S. Burn Eschar Stimulates Fibroblast and Adipose Mesenchymal Stromal Cell Proliferation and Migration but Inhibits Endothelial Cell Sprouting. *Int. J. Mol. Sci.* **2017**, *18*, 1790. [CrossRef] [PubMed]
4. Akita, S.; Hayashida, K.; Yoshimoto, H.; Fujioka, M.; Senju, C.; Morooka, S.; Nishimura, G.; Mukae, N.; Kobayashi, K.; Anraku, K.; et al. Novel Application of Cultured Epithelial Autografts (CEA) with Expanded Mesh Skin Grafting over na Artificial Dermis or Dermal Wound Bed Preparation. *Int. J. Mol. Sci.* **2018**, *19*, 57. [CrossRef] [PubMed]
5. Mostaço-Guidolin, L.; Rosin, N.L.; Hackett, T.L. Imaging Collagen in Scar Tissue: Developments in Second Harmonic Generation Microscopy for Biomedical Applications. *Int. J. Mol. Sci.* **2017**, *18*, 1772. [CrossRef] [PubMed]
6. Ogawa, R. Mechanobiology of scarring. *Wound Repair Regen.* **2011**, *19* (Suppl. 1), s2–s9. [CrossRef] [PubMed]
7. Harn, H.I.; Ogawa, R.; Hsu, C.K.; Hughes, M.W.; Tang, M.J.; Chuong, C.M. The tension biology of wound healing. *Exp. Dermatol.* **2017**. [CrossRef] [PubMed]
8. Januszyk, M.; Kwon, S.H.; Wong, V.W.; Padmanabhan, J.; Maan, Z.N.; Whittam, A.J.; Major, M.R.; Gurtner, G.C. The Role of Focal Adhesion Kinase in Keratinocyte Fibrogenic Gene Expression. *Int. J. Mol. Sci.* **2017**, *18*, 1915. [CrossRef] [PubMed]
9. Denda, M.; Sokabe, T.; Fukumi-Tominaga, T.; Tominaga, M. Effects of skin surface temperature on epidermal permeability barrier homeostasis. *J. Investig. Dermatol.* **2007**, *127*, 654–659. [CrossRef] [PubMed]
10. Huang, C.; Akaishi, S.; Ogawa, R. Mechanosignaling pathways in cutaneous scarring. *Arch. Dermatol. Res.* **2012**, *304*, 589–597. [CrossRef] [PubMed]
11. Huang, C.; Holfeld, J.; Schaden, W.; Orgill, D.; Ogawa, R. Mechanotherapy: Revisiting physical therapy and recruiting mechanobiology for a new era in medicine. *Trends Mol. Med.* **2013**, *19*, 555–564. [CrossRef] [PubMed]
12. Park, C.W.; Kim, H.J.; Choi, Y.W.; Chung, B.Y.; Woo, S.Y.; Song, D.K.; Kim, H.O. TRPV3 Channel in Keratinocytes in Scars with Post-Burn Pruritus. *Int. J. Mol. Sci.* **2017**, *18*, 2425. [CrossRef] [PubMed]
13. Maeda, M.; Kojima, S.; Sugiyama, T.; Jin, D.; Takai, S.; Oku, H.; Kohmoto, R.; Ueki, M.; Ikeda, T. Effects of Gelatin Hydrogel Containing Anti-Transforming Growth Factor-β Antibody in a Canine Filtration Surgery Model. *Int. J. Mol. Sci.* **2017**, *18*, 985. [CrossRef] [PubMed]
14. Ogawa, R. Keloid and Hypertrophic Scars Are the Result of Chronic Inflammation in the Reticular Dermis. *Int. J. Mol. Sci.* **2017**, *18*, 606. [CrossRef] [PubMed]
15. Kim, J.; Park, J.C.; Lee, M.H.; Yang, C.E.; Lee, J.H.; Lee, W.J. High-Mobility Group Box 1 Mediates Fibroblast Activity via RAGE-MAPK and NF-κB Signaling in Keloid Scar Formation. *Int. J. Mol. Sci.* **2018**, *19*, 76. [CrossRef] [PubMed]

16. Yamawaki, S.; Naitoh, M.; Kubota, H.; Aya, R.; Katayama, Y.; Ishiko, T.; Tamura, T.; Yoshikawa, K.; Enoshiri, T.; Ikeda, M.; et al. HtrA1 Is Specifically Up-Regulated in Active Keloid Lesions and Stimulates Keloid Development. *Int. J. Mol. Sci.* **2018**, *19*, 1275. [CrossRef] [PubMed]
17. Ogawa, R. The most current algorithms for the treatment and prevention of hypertrophic scars and keloids. *Plast. Reconstr. Surg.* **2010**, *125*, 557–568. [CrossRef] [PubMed]
18. Ogawa, R.; Akaishi, S.; Huang, C.; Dohi, T.; Aoki, M.; Omori, Y.; Koike, S.; Kobe, K.; Akimoto, M.; Hyakusoku, H. Clinical applications of basic research that shows reducing skin tension could prevent and treat abnormal scarring: The importance of fascial/subcutaneous tensile reduction sutures and flap surgery for keloid and hypertrophic scar reconstruction. *J. Nippon Med. Sch.* **2011**, *78*, 68–76. [CrossRef] [PubMed]
19. Ogawa, R.; Akaishi, S.; Kuribayashi, S.; Miyashita, T. Keloids and Hypertrophic Scars Can Now Be Cured Completely: Recent Progress in Our Understanding of the Pathogenesis of Keloids and Hypertrophic Scars and the Most Promising Current Therapeutic Strategy. *J. Nippon Med. Sch.* **2016**, *83*, 46–53. [CrossRef] [PubMed]
20. Lee, H.J.; Jang, Y.J. Recent Understandings of Biology, Prophylaxis and Treatment Strategies for Hypertrophic Scars and Keloids. *Int. J. Mol. Sci.* **2018**, *19*, 711. [CrossRef] [PubMed]
21. Park, T.H.; Cho, H.J.; Lee, J.W.; Kim, C.W.; Chong, Y.; Chang, C.H.; Park, K.S. Could −79 °C Spray-Type Cryotherapy Be an Effective Monotherapy for the Treatment of Keloid? *Int. J. Mol. Sci.* **2017**, *18*, 2536. [CrossRef] [PubMed]
22. Cui, H.S.; Hong, A.R.; Kim, J.B.; Yu, J.H.; Cho, Y.S.; Joo, S.Y.; Seo, C.H. Extracorporeal Shock Wave Therapy Alters the Expression of Fibrosis-Related Molecules in Fibroblast Derived from Human Hypertrophic Scar. *Int. J. Mol. Sci.* **2018**, *19*, 124. [CrossRef] [PubMed]

International Journal of
Molecular Sciences

MDPI

Article

Effects of Gelatin Hydrogel Containing Anti-Transforming Growth Factor-β Antibody in a Canine Filtration Surgery Model

Michiko Maeda [1], Shota Kojima [1,*], Tetsuya Sugiyama [1,2], Denan Jin [3], Shinji Takai [3], Hidehiro Oku [1], Ryohsuke Kohmoto [1], Mari Ueki [1] and Tsunehiko Ikeda [1]

[1]   Department of Ophthalmology, Osaka Medical College, Takatsuki-City, Osaka 569-8686, Japan;
    opt182@osaka-med.ac.jp (M.M.); tsugiyama@osaka-med.ac.jp (T.S.); opt025@osaka-med.ac.jp (H.O.);
    ryousuke0218@hotmail.co.jp (R.K.); opt089@osaka-med.ac.jp (M.U.); tikeda@osaka-med.ac.jp (T.I.)
[2]   Nakano Eye Clinic of Kyoto Medical Cooperative, Kyoto 604-8404, Japan
[3]   Department of Innovative Medicine, Osaka Medical College, Takatsuki-City, Osaka 569-8686, Japan;
    pha012@osaka-med.ac.jp (D.J.); pha010@osaka-med.ac.jp (S.T.)
*   Correspondence: shota@osaka-med.ac.jp; Tel.: +81-072-683-1221; Fax: +81-072-681-8195

Academic Editor: Rei Ogawa
Received: 3 April 2017; Accepted: 2 May 2017; Published: 5 May 2017

Abstract: In this present study, we investigated the effect of a controlled release of anti-transforming growth factor β (TGF-β) antibody on intraocular pressure (IOP), bleb formation, and conjunctival scarring in a canine glaucoma filtration surgery model using gelatin hydrogel (GH). Glaucoma surgery models were made in 14 eyes of 14 beagles and divided into the following two groups: (1) subconjunctival implantation of anti-TGF-β antibody-loaded GH (GH-TGF-β group, $n = 7$), and (2) subconjunctival implantation of GH alone (GH group, $n = 7$). IOP and bleb features were then assessed in each eye at 2- and 4-weeks postoperative, followed by histological evaluation. We found that IOP was significantly reduced at 4-weeks postoperative in the two groups ($p < 0.05$) and that IOP in the GH-TGF-β-group eyes was significantly lower than that in the GH-group eyes ($p = 0.006$). In addition, the bleb score at 4-weeks postoperative was significantly higher in the GH-TGF-β group than in the GH group ($p < 0.05$), and the densities of fibroblasts, proliferative-cell nuclear antigen (PCNA)-positive cells, mast cells, and TGF-β-positive cells were significantly lower in the GH-TGF-β group than in the GH group. The findings of this study suggest that, compared with the GH-group eyes, implantation of anti-TGF-β antibody-loaded GH maintains IOP reduction and bleb formation by suppressing conjunctival scarring due to the proliferation of fibroblasts for a longer time period via a sustained release of anti-TGF-β antibody from GH.

Keywords: trabeculectomy; glaucoma; gelatin hydrogel; transforming growth factor-β; beagles

1. Introduction

Glaucoma filtration surgery (i.e., trabeculectomy) is a primary treatment for glaucoma that results in decreased intraocular pressure (IOP) by draining the aqueous humor to the subconjunctival space and forming a bleb. Reportedly, the most common cause of unsuccessful trabeculectomy surgery is subconjunctival scarring of the filtration bleb, which leads to subconjunctival fibrosis [1,2]. The findings of a large prospective randomized trial showed that a single application of mitomycin C (MMC) or 5-flurouracil (5-FU) during trabeculectomy surgery greatly improves the surgical results; i.e., the prolonged bleb persistence and IOP reduction via strong suppression of fibroblast proliferation [3]. However, their application also increases the risk of complications such as a thin bleb, bleb infection, and infectious endophthalmitis at the late phase [4–6].

In this present study, we investigated transforming growth factor β (TGF-β), which is known to have three isoform types in humans; i.e., β1, β2, and β3. Isoforms β1 and β2 are known to greatly stimulate the dermal scarring response [7,8]. β2 is the mainly expressed ocular isoform, and is identified in both normal and diseased eyes [9,10]. The conjunctival scarring response in trabeculectomy surgery is thought to be affected by the passage of the aqueous humor including growth factors such as TGF-β, and subconjunctival scarring post glaucoma surgery is strongly affected by cytokines (especially TGF-β in the aqueous humor) [11,12]. Compared with other growth factors, TGF-β2 is reportedly dominant in the aqueous humor of glaucoma patients [13,14]. The TGF-β family is the main stimulator leading to conjunctival scarring post trabeculectomy, and various cells, such as fibroblasts and macrophages, can secrete them [15]. It was previously reported that TGF-β2 could increase α-smooth muscle actin (α-SMA) expression and the transdifferentiation of fibroblasts in conjunctiva to myofibroblasts [16]. In another previous study, the authors' findings revealed that bleb failure post trabeculectomy primarily occurred due to the excessive accumulation of collagen in the subconjunctival space, and that high activity of TGF-β was associated with scarring [17]. Numerous studies have reported that the expression of TGF-β activates the proliferation by human Tenon's fibrosis, and excessive production of granulation tissue constituents leading to scar formation [18–20]. In addition, several studies have reported that TGF-β inhibitors may effectively reduce scarring by reducing TGF-β activity via neutralization with antibodies [21,22]. Subconjunctival injections of anti-TGF-β antibody, as a drug substituting for MMC, were performed in a clinical trial for the suppression of fibroblast proliferation post trabeculectomy, however, the outcome was reportedly unsuccessful [23].

Various drug delivery systems (DDSs) have been tested for sustained drug release, since it is important to prevent scarring over an extended period following glaucoma surgery. Several previous studies have focused on subconjunctivally implanting DDSs to provide a sustained release of antiproliferative drugs over an extended time period post glaucoma surgery [24–26]. Most of those studies reported that these DDSs maintained IOP reduction and prolonged bleb persistence to the same degree as the conventional application of MMC and 5-FU, while significantly reducing their toxicity. However, most of those DDSs have yet to obtain successive results in the treatment of glaucoma patients [25].

Gelatin hydrogel (GH), a biodegradable material developed in Japan, has reportedly been used as a DDS for bioactive proteins in other medical fields [27]. Various growth factors gradually released from GH have been effective for therapy of various tissues [28,29]. In addition, GH has been applied to clinical therapies, such as for severe skin lesions complicating autoimmune vasculitis syndromes, peripheral arterial disease, and severe ischemic limb pain, and was found to be both safe and effective [30,31]. In the field of ophthalmology, GH impregnated with basic fibroblast growth factor has reportedly been used to induce experimental models of subretinal or corneal neovascularization [32,33].

We previously reported the possibility of using GH containing chymase inhibitor and GH containing MMC for longer-term maintenance of filtering blebs and IOP reduction by the prolonged suppression of subconjunctival scarring [34,35]. In this present study, we investigated the effect of a sustained release of anti-TGF-β antibody from GH in a canine glaucoma surgery model for IOP reduction and the effect on tissue in comparison with the application of GH alone.

2. Results

2.1. Verification of Anti-TGF-β Antibody in GH

Goat anti-Chicken IgY (H + L) secondary antibody was utilized to detect anti-TGF-β1-2 antibody. GH soaked overnight in phosphate-buffered saline (PBS) did not show a positive staining image by immunostaining, however, we were able to verify a wide range of positive staining images at sections of sliced GH with anti-TGF-β antibody overnight (Figure 1).

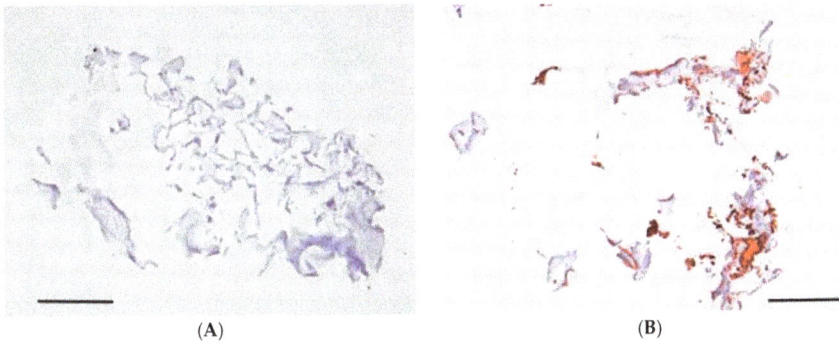

Figure 1. Gelatin hydrogel (GH) containing anti-transforming growth factor β (TGF-β) antibody. GH soaked overnight in phosphate-buffered saline (PBS) (**A**) did not show a positive staining image by immunostaining, however, we were able to verify a wide range of positive staining images (red) at sections of sliced GH with anti-TGF-β antibody (**B**) overnight. Scale bars: 500 μm.

2.2. IOP Change

The initial IOP values (mean ± SD) were 15.9 ± 0.7 mmHg in the GH containing anti-TGF-β antibody group (GH-TGF-β group) and 15.5 ± 0.8 mmHg in the GH alone group (GH group). The IOP values at 2-weeks postoperative were 8.1 ± 0.4 mmHg in the GH-TGF-β group and 8.0 ± 0.4 mmHg in the GH group. The IOP values at 4-weeks postoperative were 9.4 ± 0.7 mmHg in the GH-TGF-β group and 12.9 ± 0.7 mmHg in the GH group. In the eyes in both groups, IOP was found to be significantly reduced at 2- and 4-weeks postoperative ($p < 0.05$, unpaired t-test, Figure 2). Although there was no significant difference in IOP between the eyes in both groups at 2-weeks postoperative, IOP at 4-weeks postoperative was significantly lower in the GH-TGF-β group than in the GH group ($p < 0.05$, unpaired t-test). At 4-weeks postoperative, IOP once again began to increase in the GH group, however, IOP reduction was maintained in the GH-TGF-β group ($p < 0.05$, repeated measures ANOVA).

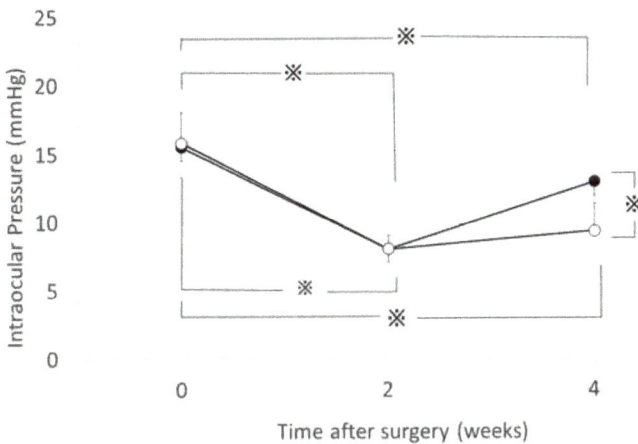

Figure 2. The effects to intraocular pressure (IOP) change by GH containing anti-TGF-β antibody. IOP changes in the GH-TGF-β group (○) and in the GH group (●). Data are shown as the mean ± SD of 14 beagles. (※ $p < 0.05$, unpaired t-test). At 4-weeks postoperative, IOP once again began to increase in the GH group, however, IOP reduction was maintained in the GH-TGF-β group ($p < 0.05$, repeated measures ANOVA).

2.3. Bleb Score

The bleb scores at 4-weeks postoperative were 3.7 ± 0.2 (mean ± SD) in the GH-TGF-β group and 2.7 ± 0.4 in the GH group; i.e., the bleb score was significantly higher in the GH-TGF-β group than in the GH group ($p < 0.05$, Mann–Whitney U-test, Figure 3).

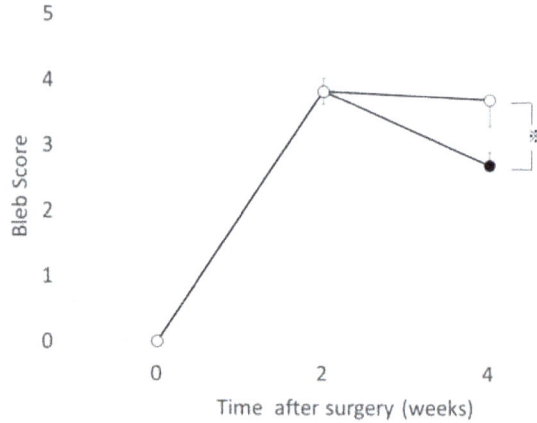

Figure 3. Comparison of bleb scores. Bleb score changes in the GH-TGF-β (○) group and in the GH group (●). Data are shown as the mean ± SD for 14 beagles. The bleb score at 4-weeks postoperative was significantly higher in the GH-TGF-β group than in the GH group (※ $p < 0.05$, Mann–Whitney U-test).

2.4. Subconjunctival/Scleral Area Ratio

As shown in Figure 4, the subconjunctival area in the GH-TGF-β group was less thickened compared with that in the GH group. The ratio of subconjunctival area to scleral area was significantly lower in the GH-TGF-β-group eyes than in the GH-group eyes ($p = 0.001$, unpaired t-test, Table 1).

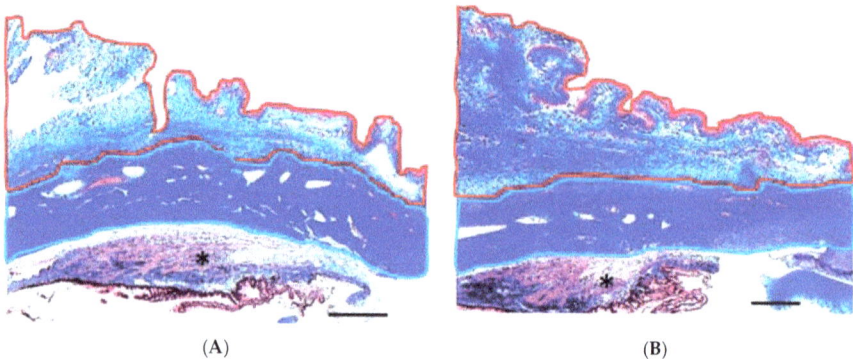

(A) (B)

Figure 4. Subconjunctival/scleral area ratio. Representative photomicrographs of the sections, posterior to the sclerectomy area, obtained from the eyes treated in the GH-TGF-β group (A) and GH group (B) at 4-weeks postoperative and stained with azan stain. Collagen fiber is stained with blue. The subconjunctival area and the scleral area are surrounded by red and light-blue lines, respectively. *: ciliary body. Scale bars: 1 mm.

Table 1. Compressions of the ratio of the conjunctival area to the scleral area, and densities of fibroblasts, TGF-β-positive cells, proliferative-cell nuclear antigen (PCNA)-positive cells, and mast cells between eyes in the GH-TGF-β group and the GH group. Data are shown as the mean ± SD for 14 beagles.

Indexes	GH-TGF-β Group	GH Group	*p*-Value (Student's *t*-Test)
Ratio of the conjunctival area to the scleral area	1.0 ± 0.1	2.4 ± 0.1	0.001
Density of fibroblasts, per mm^2	27.8 ± 8.6	67.6 ± 18.7	0.01
Density of TGF-β-positive cells, per mm^2	9.8 ± 1.5	18.2 ± 3.3	0.04
Density of PCNA-positive cells, per mm^2	4.2 ± 3.2	14.4 ± 6.0	0.03
Density of mast cells, per mm^2	7.2 ± 1.6	13.8 ± 2.0	0.01

2.5. Vimentin-Positive Cells

A lower number of vimentin-positive cells (fibroblasts stained with anti-vimentin antibody) was found in the GH-TGF-β-group eyes than in the GH-group eyes (Figure 5). The densities of fibroblasts in the lesion were significantly higher in the GH group compared with those in the GH-TGF-β group (*p* = 0.01, Student's *t*-test, Table 1).

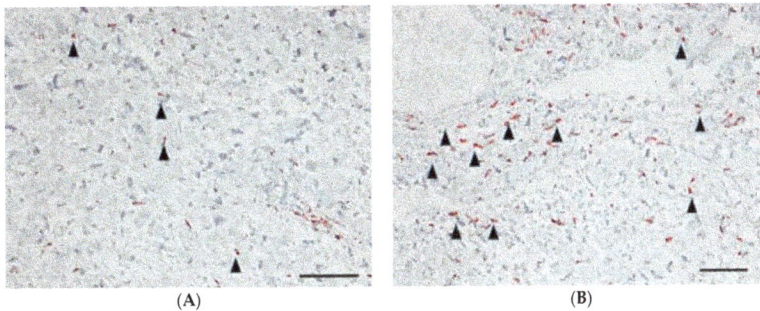

(A) (B)

Figure 5. Vimentin-positive cells in the subconjunctival lesion. Representative immunohistochemical staining images of the section for vimentin in the eyes in the GH-TGF-β group (**A**) and those in the GH group (**B**). Vimentin-positive cells are indicated by black arrowheads. Scale bars: 100 μm.

2.6. TGF-β Antibody-Positive Cells, Proliferative Cell Nuclear Antigen (PCNA)-Positive Cells, and Mast Cells

The numbers of TGF-β-positive cells, PCNA-positive cells, and mast cells were also significantly lower in the GH-TGF-β group compared to the GH group (*p* = 0.04, *p* = 0.03, and *p* = 0.01, respectively, Student's *t*-test; Figures 6–8, Table 1).

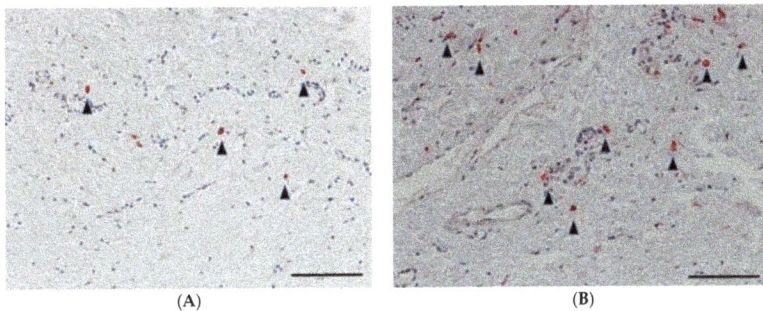

(A) (B)

Figure 6. TGF-β-positive cells in the subconjunctival lesion. Representative immunohistochemical staining images of the section for TGF-β antibody-positive cells in the GH-TGF-β-group eyes (**A**) and in the GH-group eyes (**B**). TGF-β-positive cells are indicated by black arrowheads. Scale bars: 100 μm.

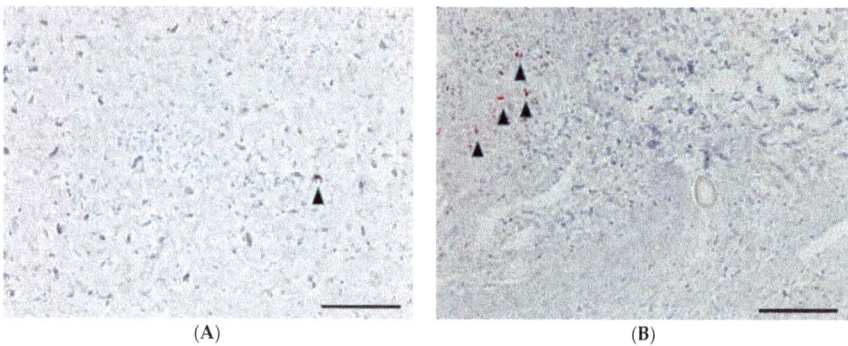

Figure 7. PCNA-positive cells in the subconjunctival lesion. Representative immunohistochemical staining images of the section for PCNA-positive cells in the GH-TGF-β-group eyes (**A**) and in the GH-group eyes (**B**). PCNA-positive cells are indicated by black arrowheads. Scale bars: 100 μm.

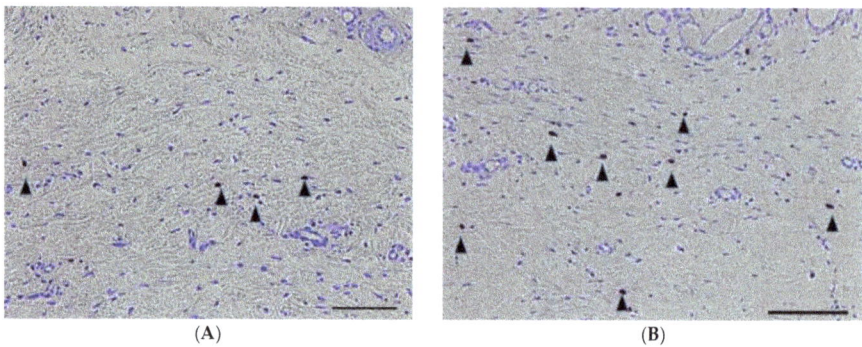

Figure 8. Mast cells in the subconjunctival lesion. Representative photomicrographs of the sections obtained from eyes in the GH-TGF-β group (**A**) and the GH group (**B**) and stained with Toluidine blue stain. Mast cells are indicated by black arrowheads. Scale bars: 100 μm.

3. Discussion

To the best of our knowledge, there have been no previous studies regarding the use of GH or other DDSs to release anti-TGF-β antibody in glaucoma surgery. Thus, this is the first experimental study regarding the effects of extended released anti-TGF-β antibody on wound healing in an experimental glaucoma filtration surgery model.

As mentioned in the Materials and Methods section, we prepared GH containing anti-TGF-β antibody. To confirm that the GH contains anti-TGF-β antibody, immunostaining was performed. We verified the existence of anti-TGF-β antibody activity in GH, indicating that GH embedded underneath the conjunctiva can release the anti-TGF-β antibody.

In this present study, we used our previously described [34,35] simple sclerotomy as a filtration surgery model to evaluate the effects of drugs or DDSs on IOP, bleb formation, and histological changes. In this model, the scleral flap and suturing flap associated with conventional trabeculectomy were not made. In order to precisely evaluate the effects of drugs in the surgery model, the same outward aqueous flow is required in all experimental eyes. However, controlling suture tightness to obtain the same outward aqueous flow is very difficult. Therefore, in this present study, we used the simple sclerotomy, as we deemed it to be the most appropriate filtration surgery model.

Our findings showed that IOP was significantly reduced at 2- and 4-weeks postoperative in both the GH group and the GH-TGF-β group. However, IOP once again began to increase at 4-weeks postoperative in the GH group, while IOP reduction was maintained in the GH-TGF-β group. This IOP change in the GH group is similar to the findings in our previously reported experiment [34,35] using the same glaucoma filtration surgery model and DDS, whereas IOP in the GH-TGF-β group was lower at 4-weeks postoperative, thus suggesting that anti-TGF-β antibody released from GH maintains IOP reduction. The bleb score at 4-weeks postoperative was significantly higher in the GH-TGF-β group than in the GH group. Together with abovementioned results, anti-TGF-β antibody from GH maintained the filtration bleb, and resulted in prolonged IOP reduction. To investigate the mechanism of the maintained bleb formation, we performed histological experiments. The ratio of subconjunctival area to scleral area was significantly lower in the GH-TGF-β-group eyes than in the GH-group eyes, thus suggesting that the above-described bleb formation in the GH-TGF-β group resulted from the inhibition of cell proliferation post glaucoma surgery in the bleb. To confirm this hypothesis, immunohistological analysis was performed. The densities of fibroblasts in the lesion were significantly higher in the GH group than in the GH-TGF-β group. The numbers of TGF-β-positive cells, PCNA-positive cells, and mast cells were also significantly lower in the GH-TGF-β group than in the GH group. Those results thus verified our hypothesis.

The findings of a previous clinical study showed no significant IOP reduction via the injection of anti-TGF-β antibody post filtration surgery [23]. The difference between the results in that study and those in this present study is whether or not a DDS was used. The injected antibody alone was washed out in the early stage due to the effect of filtration in trabeculectomy in these injection methods, whereas the sustained release of the antibody from GH was more effective to maintain bleb formation and IOP reduction. In a previous study [29], it was reported that the sustained release of TGF-β from GH enhanced the activity of bone regeneration. In that study, the authors employed a rabbit model with a calvarial defect and applied the GH containing TGF-β to the rabbit skull defect. As a control, PBS with TGF-β was employed. The authors then compared the bone mineral density (BMD) at the skull defect of the rabbit after treatment with GH containing TGF-β and PBS with TGF-β. In that study, it was described that GH containing TGF-β enhanced the BMD of the skull defect to a significantly higher extent than PBS with TGF-β. The authors also compared the release of TGF-β via subcutaneous implantation of GH containing TGF-β, and injection with TGF-β into the back of a mouse. The findings illustrated that TGF-β was retained by the implantation of GH containing TGF-β for longer time periods than TGF-β injection, and that free TGF-β disappeared from the injected site within one day. These findings indicated that the sustained release of TGF-β from GH was necessary to effectively enhance its osteoinductive function.

In the present study, we utilized GH as a DDS to spontaneously release anti-TGF-β antibody to obtain an extensive effect of anti-fibrosis in the subconjunctival area. Previous studies have shown that the controlled release of drugs over a time range of five days to three months was possible via the use of GH [36], and that the controlled release was effective for regenerative therapy of various tissues [29].

We previously reported the possibility of GH application as a new DDS to obtain a longer-term maintenance of filtering blebs [34,35]. In addition, the findings in those two reports demonstrated that the implantation of MMC-loaded GH had almost the same effects on IOP and bleb formation as the application of MMC alone in a canine filtration surgery model, and that GH containing chymase inhibitor made the period of filtration bleb formation and IOP reduction longer by decreasing cell proliferation. Taken together, this DDS is worthy of further investigation to improve the postoperative success rate of glaucoma surgery.

It should be noted that there were several limitations in the present study. First, the effect of the anti-TGF-β antibody might be different in the human eye. Thus, the toxicity of the antibody and GH DDS should be verified in primates in the future. Second, longer-term observation of the effects and toxicities of anti-TGF-β antibody might be needed. Although a 4-weeks observation period is effective for examining strong and dynamic scarring reaction in the early stage, a 3-month or greater observation period is much more informative to predict the long-term outcome and is necessary to determine whether this method can be clinically utilized. Third, conventional MMC application was not employed as a negative control, so that data is required in a future study. Fourth, since the simple sclerectomy used in the study might have a different effect on TGF-β expression around the flap site, careful interpretation from this experiment is necessary. In addition, the most effective dosage of anti-TGF-β antibody for use in GH for glaucoma surgery should be investigated.

4. Materials and Methods

4.1. Verification of Anti-TGF-β Antibody in GH

Anti-TGF-β1-2 antibody (Polyclonal Chicken IgY) was purchased from R&D Systems, Inc. (Minneapolis, MN, USA). Preparation of the GH containing the anti-TGF-β antibody was as follows: anti-TGF-β neutralizing antibody solution was produced by diluting with physiological saline, and a $5 \times 5 \times 1.5$ mm block of freeze-dried GH (MedGel PI5®; Wako Chemical, Tokyo, Japan) was then soaked in the 0.1% anti-TGF-β antibody solution overnight at 4 °C. The GH preparation method used in this present study was the same as that previously described [34,35].

In order to verify that the anti-TGF-β antibody was trapped and existent in the GH, we also fixed the GH with Carnoy's Solution (Muto Pure Chemicals Co., Ltd., Tokyo, Japan) to prepare a paraffin block. Furthermore, the GH was soaked in physiological saline overnight and compared as a negative control. All of the GH sections obtained from the above-described paraffin blocks were incubated with goat anti-Chicken IgY (H + L) secondary antibody, biotin conjugate (Thermo Fisher Scientific, Inc., Waltham, MA, USA) for 30 min at room temperature, followed by incubation with avidin-biotin-peroxidase complex (LSAB 2 Kit/HRP; Dako Japan, Kyoto, Japan) for 30 min to identify TGF-β-positive cells.

4.2. Animals and IOP Measurement

This experimental protocol was approved by the Committee of Animal Use and Care of Osaka Medical College (No. 28008). This study involved 14 eyes of 14 beagles purchased from Japan SLC, Inc. (Hamamatsu, Japan). The beagles were fed regular canine food, had constant free access to tap water, and were housed in an air-conditioned room at approximately 23 °C and 60% humidity with a 12-h light–dark cycle. All of the animal experiments were conducted in accordance with the ARVO Statement for Use of Animals in Ophthalmic and Vision Research. The IOP measurements were gauged via the use of a calibrated pneumatonometer (Model 30 Classic; Medtronic Solan, Jacksonville, FL, USA) under general anesthesia with intravenous injection of pentobarbital sodium (35 mg per kg body weight) in a front-facing position.

4.3. Glaucoma Filtration Surgery Model

For the glaucoma filtration surgery model, the beagles were anesthetized with pentobarbital sodium as described above. Briefly, a control suture was first fixed to the cornea using 8-0 Vicryl® (Ethicon US, LLC., Dallas, TX, USA) suture. Next, a 10-mm fornix-based flap of conjunctiva and the Tenon's capsule (5 mm in length) was made as previously described [34,35], and hemostasis was then performed. After a 3×1 mm scleral portion was removed at the limbus, peripheral iridectomy was performed. The conjunctiva was closed using a 10-0 nylon suture. After surgery, the appropriate amount of 3 mg/g ofloxacin was applied to the eye.

4.4. Experiment Protocol

At the end of the glaucoma filtration surgery described above, a 5×5 mm block of GH-TGF-β group ($n = 7$) or GH group ($n = 7$) was surgically implanted under the conjunctiva before closing the conjunctiva. In each dog, IOP and bleb score was assessed every 2 weeks for 1-month postoperative. After the final measurement, a dog was killed with a lethal dose of KCl intracardially injected and ophthalmectomy was performed, followed by soaking in 4 °C saline solution. Then, we identified the bleb area by a marked 10-0 nylon suture and excised the area by 10 × 5 mm including conjunctiva, subconjunctival tissue, and sclera and performed the following histological examination.

4.5. Bleb Scores

Blebs were examined via slit-lamp microscopy and graded according to the definition previously reported by Perkins et al. [37], reflecting increasing bleb height and size as follows: Score 1: minimally high conjunctiva thickening without swelling; Score 2: mild swelling present; Score 3: elevated bleb covering an area equivalent to 2–3 clock hours of the eye; and Score 4: greatly elevated bleb covering an area equivalent to more than 4 clock hours of the eye. A score of 0 indicated no observed bleb.

4.6. Histological Examination

Conjunctival and scleral tissue specimens were prepared for histologic analysis after fixation for 24 h with Carnoy Solution (Muto Pure Chemicals Co., Ltd., Tokyo, Japan) and embedded in paraffin. Next, 5-μm-thick sections were cut and mounted on silanized slides (Dako Japan), and then deparaffinized with xylene and a series of graded ethanol. The change in thickness of the conjunctiva and subconjunctival tissue was investigated via the ratio of subconjunctival area to scleral area stained with Hematoxylin-Eosin and Azan-Mallory. Mast cells were stained with Toluidine Blue for identification. The ratio of subconjunctival area to scleral area was then calculated (MacSCOPE Ver 2.2; Mitani Corporation, Fukui, Japan). To retrieve the antigen, sections were pretreated with 10 mM citrate buffer (pH 6.0) and autoclaved for 5 min at 120 °C before immunohistochemical staining. The sections were then soaked in absolute methanol containing 3% hydrogen peroxide for 5 min at room temperature to remove endogenous peroxidase activity. To suppress nonspecific binding, the sections were incubated with Serum-Free Protein Block (X0909; Dako Japan) for 5 min. To identify the PCNA-positive cells, the sections were incubated with mouse monoclonal antibody against PCNA (PC10, M0879; Dako Japan) for 15 h at 4 °C. Then, the slides were incubated with biotin-conjugated secondary antibody (LSAB 2 Kit/HRP; Dako Japan) for 30 min after being washed in PBS. Thereafter, those sections were incubated with avidin-biotin-peroxidase complex (LSAB 2 Kit/HRP; Dako Japan) for 30 min, washed with PBS, and then incubated with 0.05% 3,3-diaminobenzidine. The slides were then washed in running water, counterstained with hematoxylin, and mounted with cover glasses. To identify the TGF-β-positive cells, the sections were incubated with Polyclonal Chicken IgY anti-TGF-β1-2 antibody (R&D Systems, Inc.) for 15 h at 4 °C. To identify the fibroblasts, monoclonal mouse anti-vimentin antibody (M0725; Dako Japan) was also used in this present study.

We counted the PCNA-positive cells, fibroblasts, mast cells, and TGF-β-positive cells at the sites where they accumulated in the subconjunctival lesions, posterior to the sclerectomy area, by use of a light microscope (number per ×100 fields). The average number of each type of cell in five randomly selected fields of implanted GH was then calculated.

4.7. Masked Manner

All measurements were performed by investigators (MM and SK) who were masked from identifying which eye or tissue was in the GH-TGF-β or GH group.

4.8. Statistical Analysis

Each measurement was expressed as the mean ± SD. Statistical comparisons for repeated measurements used repeated-measures ANOVA, followed by other tests. Bleb scores were statistically analyzed via the Mann–Whitney U-test. IOP change and subconjunctival/scleral area ratio were evaluated via the unpaired t-test. Other parameters were evaluated via the Student's t-test. Differences were considered statistically significant at a p-value of <0.05.

5. Conclusions

In conclusion, the findings of this present study demonstrated that implantation of an anti-TGF-β antibody-loaded GH was more effective for the maintenance of IOP reduction and bleb formation by the sustained release of anti-TGF-β antibody. Our findings also demonstrated that it is possible to suppress the scarring effects and maintain IOP reduction and filtration bleb formation longer than is possible with only a subconjunctival injection of anti-TGF-β antibody.

Acknowledgments: The authors wish to thank John Bush for reviewing the manuscript.

Author Contributions: Tetsuya Sugiyama, Shota Kojima, Shinji Takai, Denan Jin, and Michiko Maeda conceived and designed the experiments; Michiko Maeda and Shota Kojima performed the experiments; Michiko Maeda, Shinji Takai, and Denan Jin analyzed the data; Shinji Takai and Denan Jin contributed analysis tools; Michiko Maeda wrote the paper; Tetsuya Sugiyama, Shota Kojima, Hidehiro Oku, Shinji Takai, Denan Jin, Ryohsuke Kohmoto, Mari Ueki, and Tsunehiko Ikeda edited the paper.

Conflicts of Interest: The authors declare no conflict of interest.

References

1. Skuta, G.L.; Parrish, R.K. Wound healing in glaucoma filtering surgery. *Surv. Ophthalmol.* **1987**, *32*, 149–170. [CrossRef]
2. Tripathi, R.C.; Li, J.; Chalam, K.V.; Tripathi, B.J. Expression of growth factor mRNAs by human Tenon's capsule fibroblasts. *Exp. Eye Res.* **1996**, *63*, 339–346. [PubMed]
3. Lama, P.J.; Fechtner, R.D. Antifibrotics and wound healing in glaucoma surgery. *Surv. Ophthalmol.* **2003**, *48*, 314–346. [CrossRef]
4. Mochizuki, K.; Jikihara, S.; Ando, Y.; Hori, N.; Yamamoto, T.; Kitazawa, Y. Incidence of delayed onset infection after trabeculectomy with adjunctive mitomycin C or 5-fluorouracil treatment. *Br. J. Ophthalmol.* **1997**, *81*, 877–883. [CrossRef] [PubMed]
5. Higginbotham, E.J.; Stevens, R.K.; Musch, D.C.; Karp, K.O.; Lichter, P.R.; Bergstrom, T.J.; Skuta, G.L. Bleb-related endophthalmitis after trabeculectomy with mitomycin C. *Ophthalmology* **1996**, *103*, 650–656. [PubMed]
6. Greenfield, D.S.; Suner, I.J.; Miller, M.P.; Kangas, T.A.; Palmberg, P.F.; Flynn, H.W., Jr. Endophthalmitis after filtering surgery with mitomycin. *Arch. Ophthalmol.* **1996**, *114*, 943–949. [PubMed]
7. Shah, M.; Foreman, D.M.; Ferguson, M.W. Neutralisation of TGF-β1 and TGF-β2 or exogenous addition of TGF-β3 to cutaneous rat wounds reduces scarring. *J. Cell Sci.* **1995**, *108*, 985–1002. [PubMed]
8. Levine, J.H.; Moses, H.L.; Gold, L.I.; Nanney, L.B. Spatial and temporal patterns of immunoreactive transforming growth factor-β1, -β2 and -β3 during excisional wound repair. *Am. J. Pathol.* **1993**, *143*, 368–380. [PubMed]
9. Lutty, G.A.; Merfes, C.; Threlkeld, A.B.; Crone, S.; Mcleod, D.S. Heterogeneity in localization of isoforms of TGF-β in human retina, vitreous and choroid. *Investig. Ophthalmol. Vis. Sci.* **1993**, *34*, 477–487.
10. Pasquale, L.R.; Dorman-Pease, M.E.; Lutty, G.A.; Quigley, H.A.; Jampel, H.D. Immunolocalisation of TGF-β1, TGF-β2 and TGF-β3 in the anterior segment of the human eye. *Investig. Ophthalmol. Vis. Sci.* **1993**, *34*, 23–30.
11. Jampel, H.D.; Roche, N.; Stark, W.J.; Roberts, A.B. Transforming growth factor-β in human aqueous humor. *Curr. Eye Res.* **1990**, *9*, 963–969. [CrossRef] [PubMed]
12. Cordeiro, M.F.; Reichel, M.B.; Gay, J.A.; D'Esposita, F.; Alexander, R.A.; Khaw, P.T. Transforming growth factor-β1, -β2 and -β3 in vivo: Effects on normal and mitomycin-C modulated conjunctival scarring. *Investig. Ophthalmol. Vis. Sci.* **1999**, *40*, 1975–1982.
13. Tripathi, R.C.; Ki, J.; Chan, W.F. Aqueous humor in glaucomatous eyes contains an increased level of TGF2. *Exp. Eye Res.* **1994**, *59*, 723–727. [CrossRef] [PubMed]
14. Picht, G.; Welge-Luessen, U.; Grehn, F.; Lütjen-Drecoll, E. Transforming growth factor β2 levels in the aqueous humor in different types of glaucoma and the relation to filtering bleb development. *Graefe's Arch. Clin. Exp. Ophthalmol.* **2001**, *239*, 199–207. [CrossRef]
15. Cordeiro, M.F. Beyond mitomycin: TGF-β and wound healing. *Prog. Retin. Eye Res.* **2002**, *21*, 75–89. [CrossRef]
16. Zhu, X.; Li, L.; Zou, L.; Zhu, X.; Xian, G.; Li, H.; Tan, Y.; Xie, L. A novel aptamer targeting TGF-β receptor II inhibits transdifferentiation of human tenon's fibroblasts into myofibroblast. *Investig. Ophthalmol. Vis. Sci.* **2012**, *53*, 6897–6903. [CrossRef] [PubMed]

17. Zhu, X.; Xu, D.; Zhu, X.; Li, L.; Li, H.; Gou, F.; Chen, X.; Tan, Y.; Xie, L. Evaluation of chitosan/aptamer targeting TGF-β receptor II thermo-sensitive gel for scarring in rat glaucoma filtration surgery. *Investig. Ophthalmol. Vis. Sci.* **2015**, *56*, 5465–5476. [CrossRef] [PubMed]

18. Meyer-Ter-Vehn, T.; Sieprath, S.; Katzenberger, B.; Gebhardt, S.; Grehn, F.; Schlunck, G. Contractility as a prerequisite for TGF-β-induced myofibroblast transdifferentiation in human tenon fibroblasts. *Investig. Ophthalmol. Vis. Sci.* **2006**, *47*, 4895–4904. [CrossRef] [PubMed]

19. Cordeiro, M.F. Role of transforming growth factor β in conjunctival scarring. *Clin. Sci.* **2003**, *104*, 181–187. [CrossRef] [PubMed]

20. Branton, M.H.; Kopp, J.B. TGF-β and fibrosis. *Microbes Infect.* **1999**, *1*, 1349–1365. [CrossRef]

21. Freedman, J. TGF-β2 antibody in trabeculectomy. *Ophthalmology* **2009**, *116*, 166. [CrossRef] [PubMed]

22. Mead, A.L.; Wong, T.T.L.; Cordeiro, M.F.; Anderson, I.K.; Khaw, P.T. Evaluation of anti-TGF-β2 antibody as a new postoperative anti-scarring agent in glaucoma surgery. *Investig. Ophthalmol. Vis. Sci.* **2003**, *44*, 3394–3401. [CrossRef]

23. CAT-152 0102 Trabeculectomy Study Group; Khaw, P.; Grehn, F.; Hollo, G.; Overton, B.; Wilson, R.; Vogel, R.; Smith, Z. A phase III study of subconjunctival human anti-transforming growth factor β2 monoclonal antibody (CAT-152) to prevent scarring after first-time trabeculectomy. *Ophthalmology* **2007**, *114*, 1822–1830. [PubMed]

24. Blandford, D.L.; Smith, T.J.; Brown, J.D.; Pearson, P.A.; Ashton, P. Subconjunctival sustained release 5-fluorouracil. *Investig. Ophthalmol. Vis. Sci.* **1992**, *33*, 3430–3435.

25. Min, J.K.; Kee, C.W.; Sohn, S.; Lee, H.J.; Woo, J.M.; Yim, J.H. Surgical outcome of mitomycin C-soaked collagen matrix implant in trabeculectomy. *J. Glaucoma* **2013**, *22*, 456–462. [CrossRef] [PubMed]

26. Yan, Z.C.; Bai, Y.J.; Tian, Z.; Hu, H.Y.; You, X.H.; Lin, J.X.; Liu, S.R.; Zhuo, Y.H.; Luo, R.J. Anti-proliferation effects of Sirolimus sustained delivery film in rabbit glaucoma filtration surgery. *Mol. Vis.* **2011**, *17*, 2495–2506. [PubMed]

27. Tabata, Y. Biomaterial technology for tissue engineering applications. *J. R. Soc. Interface* **2009**, *6*, S311–S324. [CrossRef] [PubMed]

28. Tabata, Y.; Hijikata, S.; Ikada, Y. Enhanced vascularization and tissue granulation by basic fibroblast growth factor impregnated in gelatin hydrogels. *J. Control. Release* **1994**, *31*, 189–199. [CrossRef]

29. Yamamoto, M.; Tabata, Y.; Hong, L.; Miyamoto, S.; Hashimoto, N.; Ikada, Y. Bone regeneration by transforming growth factor β1 released from a biodegradable hydrogel. *J. Control. Release* **2000**, *64*, 133–142. [CrossRef]

30. Kawanaka, H.; Takagi, G.; Miyamoto, M.; Tara, S.; Takagi, I.; Takano, H.; Yasutake, M.; Tabata, Y.; Mizuno, K. Therapeutic angiogenesis by controlled-release fibroblast growth factor in a patient with Chung-Strauss syndrome complicated by an intractable ischemic leg ulcer. *Am. J. Med. Sci.* **2009**, *338*, 341–342. [CrossRef] [PubMed]

31. Hashimoto, T.; Koyama, H.; Miyata, T.; Hosaka, A.; Tabata, Y.; Takato, T.; Nagawa, H. Selective and sustained delivery of basic fibroblast growth factor (bFGF) for treatment of peripheral arterial disease: Results of a phase I trial. *Eur. J. Vasc. Endovasc. Surg.* **2009**, *38*, 71–75. [CrossRef] [PubMed]

32. Kimura, H.; Sakamoto, T.; Hinton, D.R.; Spee, C.; Ogura, Y.; Tabata, Y.; Ikada, Y.; Ryan, S.J. A new model of subretinal neovascularization in the rabbit. *Investig. Ophthalmol. Vis. Sci.* **1995**, *36*, 2110–2119.

33. Yang, C.F.; Yasukawa, T.; Kimura, H.; Miyamoto, H.; Honda, Y.; Tabata, Y.; Ikada, Y.; Ogura, Y. Experimental corneal neovascularization by basic fibroblast growth factor incorporated into gelatin hydrogel. *Ophthalmic Res.* **2000**, *32*, 19–24. [CrossRef] [PubMed]

34. Kojima, S.; Sugiyama, T.; Takai, S.; Jin, D.; Shibata, M.; Oku, H.; Tabata, Y.; Ikeda, T. Effects of gelatin hydrogel containing chymase inhibitor on scarring in a canine filtration surgery model. *Investig. Ophthalmol. Vis. Sci.* **2011**, *52*, 7672–7680. [CrossRef] [PubMed]

35. Kojima, S.; Sugiyama, T.; Takai, S.; Jin, D.; Ueki, M.; Oku, H.; Tabata, Y.; Ikeda, T. Effects of gelatin hydrogel loading mitomycin C on conjunctival scarring in a canine filtration surgery model. *Investig. Ophthalmol. Vis. Sci.* **2015**, *56*, 2601–2605. [CrossRef] [PubMed]

36. Ikada, Y.; Tabata, Y. Protein release from gelatin matrices. *Adv. Drug Deliv. Rev.* **1998**, *31*, 287–301. [PubMed]

37. Perkins, T.W.; Faha, B.; Ni, M.; Kiland, J.A.; Poulsen, G.L.; Antelman, D.; Atencio, I.; Shinoda, J.; Sinha, D.; Brumback, L.; et al. Adenovirus-mediated gene therapy using human p21WAF-1/Cip-1 to prevent wound healing in a rabbit model of glaucoma filtration surgery. *Arch. Ophthalmol.* **2002**, *120*, 941–949. [CrossRef] [PubMed]

International Journal of
Molecular Sciences

MDPI

Article

Scar Prevention and Enhanced Wound Healing Induced by Polydeoxyribonucleotide in a Rat Incisional Wound-Healing Model

Woonhyeok Jeong [1], Chae Eun Yang [2], Tai Suk Roh [3], Jun Hyung Kim [1], Ju Hee Lee [4],* and Won Jai Lee [2],*

[1] Department of Plastic and Reconstructive Surgery, School of Medicine & Institute for Medical Science, Keimyung University, Dongsan Medical Center, Daegu 41931, Korea; psjeong0918@gmail.com (W.J.); med69@dsmc.or.kr (J.H.K.)
[2] Department of Plastic and Reconstructive Surgery, Institute for Human Tissue Restoration, Yonsei University Health System, Severance Hospital, Seoul 03722, Korea; cheniya@yuhs.ac
[3] Department of Plastic and Reconstructive Surgery, Institute for Human Tissue Restoration, Gangnam Yonsei University Health System, Severance Hospital, Seoul 06273, Korea; tsroh@yuhs.ac
[4] Department of Dermatology and Cutaneous Biology Research Institute, Severance Hospital, Yonsei University College of Medicine, Seoul 03722, Korea
* Correspondence: juhee@yuhs.ac (J.H.L.); pswjlee@yuhs.ac (W.J.L.);
   Tel.: +82-2-2228-2080 (J.H.L.); +82-2-2228-2219 (W.J.L.)

Received: 23 June 2017; Accepted: 1 August 2017; Published: 3 August 2017

Abstract: High-mobility group box protein-1 (HMGB-1) plays a central role in the inflammatory network, and uncontrolled chronic inflammation can lead to excessive scarring. The aim of this study was to evaluate the anti-inflammatory effects of polydeoxyribonucleotide (PDRN) on scar formation. Sprague-Dawley rats ($n = 30$) underwent dorsal excision of the skin, followed by skin repair. PDRN (8 mg/kg) was administered via intraperitoneal injection for three (PDRN-3 group, $n = 8$) or seven (PDRN-7 group, $n = 8$) days, and HMGB-1 was administered via intradermal injection in addition to PDRN treatment for three days (PDRN-3+HMGB-1 group; $n = 6$). The scar-reducing effects of PDRN were evaluated in the internal scar area and by inflammatory cell counts using histology and immunohistochemistry. Western blot, immunohistochemistry and immunofluorescence assays were performed to observe changes in type I and type III collagen and the expression of HMGB-1 and CD45. Treatment with PDRN significantly reduced the scar area, inflammatory cell infiltration and the number of CD45-positive cells. In addition, the increased expression of HMGB-1 observed in the sham group was significantly reduced after treatment with PDRN. Rats administered HMGB-1 in addition to PDRN exhibited scar areas with inflammatory cell infiltration similar to the sham group, and the collagen synthesis effects of PDRN were reversed. In summary, PDRN exerts anti-inflammatory and collagen synthesis effects via HMGB-1 suppression, preventing scar formation. Thus, we believe that the anti-inflammatory and collagen synthesis effects of PDRN resulted in faster wound healing and decreased scar formation.

Keywords: polydeoxyribonucleotide; cicatrix; inflammation; wounds and injuries; rats

1. Introduction

Inflammation is an inevitable first step in the process of wound healing and is closely related to scar formation. However, continuous and chronic inflammation stimulates the secretion of pro-inflammatory cytokines and causes excessive scarring [1–4]. In one prior study involving a scarless fetal wound-healing model, scar formation was caused by the injection of mast cells [5]. In addition, despite the observation that neutrophil depletion did not alter wound-breaking strength or collagen deposition, neutrophil depletion resulted in wounds that healed in a more organized fashion compared with normal wounds [6]. Furthermore, another previous study revealed that macrophage depletion also reduced scar formation [7]. Hence, we hypothesized that the inhibition of inflammatory cell infiltration could be a factor in reducing scar formation.

The sustained infiltration of immune cells during prolonged and intense inflammation contributes to the continuous growth of keloid lesions [8,9]. Moreover, keloid growth involves an abnormal response to inflammation [4,10,11]. In particular, extracellular high-mobility group box protein-1 (HMGB-1) plays a central role in the inflammatory network, as it is induced by a number of cytokines and can in turn induce a series of inflammatory reactions [12]. HMGB-1 can stimulate inflammation by binding to several receptors and acts as a potent inflammatory cytokine [13]. Although there have been few studies on the relationship between HMGB-1 and fibrosis or scarring, the serum level of HMGB-1 has been positively correlated with skin thickness in systemic sclerosis [14]. Furthermore, icariin, which is used to treat erectile dysfunction, has been shown to reduce liver fibrosis in a thioacetamide-induced liver fibrosis model by antagonizing the increase in HMGB-1 in addition to other mechanisms [15].

Polydeoxyribonucleotide (PDRN) is composed of a mixture of nucleotides extracted from trout sperm. PDRN exerts anti-inflammatory effects by inhibiting mast cell degranulation and inflammatory cytokines [16,17]. A previous study reported that PDRN administration reduced pro-inflammatory mediators, such as tumor necrosis factor alpha (TNF-α), interleukin 6 (IL-6), and HMGB-1 [18]. Although it could be hypothesized that PDRN may reduce scarring by down-regulating inflammatory reactions and HMGB-1, no previous studies have investigated the relationship between PDRN and scarring. Accordingly, the aim of this study was to evaluate the anti-inflammatory effects of PDRN, including reduced infiltration of inflammatory cells and HMGB-1 expression, on scar formation via short-duration administration.

2. Results

2.1. Polydeoxyribonucleotide Decreases Scar Size in Incisional Scar Tissue in Rats

On Day 7 of the postoperative period, all groups exhibited complete re-epithelialization and the formation of granulation tissue, as demonstrated by hematoxylin and eosin (H&E) and Masson's trichrome (M-T) staining (Figure 1A). Although the sham group displayed active inflammation with abundant inflammatory cells and fewer collagen fibers on Day 14 of the postoperative period, the PDRN-3 and PDRN-7 groups exhibited lower inflammatory cell infiltration with collagen fibers in the scar area in the H&E- and M-T-stained tissues (Figures 1B and 2).

To estimate the scar area and the degree of granulation tissue formation, only the boundary of the scar area that covered below the epidermis and above the panniculus carnosus was measured. In each wound, the scar and/or granulation tissue areas were estimated from two H&E-stained tissue sections representing different areas of the same wound. Each measurement is shown as the mean ± SEM. In the quantitative analysis of the scar area, the scar sizes of the sham, PDRN-3, and PDRN-7 groups were 51,272 ± 5793 μm^2, 13,201 ± 2243 μm^2, and 21,329 ± 1518 μm^2, respectively, on Day 7 of the postoperative period (* $p < 0.05$, *** $p < 0.001$; Figure 1C), while, on Day 14 of the postoperative period, the scar sizes decreased to 35,368 ± 3511 μm^2, 12,304 ± 1842 μm^2, and 13,291 ± 1076 μm^2, respectively (** $p < 0.01$, *** $p < 0.001$; Figure 1C). These results indicated that PDRN administration reduced the scar size compared with the sham group.

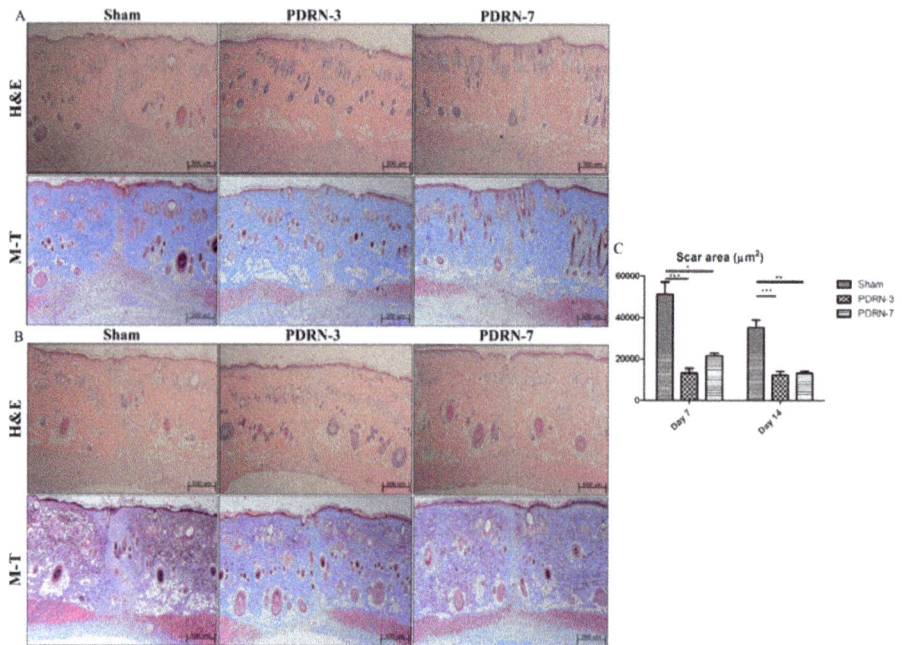

Figure 1. H&E- and M-T-stained tissues from the sham, Polydeoxyribonucleotide (PDRN)-3, and PDRN-7 groups on Days 7 and 14 (magnification, 40×). (**A**) All groups exhibited complete re-epithelialization and the formation of granulation tissue on Day 7. However, the PDRN-3 group showed the narrowest granulation tissue area among all groups. (**B**) Hematoxylin and eosin (H&E)- and Masson's trichrome (M-T)-stained tissues from the sham, PDRN-3, and PDRN-7 groups on Day 14 (magnification, 40×). The sham group continued to show a wide granulation tissue area with inflammation. However, the PDRN-3 and PDRN-7 groups showed more collagen deposition within narrower scar areas, as demonstrated by M-T staining. (**C**) Quantitative analyses of the scar areas in each treatment group. The scar areas were significantly narrower in the PDRN-3 and PDRN-7 groups than in the sham group on Days 7 and 14. However, no significant difference in scar size was observed between the PDRN-3 and PDRN-7 groups on Day 7 or 14 (* $p < 0.05$, ** $p < 0.01$, *** $p < 0.001$).

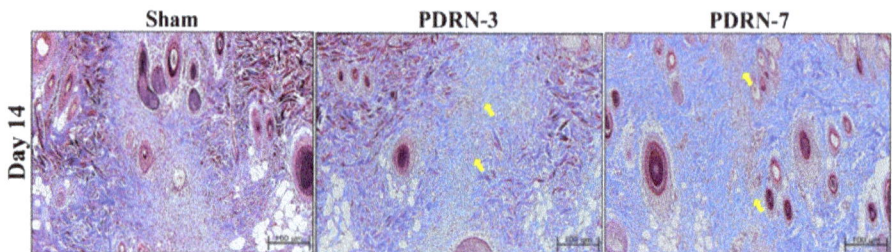

Figure 2. The sham group still exhibited granulation tissue and fewer collagen fibers with inflammatory cell infiltration on Day 14. However, the PDRN-treated groups demonstrated reduced inflammatory cell infiltration and collagen fibers (arrows) within the scar area.

2.2. Polydeoxyribonucleotide Decreases Inflammatory Cell Infiltration in Incisional Scar Tissue in Rats

To observe inflammatory cell infiltration within the scar tissue, we performed CD45 immunofluorescence staining on tissue collected on postoperative Day 7. More CD45-expressing leukocytes were detected in the sham group than in the PDRN-3 and PDRN-7 groups (Figure 3A). The number of inflammatory cells was calculated from four serial H&E-stained tissue sections from within the dermis of the scar area (Figure 3B). On Day 7 of the postoperative period, the mean numbers of inflammatory cells within the scar tissue were 21.16 ± 2.49, 13.47 ± 1.77, and 14.31 ± 2.28 in the sham, PDRN-3, and PDRN-7 groups, respectively. On Day 14 of the postoperative period, the mean numbers of inflammatory cells within the scar tissue were 15.34 ± 1.81, 7.53 ± 1.02, and 8.00 ± 1.12 in the sham, PDRN-3, and PDRN-7 groups, respectively. The numbers of inflammatory cells on Days 7 and 14 were significantly lower in the PDRN-3 and PDRN-7 groups than the sham group (* $p < 0.05$, *** $p < 0.001$; Figure 3C).

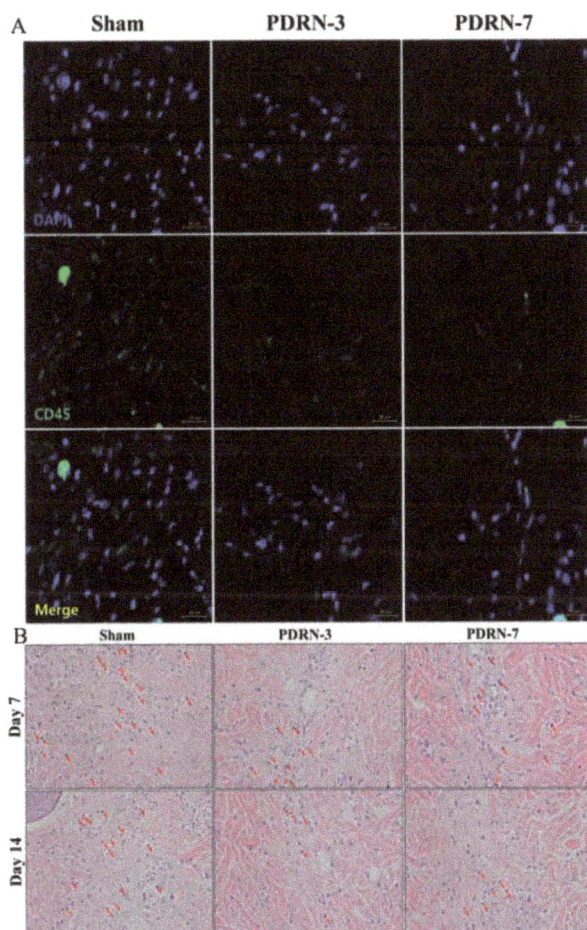

Figure 3. *Cont.*

C

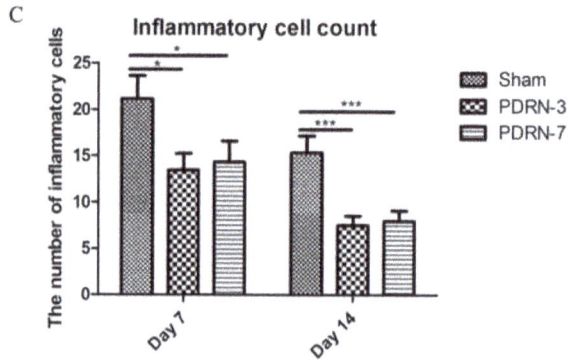

Figure 3. (**A**) Immunofluorescence analysis of CD45-expressing leukocytes on Day 7 (magnification, 400×). The sham group exhibited abundant cellular infiltration, and the majority of the cells were CD45-positive (green) leukocytes. The PDRN-3 and PDRN-7 groups exhibited diminished cellular infiltration and CD45-positive leukocytes. (**B**) Inflammatory cell infiltration in the scar area, as demonstrated by H&E staining (magnification, 400×). On Days 7 and 14, the sham group exhibited significant inflammatory cell infiltration into the scar area compared with the PDRN-3 and PDRN-7 groups. (**C**) Comparison of inflammatory cell counts. The infiltration of inflammatory cells in the scar area was lower in the PDRN-treated groups than in the sham group on Days 7 and 14 (* $p < 0.05$, *** $p < 0.001$).

2.3. Polydeoxyribonucleotide Decreases HMGB-1 Expression in Incisional Scar Tissue in Rats

Inflammatory cells were clearly observed within the scar tissue following staining for high mobility group box-1 (HMGB-1). On Day 7 of the postoperative period, increased HMGB-1 protein expression was observed in the sham group, whereas the PDRN-3 and PDRN-7 groups exhibited decreased HMGB-1 protein expression within the narrow scar areas. On Day 14 of the postoperative period, the sham group continued to exhibit high HMGB-1 expression in the wide scar areas, whereas the PDRN-3 and PDRN-7 groups showed markedly decreased HMGB-1 expression and only a small number of inflammatory cells (Figure 4A). Semi-quantitative analysis indicated that, on Days 7 and 14, the PDRN-3 and PDRN-7 groups exhibited significantly lower HMGB-1 protein expression than the sham group (* $p < 0.05$, ** $p < 0.01$, *** $p < 0.001$; Figure 4B).

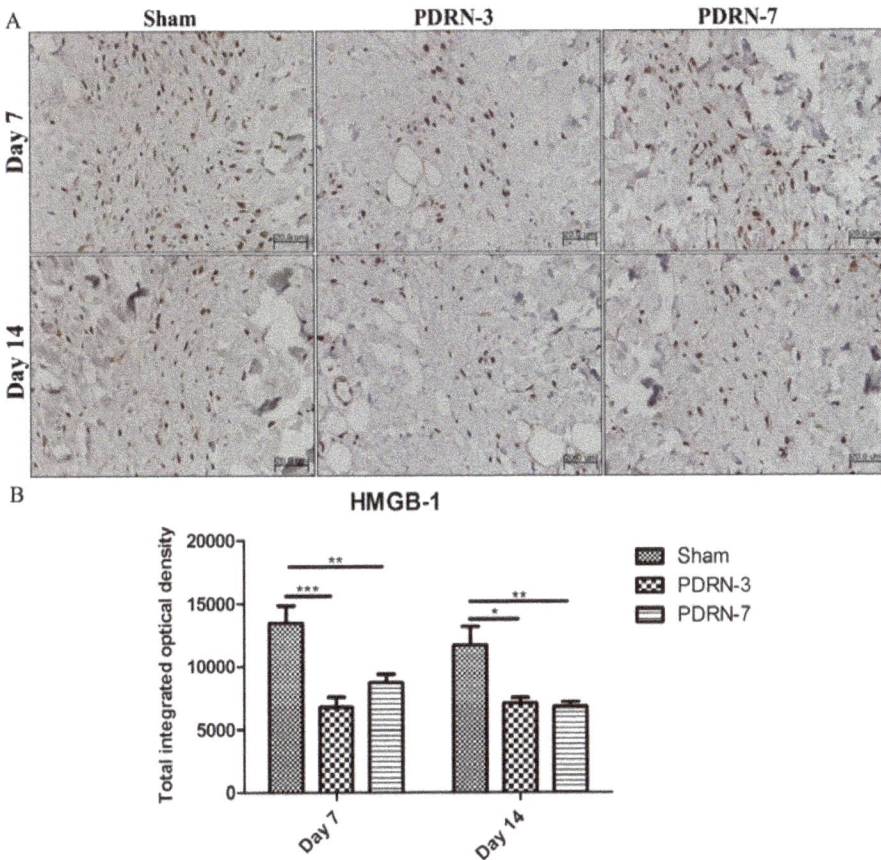

Figure 4. (**A**) Immunohistochemistry of high mobility group box-1 (HMGB-1) (magnification, 400×). All groups exhibited extracellular expression of HMGB-1, though the PDRN-3 and PDRN-7 groups displayed weaker expression within the narrow scar areas on Day 7. On Day 14, the sham group continued to show high extracellular expression of HMGB-1 in the wide scar areas, whereas extracellular expression of HMGB-1 was absent in the PDRN-3 and PDRN-7 groups. (**B**) Semi-quantitative analysis of HMGB-1 expression levels. The sham group showed significantly higher levels of HMGB-1 expression than the PDRN-3 and PDRN-7 groups on Days 7 and 14 (* $p < 0.05$, ** $p < 0.01$, *** $p < 0.001$).

2.4. HMGB-1 Administration Reverses the Anti-Inflammatory and Collagen Synthesis Effects of PDRN

We next examined whether HMGB-1 administration could reverse the effects of PDRN. On Day 7 of the postoperative period, the sham and PDRN-3 + HMGB-1 groups exhibited wider granulation tissue areas than did the PDRN-3 group (Figure 5A). The sham and PDRN-3 + HMGB-1 groups continued to show higher inflammation with wider scar areas on Day 14 of the postoperative period (Figure 5B). Quantitative analysis of the scar area indicated that the scar sizes in the PDRN-3 + HMGB-1 group were 44,688 ± 3573 µm^2 and 34,593 ± 2751 µm^2 on Days 7 and 14, respectively. Furthermore, the scar sizes in the PDRN-3 + HMGB-1 group were similar to those of the sham group and significantly wider than those of the PDRN-3 group on Days 7 and 14 (* $p < 0.05$, ** $p < 0.01$; Figure 5C). Administration of HMGB-1 to the PDRN-3 group resulted in enhanced inflammation (Figure 5D). The mean numbers of inflammatory cells in the PDRN-3 + HMGB-1 group were 18.07 ± 2.10 and 11.75 ± 1.27 on Days 7 and 14, respectively (* $p < 0.05$, *** $p < 0.001$; Figure 5E).

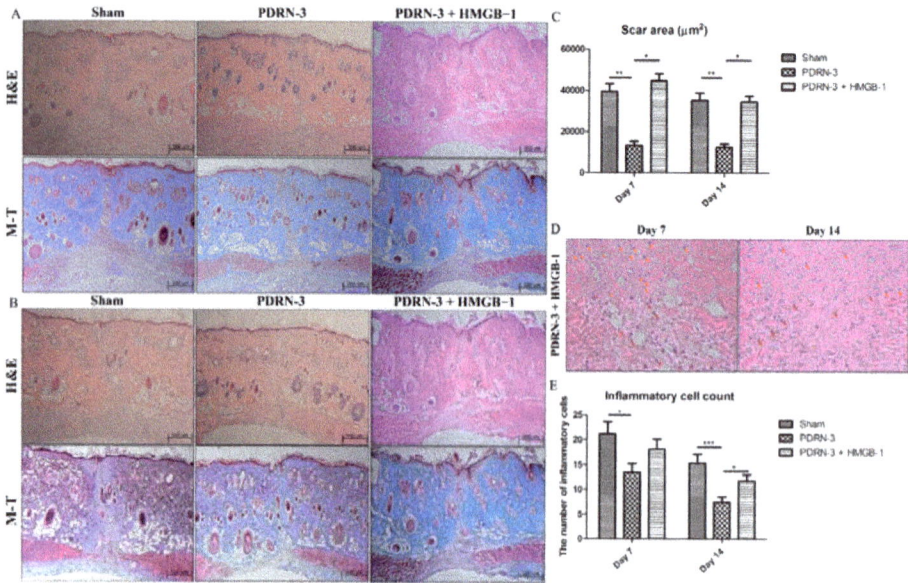

Figure 5. (**A**) H&E- and M-T-stained tissues from the sham, PDRN 3 and PDRN-3 + HMGB-1 groups on Day 7 (magnification, 40×). The PDRN-3 + HMGB-1 group exhibited wider scars compared with the PDRN 3 group. (**B**) H&E- and M-T-stained tissues from the sham, PDRN 3 and PDRN-3 + HMGB-1 groups on Day 14 (magnification, 40×). Although the rats in the PDRN-3 + HMGB-1 group were treated with PDRN, this group exhibited wide scars similar to the sham group on Day 14. The scar-narrowing effect of PDRN was reversed by administration of HMGB-1. (**C**) The scar areas in the PDRN-3 + HMGB-1 group were significantly wider than those in the PDRN-3 group. Additional administration of HMGB-1 reversed the scar-narrowing effect of PDRN (* $p < 0.05$, ** $p < 0.01$). (**D**) Inflammatory cell infiltration (arrow) in the PDRN-3 + HMGB-1 group, as demonstrated by H&E staining (magnification, 400×). Many inflammatory cells had infiltrated the scar area on Days 7 and 14. (**E**) The total inflammatory cell count was significantly higher in the PDRN-3 + HMGB-1 group than in the PDRN-3 group on Day 14 (* $p < 0.05$, *** $p < 0.001$). This result indicated that the anti-inflammatory effect of PDRN was reversed by HMGB-1 administration.

The synthesis of type I and type II collagen was analyzed by Western blots in the sham, PDRN-3, and PDRN-3 + HMGB-1 groups. On Day 7 of the postoperative period, type I collagen in the PDRN-3 and PDRN-3 + HMGB-1 groups increased by 1.36 ± 0.07-fold and 1.08 ± 0.03-fold, respectively, compared with the sham group (* $p < 0.05$, ** $p < 0.01$; Figure 6A). Type III collagen in the PDRN-3 and PDRN-3 + HMGB-1 groups also increased by 3.07 ± 0.31-fold and 1.35 ± 0.03-fold, respectively, compared with the sham group (** $p < 0.01$, *** $p < 0.001$; Figure 6A). On Day 14 of the postoperative period, type I collagen in the PDRN-3 and PDRN-3 + HMGB-1 groups increased by 1.43 ± 0.03-fold and 0.75 ± 0.02-fold, respectively, compared with the sham group (*** $p < 0.001$; Figure 6B). Type III collagen in the PDRN-3 and PDRN-3 + HMGB-1 groups also increased by 1.38 ± 0.11-fold and 0.88 ± 0.01-fold, respectively, compared with the sham group (* $p < 0.05$, ** $p < 0.01$; Figure 6B). These results indicated that PDRN administration stimulated wound healing by reducing inflammation and increasing collagen synthesis. Furthermore, the effect of PDRN was reversed by HMGB-1 administration.

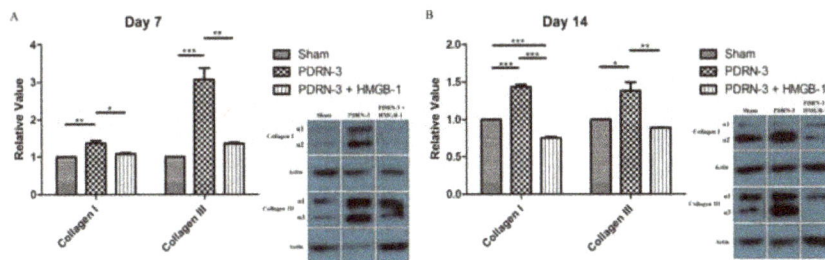

Figure 6. (**A**) Western blots for type I and type III collagen on Day 7. The expression of type I and III collagen was significantly higher in the PDRN-3 group compared with the sham and PDRN-3 + HMGB-1 group (* $p < 0.05$, ** $p < 0.01$, *** $p < 0.001$). (**B**) Western blots for type I and type III collagen on Day 14. The PDRN-3 group also demonstrated higher collagen synthesis compared with the other groups. PDRN accelerated wound healing by promoting early collagen synthesis. This effect of PDRN was reversed by HMGB-1 administration (* $p < 0.05$, ** $p < 0.01$, *** $p < 0.001$).

3. Discussion

As an A2AR agonist, PDRN exerts angiogenic effects via vascular endothelial growth factor (VEGF) augmentation [19,20] and tissue-repair effects via fibroblast stimulation [21,22]. Additionally, the activation of A2AR has an anti-inflammatory effect due to the inhibition of several pro-inflammatory mediators [16,18,23]. Previous investigations found that the injection of PDRN until the proliferative phase of wound healing resulted in a fibroplasia effect [20,22,24]. Although a fibroplasia effect from the prolonged injection of PDRN is beneficial to the compromised wound, it is not beneficial with respect to scar formation. Scar formation as the final result of wound healing is due to temporary overlaps of three phases: inflammatory, proliferative, and remodeling. In normal wound healing, there is an influx of inflammatory cells to the wound site until Days 4–6, followed by a proliferative phase during which inflammatory cells are replaced with fibroblasts. The transition between inflammation and proliferation is important because abnormal inflammatory prolongation results in excessive scarring [25]. Thus, we suspected that PDRN, which has both anti-inflammatory and collagen synthesis effects, could be beneficial in scar formation when administered during the inflammatory phase. Because uncontrolled and prolonged inflammation of the dermis produces pathologic scars, reduced inflammation and faster wound healing could have a beneficial effect on scar formation [4]. In this regard, we hypothesized that intensive administration of PDRN during the inflammatory phase for approximately three to seven days post-wounding could reduce inflammation and promote progression to the proliferative phase and early collagen synthesis.

Minimizing inflammation is thought to be associated with reducing scar formation [26,27]. Continuous and histologically localized inflammation of the reticular layer of the dermis produces pathologic scars [4]. Similarly, it has been shown that in surgical wounds, dermal inflammation that persists for 1–2 weeks can result in aberrant scarring and eventually pathologic scars [4]. Thus, we attempted to determine the degree of inflammation after PDRN administration. CD45, which is also referred to as common leukocyte antigen, is a ubiquitous membrane glycoprotein expressed in all hematopoietic cells, except mature erythrocytes [28]. The degree of cellular infiltration was significantly greater in the sham group, and the majority of these cells were CD45+ leukocytes. To objectively analyze the infiltration of inflammatory cells, we counted cell numbers in H&E-stained tissues. The scar areas in the sham group were more abundantly infiltrated with inflammatory cells than those in the PDRN-treated groups. Hence, our results indicated that the administration of PDRN could decrease inflammation, which may be a factor in excessive scar formation.

HMGB-1 is a ubiquitous nuclear protein that exists in eukaryotic cells [29]. Extracellular HMGB-1 regulates the synthesis of monocyte-derived pro-inflammatory cytokines such as TNF-α and IL-1 [30].

Extracellular HMGB-1 secreted from necrotic and inflammatory cells triggers inflammation by inducing inflammatory cell chemotaxis, which in turn initiates the production of pro-inflammatory cytokines by other inflammatory cells [12,31]. In previous investigations, administration of PDRN down-regulated the expression of the inflammatory cytokine HMGB-1 in arthritis and periodontitis models [16,18]. Although the relationship between HMGB-1 and scar formation remains unclear, HMGB-1 induced scar formation when applied to early embryonic murine skin wounds [32]. Thus, we hypothesized that PDRN may down-regulate inflammation and scar formation in surgical wounds by suppressing HMGB-1. On Day 7, extracellular HMGB-1 expression was widespread throughout the granulation tissue in all treatment groups, particularly in the sham group. On Day 14, extracellular HMGB-1 expression remained apparent in the sham group but was decreased in the PDRN treatment groups. Thus, our results implied that PDRN administration reduced HMGB-1 as a potent inflammatory mediator. To confirm the role of HMGB-1 in PDRN action, we administered HMGB-1 to the PDRN-3 group. This additional HMGB-1 administration counteracted the effects of PDRN, resulting in a wide scar area. Furthermore, collagen synthesis was also significantly suppressed by the administration of HMGB-1 to PDRN-treated rats. Inflammatory cell infiltration was also increased on Day 14 in PDRN-treated rats that were administered HMGB-1. Therefore, HMGB-1 could reverse the collagen synthesis and anti-inflammatory effects of PDRN. These results supported our hypothesis that PDRN exerts its anti-inflammatory and collagen synthesis effects via HMGB-1 suppression.

Histologic analyses of tissue samples collected on Day 7 showed that all groups were in the early proliferative phase, as indicated by the formation of granulation tissue; however, substantial inflammatory cell infiltration was observed in the sham group, which continued to exhibit granulation tissue and inflammatory cell infiltration on Day 14. Furthermore, the scar widths were significantly narrower in the PDRN-treated groups than in the sham group. PDRN could reduce the granulation tissue that serves as potential scar tissue during the early wound-healing phase. Thus, PDRN prevents scar formation via the promotion of fast wound healing by suppressing inflammation and enhancing collagen synthesis.

In summary, we concluded that faster wound healing and decreased scar formation were induced by the anti-inflammatory and collagen synthesis effects of PDRN. However, the short experimental period could be a potential limitation of our study. Nonetheless, it was obvious that the reduced formation of granulation tissue and decreased infiltration of inflammatory cells combined with faster wound healing following PDRN administration could improve the characteristics and sizes of scars.

4. Materials and Methods

4.1. Animal Model

Twenty-four male Sprague-Dawley (SD) rats were used to study incisional wounds. All animal protocols used in this study were approved by the Yonsei University Institutional Animal Care and Use Committee (16 April 2014). General anesthesia was induced via intraperitoneal injection of a zolazepam tiletamine mixture (30 mg/kg, Zoletil®; Virbac, Carros, France) and xylazine (10 mg/kg, Rompun®; Bayer, Leverkusen, Germany). A 6×1 cm^2 rectangular design was made to excise the skin and the panniculus carnosus muscle. The skin and panniculus carnosus muscle were excised, and only the skin layer was closed to maximize tension stress by leaving the muscle layer unrepaired. After surgery, the rats were randomly assigned to one of three treatment groups: sham ($n = 8$), PDRN-3 ($n = 8$), and PDRN-7 ($n = 8$). The sham group was injected with 1 mL of normal saline for seven days, whereas the PDRN-3 and PDRN-7 groups were administered PDRN via intraperitoneal injection (8 mg/kg, Placentex Integro®, Mastelli SRL, Sanremo, Italy) for three and seven days, respectively.

Another experiment was performed to clarify whether the effects of PDRN on scar diminishing and inflammation were mediated by HMGB-1. To determine whether HMGB-1 administration increased scar formation and inflammatory cell infiltration, 400 µg of HMGB-1 (HMGBiotech, Milan, Italy) diluted in 500 µL of normal saline was administered on a central 2-cm area of the incisional wound via intradermal injection for three days followed by PDRN administration as described for the PDRN-3 group (PDRN-3 + HMGB-1 group; $n = 6$). HMGB-1 was administered before intraperitoneal injection of PDRN. Other experimental protocols were the same as described above.

4.2. Histologic Analysis

Four SD rats in each group were euthanized on Days 7 and 14, and tissue biopsies were performed to evaluate inflammatory cell counts and scar areas. Tissue samples (10 mm thick) were obtained from the middle region of the wound, where there was maximal tension. All tissues were fixed in 10% neutral buffered formalin, embedded in a paraffin block, and stained with H&E and M-T stain.

The H&E- and M-T-stained tissues were examined under a light microscope at $40\times$ to estimate the scar areas and the degree of tissue granulation. The scar area was estimated using ImageJ® software version 1.49 (National Institutes of Health, Bethesda, MD, USA). In each wound, the scar and/or granulation tissue areas were obtained from two tissue sections representing different areas of the same wound. The mean scar and/or granulation tissue areas for each wound were then converted from pixel numbers to square micrometers that were calculated using the ratio of pixel numbers to the scale bar.

The H&E-stained tissues were also examined using a light microscope at $400\times$ to evaluate the degree of inflammatory cell infiltration in the scar tissue. The inflammatory cells were counted in four serial sections of tissue from within the dermis of the scar area. The numbers of inflammatory cells were calculated for each wound from two tissue sections representing different areas of the same wound, and then the mean number of inflammatory cells was obtained.

4.3. Immunohistochemistry for HMGB-1

Tissues obtained from the middle region of the wound were fixed with 10% formaldehyde and embedded in a paraffin block. Tissue sections were pretreated with a 3% hydrogen peroxide solution for 10 min to block endogenous peroxidase activity and then treated with a protein blocking serum-free reagent (X0909 DAKO, Carpinteria, CA, USA) for 30 min to prevent non-specific reactions. The sections were incubated at 4 °C overnight with primary antibodies (HMGB-1, Abcam, Cambridge, MA, USA) and then incubated at room temperature for 20 min with secondary antibodies from the DAKO Envision Kit (DAKO, Carpinteria, CA, USA). The expression of HMGB-1 in the scar area was semi-quantitatively analyzed using the MetaMorph® image analysis software version 7.8 (Universal Imaging, West Chester, PA, USA).

4.4. Immunofluorescence Assay for CD45

For immunofluorescence microscopy, the samples were blocked with 1% bovine serum albumin (BSA) followed by incubation with anti-CD45 (Life Technologies Co., Carlsbad, CA, USA) overnight at 4 °C. The next day, the cells were washed in phosphate-buffered saline and incubated with an Alexa Flour 488-conjugated goat anti-rabbit immunoglobulin G secondary antibody for 60 min at room temperature. The final antibody treatment also contained tetramethylrhodamine isothiocyanate–conjugated phalloidin and 4′,6-diamidino-2-phenylindole stain (DAPI; both at 1 g/mL; Sigma, St. Louis, MO, USA) for nuclear staining. The slides were mounted in Vectashield® HardSet Mounting Medium with DAPI (Vector Laboratories, Burlingame, CA, USA), and the cells were viewed under a confocal laser scanning microscope (LSM700; Carl Zeiss MicroImaging, Thornwood, NY, USA)

4.5. Western Blots for Type I and Type III Collagen

The wound homogenates were analyzed via Western blotting. The protein concentration was quantified using a bicinchoninic acid assay (Thermo Fisher Scientific, Waltham, MA) and normalized to a standard concentration using extraction buffer. The proteins were separated by sodium dodecyl sulfate-polyacrylamide gel electrophoresis (SDS-PAGE) and then transferred to polyvinylidene difluoride membranes (Millipore, Billerica, MA, USA). The membranes were blocked for 1 h with 3% BSA in 1X TBST and then incubated overnight at 4 °C with monoclonal mouse anti-collagen type I α1 and α2 antibodies and anti-collagen type III α1 and α3 antibodies (1:1000, Abcam, Cambridge, MA, USA). The primary antibody was detected using a horseradish peroxidase-conjugated goat anti-mouse or anti-rabbit secondary antibody (1:5000, Cell Signaling, Beverly, MA, USA). The protein bands were visualized using an ECL detection kit (Thermo Fisher Scientific, Waltham, MA, USA) according to the manufacturer's instructions. Finally, immunoblot signals were analyzed using ImageJ® software version 1.49 (National Institutes of Health, Bethesda, MD, USA). The results from each group were expressed as the integrated intensity relative to the sham group, measured with the same batch.

4.6. Statistical Analysis

Each measurement is shown as the mean ± SEM. All pairwise differences between the group measurements were examined by independent and paired t-tests using standard software (SPSS for Windows v15.0; SPSS Inc., Chicago, IL, USA). Statistical significance was set at $p < 0.05$.

Acknowledgments: This work was supported by a National Research Foundation of Korea (NRF) grant funded by the Korean government (MEST; No. 2014051295, Won Jai Lee).

Author Contributions: Woonhyeok Jeong performed the experiments, analyzed the data, wrote sections of the manuscript, and edited the figures; these contributions were commensurate with those of Won Jai Lee and Ju Hee Lee. Chae Eun Yang wrote sections of the manuscript. Tai Suk Roh and Jun Hyung Kim edited the manuscript. Won Jai Lee and Ju Hee Lee conceived of and designed the study, wrote a draft of the manuscript, and approved the final version of the manuscript.

Conflicts of Interest: None of the authors have any financial arrangements or potential conflicts of interest related to this article.

References

1. Oryan, A.; Alemzadeh, E.; Moshiri, A. Biological properties and therapeutic activities of honey in wound healing: A narrative review and meta-analysis. *J. Tissue Viability* **2016**, *25*, 98–118. [CrossRef] [PubMed]
2. Wang, J.; Ding, J.; Jiao, H.; Honardoust, D.; Momtazi, M.; Shankowsky, H.A.; Tredget, E.E. Human hypertrophic scar-like nude mouse model: Characterization of the molecular and cellular biology of the scar process. *Wound Repair Regen.* **2011**, *19*, 274–285. [CrossRef] [PubMed]
3. Wang, J.F.; Hori, K.; Ding, J.; Huang, Y.; Kwan, P.; Ladak, A.; Tredget, E.E. Toll-like receptors expressed by dermal fibroblasts contribute to hypertrophic scarring. *J. Cell Physiol.* **2011**, *226*, 1265–1273. [CrossRef] [PubMed]
4. Ogawa, R. Keloid and hypertrophic scars are the result of chronic inflammation in the reticular dermis. *Int. J. Mol. Sci.* **2017**, *18*, 606. [CrossRef] [PubMed]
5. Wulff, B.C.; Parent, A.E.; Meleski, M.A.; DiPietro, L.A.; Schrementi, M.E.; Wilgus, T.A. Mast cells contribute to scar formation during fetal wound healing. *J. Investig. Dermatol.* **2012**, *132*, 458–465. [CrossRef] [PubMed]
6. Dovi, J.V.; He, L.K.; DiPietro, L.A. Accelerated wound closure in neutrophil-depleted mice. *J. Leukoc. Biol.* **2003**, *73*, 448–455. [CrossRef] [PubMed]
7. Lucas, T.; Waisman, A.; Ranjan, R.; Roes, J.; Krieg, T.; Müller, W.; Roers, A.; Eming, S.A. Differential roles of macrophages in diverse phases of skin repair. *J. Immunol.* **2010**, *184*, 3964–3977. [CrossRef] [PubMed]
8. Brown, J.J.; Bayat, A. Genetic susceptibility to raised dermal scarring. *Br. J. Dermatol.* **2009**, *161*, 8–18. [CrossRef] [PubMed]
9. Bran, G.M.; Goessler, U.R.; Hormann, K.; Riedel, F.; Sadick, H. Keloids: Current concepts of pathogenesis (review). *Int. J. Mol. Med.* **2009**, *24*, 283–293. [CrossRef] [PubMed]
10. Shih, B.; Garside, E.; McGrouther, D.A.; Bayat, A. Molecular dissection of abnormal wound healing processes resulting in keloid disease. *Wound Repair Regen.* **2010**, *18*, 139–153. [CrossRef] [PubMed]
11. Al-Attar, A.; Mess, S.; Thomassen, J.M.; Kauffman, C.L.; Davison, S.P. Keloid pathogenesis and treatment. *Plast. Reconstr. Surg.* **2006**, *117*, 286–300. [CrossRef] [PubMed]
12. Bianchi, M.E.; Manfredi, A.A. High-mobility group box 1 (HMGB1) protein at the crossroads between innate and adaptive immunity. *Immunol. Rev.* **2007**, *220*, 35–46. [CrossRef] [PubMed]
13. Scaffidi, P.; Misteli, T.; Bianchi, M.E. Release of chromatin protein HMGB1 by necrotic cells triggers inflammation. *Nature* **2002**, *418*, 191–195. [CrossRef] [PubMed]

14. Yoshizaki, A.; Komura, K.; Iwata, Y.; Ogawa, F.; Hara, T.; Muroi, E.; Takenaka, M.; Shimizu, K.; Hasegawa, M.; Fujimoto, M.; et al. Clinical significance of serum HMGB-1 and sRAGE levels in systemic sclerosis: Association with disease severity. *J. Clin. Immunol.* **2009**, *29*, 180–189. [CrossRef] [PubMed]

15. Algandaby, M.M.; Breikaa, R.M.; Eid, B.G.; Neamatallah, T.A.; Abdel-Naim, A.B.; Ashour, O.M. Icariin protects against thioacetamide-induced liver fibrosis in rats: Implication of anti-angiogenic and anti-autophagic properties. *Pharmacol. Rep.* **2017**, *69*, 616–624. [CrossRef] [PubMed]

16. Bitto, A.; Polito, F.; Irrera, N.; D'Ascola, A.; Avenoso, A.; Nastasi, G.; Campo, G.; Micali, A.; Bagnato, G.; Minutoli, L.; et al. Polydeoxyribonucleotide reduces cytokine production and the severity of collagen-induced arthritis by stimulation of adenosine A_{2A} receptor. *Arthritis Rheum.* **2011**, *63*, 3364–3371. [CrossRef] [PubMed]

17. Rork, T.H.; Wallace, K.L.; Kennedy, D.P.; Marshall, M.A.; Lankford, A.R.; Linden, J. Adenosine A_{2A} receptor activation reduces infarct size in the isolated, perfused mouse heart by inhibiting resident cardiac mast cell degranulation. *Am. J. Physiol. Heart Circ. Physiol.* **2008**, *295*, H1825–H1833. [CrossRef] [PubMed]

18. Bitto, A.; Oteri, G.; Pisano, M.; Polito, F.; Irrera, N.; Minutoli, L.; Squadrito, F.; Altavilla, D. Adenosine receptor stimulation by polynucleotides (PDRN) reduces inflammation in experimental periodontitis. *J. Clin. Periodontol.* **2013**, *40*, 26–32. [CrossRef] [PubMed]

19. Bitto, A.; Polito, F.; Altavilla, D.; Minutoli, L.; Migliorato, A.; Squadrito, F. Polydeoxyribonucleotide (PDRN) restores blood flow in an experimental model of peripheral artery occlusive disease. *J. Vasc. Surg.* **2008**, *48*, 1292–1300. [CrossRef] [PubMed]

20. Bitto, A.; Galeano, M.; Squadrito, F.; Minutoli, L.; Polito, F.; Dye, J.F.; Clayton, E.A.; Calò, M.; Venuti, F.S.; Vaccaro, M.; et al. Polydeoxyribonucleotide improves angiogenesis and wound healing in experimental thermal injury. *Crit. Care Med.* **2008**, *36*, 1594–1602. [CrossRef] [PubMed]

21. Sini, P.; Denti, A.; Cattarini, G.; Daglio, M.; Tira, M.E.; Balduini, C. Effect of polydeoxyribonucleotides on human fibroblasts in primary culture. *Cell Biochem. Funct.* **1999**, *17*, 107–114. [CrossRef]

22. Galeano, M.; Bitto, A.; Altavilla, D.; Minutoli, L.; Polito, F.; Calo, M.; Cascio, P.L.; d'Alcontres, F.S.; Squadrito, F. Polydeoxyribonucleotide stimulates angiogenesis and wound healing in the genetically diabetic mouse. *Wound Repair Regen.* **2008**, *16*, 208–217. [CrossRef] [PubMed]

23. Chan, E.S.; Fernandez, P.; Cronstein, B.N. Adenosine in inflammatory joint diseases. *Purinergic Signal.* **2007**, *3*, 145–152. [CrossRef] [PubMed]

24. Altavilla, D.; Squadrito, F.; Polito, F.; Irrera, N.; Calo, M.; Lo Cascio, P.; Galeano, M.; Cava, L.L.; Minutoli, L.; Marini, H.; et al. Activation of adenosine A_{2A} receptors restores the altered cell-cycle machinery during impaired wound healing in genetically diabetic mice. *Surgery* **2011**, *149*, 253–261. [CrossRef] [PubMed]

25. Walmsley, G.G.; Maan, Z.N.; Wong, V.W.; Duscher, D.; Hu, M.S.; Zielins, E.R.; Wearda, T.; Muhonen, E.; McArdle, A.; Tevlin, R.; et al. Scarless wound healing: Chasing the holy grail. *Plast. Reconstr. Surg.* **2015**, *135*, 907–917. [CrossRef] [PubMed]

26. Szpaderska, A.M.; DiPietro, L.A. Inflammation in surgical wound healing: Friend or foe? *Surgery* **2005**, *137*, 571–573. [CrossRef] [PubMed]

27. Xue, M.; Jackson, C.J. Extracellular matrix reorganization during wound healing and its impact on abnormal scarring. *Adv. Wound Care* **2015**, *4*, 119–136. [CrossRef] [PubMed]

28. Nakano, A.; Harada, T.; Morikawa, S.; Kato, Y. Expression of leukocyte common antigen (CD45) on various human leukemia/lymphoma cell lines. *Acta. Pathol. Jpn.* **1990**, *40*, 107–115. [CrossRef] [PubMed]

29. Vaccari, T.; Beltrame, M.; Ferrari, S.; Bianchi, M.E. *Hmg4*, a new member of the *Hmg1/2* gene family. *Genomics* **1998**, *49*, 247–252. [CrossRef] [PubMed]

30. Andersson, U.; Wang, H.; Palmblad, K.; Aveberger, A.C.; Bloom, O.; Erlandsson-Harris, H.; Janson, A.; Kokkola, R.; Zhang, M.; Yang, H.; et al. High mobility group 1 protein (HMG-1) stimulates proinflammatory cytokine synthesis in human monocytes. *J. Exp. Med.* **2000**, *192*, 565–570. [CrossRef] [PubMed]

31. Lee, D.E.; Trowbridge, R.M.; Ayoub, N.T.; Agrawal, D.K. High-mobility group box protein-1, matrix metalloproteinases, and vitamin D in keloids and hypertrophic scars. *Plast. Reconstr. Surg. Glob. Open* **2015**, *3*, e425. [CrossRef] [PubMed]

32. Dardenne, A.D.; Wulff, B.C.; Wilgus, T.A. The alarmin HMGB-1 influences healing outcomes in fetal skin wounds. *Wound Repair Regen.* **2013**, *21*, 282–291. [CrossRef] [PubMed]

International Journal of
Molecular Sciences

MDPI

Review

Recent Understandings of Biology, Prophylaxis and Treatment Strategies for Hypertrophic Scars and Keloids

Ho Jun Lee [1] and Yong Ju Jang [2,*]

[1] Department of Otorhinolaryngology-Head and Neck Surgery, Chuncheon Sacred Heart Hospital, College of Medicine, Hallym University, Chuncheon 24253, Korea; leehj@hallym.or.kr
[2] Department of Otolaryngology, Asan Medical Center, University of Ulsan College of Medicine, Seoul 05505, Korea
* Correspondence: jangyj@amc.seoul.kr; Tel.: +82-2-3010-3712

Received: 22 November 2017; Accepted: 8 January 2018; Published: 2 March 2018

Abstract: Hypertrophic scars and keloids are fibroproliferative disorders that may arise after any deep cutaneous injury caused by trauma, burns, surgery, etc. Hypertrophic scars and keloids are cosmetically problematic, and in combination with functional problems such as contractures and subjective symptoms including pruritus, these significantly affect patients' quality of life. There have been many studies on hypertrophic scars and keloids; but the mechanisms underlying scar formation have not yet been well established, and prophylactic and treatment strategies remain unsatisfactory. In this review, the authors introduce and summarize classical concepts surrounding wound healing and review recent understandings of the biology, prevention and treatment strategies for hypertrophic scars and keloids.

Keywords: keloid; hypertrophic scar; scar biology; scar prevention; scar treatment

---

1. Introduction

Many life situations result in injury to the skin. Physical trauma, surgical incisions, burn injuries, vaccinations, skin piercings, herpes infection and even insect bites can cause skin injury and resultant scar problems. Each year in the developed world, approximately 100 million people suffer from scar-related issues [1]. Most superficial injuries do not leave significant scars, but deep cutaneous injuries occasionally produce serious problems, hypertrophic scars and keloids [2]. Cosmetic problems, functional problems such as contractures and patients' subjective symptoms including pruritus and pain can cause hypertrophic scars and keloids to dramatically affect patients' quality of life, physical status and psychological health [3]. Hypertrophic scars and keloids are fibroproliferative disorders that result from abnormal wound healing, defined as increased or decreased regulation of certain wound healing processes. Understanding the major mechanisms underlying abnormal wound healing and correcting them will benefit numerous patients, like the wide-spread public health effects of antibiotics in the twentieth century. Many studies on hypertrophic scars and keloids have been reported, and our understanding of these conditions is improving. However, the pathophysiology remains extremely complex. In this review, we introduce and summarize the classical concepts of wound healing and review the recent biological advances in treatment, as well as the manner in which these advances translate into preventive and treatment strategies for hypertrophic scars and keloids. This review included the basic knowledge on scar biology any kind of physician should know and may be appropriate for general physicians rather than scar specialists.

2. Methods

The original articles dealing with the biology, prophylaxis and treatment strategies for hypertrophic scars and keloids were searched and reviewed. PubMed, Web of Science and Cochrane library databases were searched on the keywords: hypertrophic scar OR keloid AND biology; hypertrophic scar OR keloid AND prophylaxis; hypertrophic scar OR keloid AND treatment. Time limits were from 1 January 2010 to the present. In addition, important reference articles from the included articles were also reviewed. Several meta-analysis were also reviewed to estimate the outcome of a certain treatment modality. Duplicates, letters, reviews, hypotheses articles dealing with the fibrotic disorders on internal organs, studies dealing with specific surgical techniques and studies published in a language other than English were excluded. Figure 1 shows the flowchart of the literature search for this review.

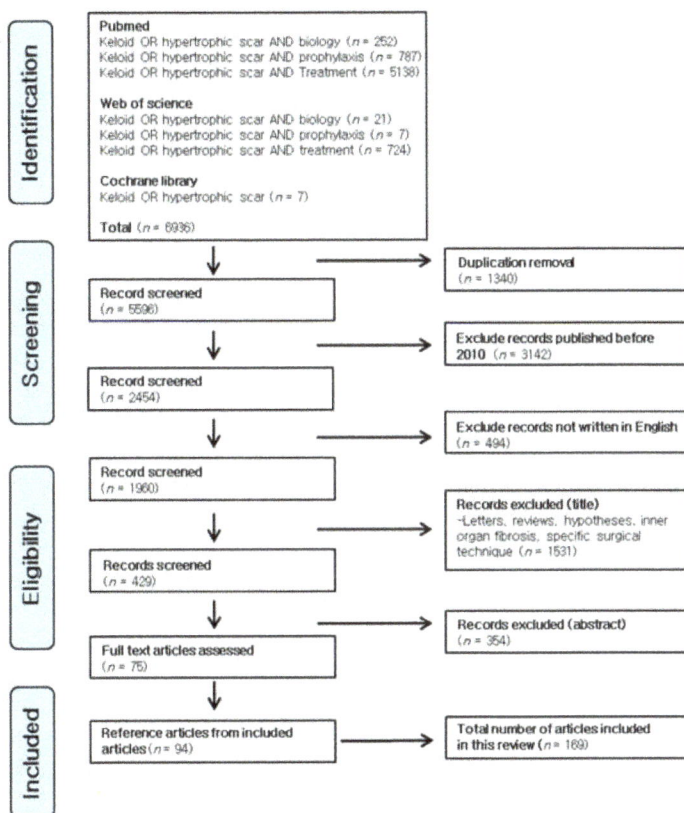

Figure 1. The flowchart of the literature search for this review.

3. Classical Concepts of Wound Healing

The classical model of wound healing involves three distinct, but overlapping phases that follow a time sequence: the inflammatory phase, the proliferative phase and the remodeling phase. The first phase of wound healing is the inflammatory phase that starts immediately after tissue injury and lasts for approximately 2–3 days after injury. Coagulation cascades, complement activation and platelet degranulation prevent further fluid and blood losses by creating platelet plugs and a fibrin matrix [4]. The immune system and inflammatory reactions are activated to prevent infection and removing devitalized tissues [5]. Neutrophils are recruited by chemotactic factors produced by platelet and bacterial degranulations [6], and monocytes are recruited and differentiated into macrophages 2–3 days after injury.

The second phase of wound healing is the proliferative phase. This phase of new tissue formation occurs approximately 2–3 days after tissue damage and may last for 3–6 weeks. Active cellular proliferation and migration characterize this phase. Keratinocytes migrate to the damaged dermis; new blood vessels grow inward within the damaged tissue; and new capillaries replace the fibrin matrix with granulation tissue via the actions of macrophages and fibroblasts. Granulation tissue forms a new substrate for keratinocyte migration. Keratinocytes proliferate and mature within granulation tissue along the wound margin, restoring the protective function of the epithelium. In the late proliferative phase, a portion of the fibroblasts differentiates into myofibroblasts in association with macrophages. Fibroblasts and myofibroblasts produce extracellular matrix (ECM), mainly in the form of collagen; this accumulated collagen forms most of the eventual scar [7]. Other constituents of ECM include elastin, hyaluronic acids and proteoglycans. Myofibroblasts, which contain actin filaments, have contractile properties and help bring the edges of the wound together over time [8]. Once wound closure is accomplished, the final remodeling phase commences. This phase is characterized by degradation of excessive tissue, transforming immature healing products into a mature form. Remodeling may last for a year or more. Excessive ECM is degraded and remodeled from type III collagen, the main component of ECM present during the early wound healing process, to mature type I collagen.

4. Important Proteins and Cytokines in the Wound Healing Processes

It is important to achieve a proper balance between these wound healing phases. Synthesis and degradation of ECM should be balanced, otherwise wound healing may be delayed or result in excessive scars. Important proteins and cytokines that influence balanced wound healing processes are summarized herein (Figure 2) and are important for understanding current investigations in keloid treatment.

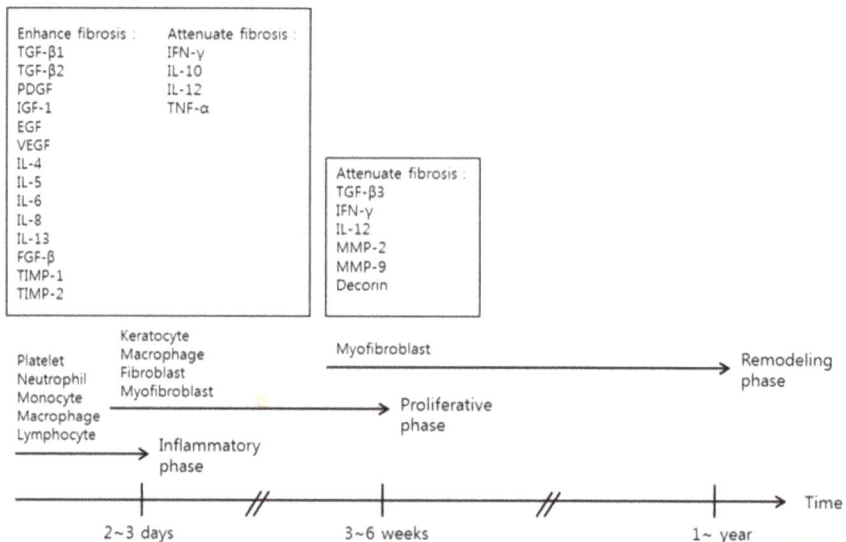

Figure 2. Important proteins and cytokines in the wound healing processes. The classical model of wound healing involves three distinct, but overlapping phases that follow a time sequence: the inflammatory, proliferative and remodeling phases. Important cells, proteins and cytokines in each phase are listed.

4.1. Inflammatory and Proliferative Phase

Prolonged and excessive inflammatory reactions result within the context of increased fibroblast activity, which in turn produces excessive ECM. In this phase, degranulation of platelets releases and activates transforming growth factor β (TGF-β), particularly TGF-β1, TGF-β2, platelet-derived growth factor (PDGF), insulin-like growth factor (IGF-1) and epidermal growth factor (EGF). Vascular endothelial growth factor (VEGF), which is produced by epidermal cells, is a positive regulator of angiogenesis. Because of this, overexpression of VEGF is related to excessive capillary formation, collagen type I production and overall scar volume increase [9]. These cytokines are not only fibrogenic growth factors, but also chemotactic agents for epithelial cells, endothelial cells, neutrophils, macrophages, mast cells and fibroblasts [4,8]. Fibroblasts originating in keloid tissues show increased receptors to these growth factors and demonstrate increased responsiveness compared with fibroblasts from normal tissues [10–13]. Tissue inhibitors of metalloproteinases (TIMPs) are endogenous inhibitors of matrix metalloproteinases (MMPs); thus, increased levels of TIMPs, especially TIMP-1 and TIMP-2, are associated with hypertrophic scar formation [9]. Tumor necrosis factor-α (TNF-α) is an inflammatory cytokine produced by monocytes and macrophages during the inflammatory phase. It has been known that this cytokine induces collagen degranulation and contributes to minimizing excessive scarring. One suggested mechanism is that TNF-α increases the MMP1/TIMP3, MMP2/TIMP3 ratios [14]. However, other studies showed that the biologic effect of TNF-α was not the same on the fibroblasts from lung and skin tissues showing tissue specificity [15] and TNF-α induced epithelial-mesenchymal transition in human skin wound healing [16]. Therefore, it is still unclear whether TNF-α would promote or attenuate scar formation.

Immune responses are also related to wound healing processes. T helper CD4 cells are thought to be major immunoregulatory cells during wound healing processes. CD4 T cells express Th1 or Th2 responses [17]. Th1 responses produce interferon-γ and interleukin (IL)-12 and are thought to be related to the attenuation of fibrogenesis. Th2 responses of CD4 cells are generally likely related to fibrogenesis. IL-4, IL-5, IL-6 and IL-13 are thought to be related to pro-fibrosis [18,19], but IL-10 is thought to be related to anti-fibrosis [20–22]. These cytokines are essential for promoting or impeding the fibroblast recruitment and proliferation, ECM deposition, angiogenesis and re-epithelialization [4]. Endothelial cytokines including IL-8, IGF-1, fibroblast growth factor (FGF)-β and heparin promote angiogenesis. Wound re-epithelialization is enhanced by EGF, TGF-α and IGF-1 [4].

4.2. Remodeling Phase

During the remodeling phase, excessive ECM is degraded, and collagen type III, an immature collagen form, is converted to mature collagen type I. TGF-β3 is considered to play a role in reducing the newly-synthesized ECM [23]. Significantly lower TGF-β3 mRNA expression was found in keloid tissues [24,25]. However, TGF-β isoforms (TGF-β1, TGF-β2 and TGF-β3) do not present its activity as isolated ligands, but are also associated with receptors and activity modulators; therefore, the mere presence or absence of TGF-β may not fully explain abnormal wound healing [26]. Members of the MMP family have major effects on ECM degradation and remodeling and mediate the degradation of type III and type I collagens, the major components of ECM [27,28]. MMP-2 and MMP-9 are active during the remodeling phase. MMP-9 is involved in degradation of type IV and V collagens, fibronectin and elastin. MMP-2 plays an important role in ECM remodeling by degrading denatured collagen [29,30]. MMPs have a downregulating effect on inflammation by decreasing and antagonizing chemokines [31,32]. Immunity, cell migration and angiogenesis are also influenced by MMPs [33]. MMP activities are regulated by TIMPs. Decorin is a proteoglycan component of dermal connective tissue that binds to type I collagen fibrils and influences TGF-β [34]. This protein is decreased in keloids and hypertrophic scars [35]. By binding and neutralizing TGF-β, decorin decreases the stimulatory effects of TGF-β on collagen, fibronectin and glycosaminoglycan synthesis [17]. Decorin also inhibits angiogenesis by interacting with VEGF receptors (VEGFR2) and by inhibiting hepatocyte growth factors and PDGF [36]. Decorin's antifibrotic properties are receiving attention as a future therapeutic agent [37,38].

5. Recent Findings of Scar Biology

Here, we introduce some recent findings on scar biology. These consist of some factors that influence pro-fibrotic or anti-fibrotic pathways.

5.1. Hypoxia

Oxygen has long been known to be an important factor in wound healing [39,40]. There have been many reports suggesting that a hypoxic environment is associated with keloid formation [41,42]. Zhao et al. measured the quantity of hypoxia inducible factor (HIF)-1α in keloid and normal tissues and reported that keloid tissues are relatively hypoxic tissues compared to normoxic tissues, and hypoxia induces a pro-fibrotic state in dermal fibroblasts via the TGF-β1/SMAD3 pathway [43].

5.2. Periostin

Periostin is a secreted extracellular matrix (ECM) protein, which was originally identified in osteoblast, periodontal ligament and periosteum [44]. This matricellular protein is expressed in the basement membrane, dermis and hair follicle [45]. Periostin is induced by TGF-β in human dermal fibroblast and has an important role in wound healing and scar pathogenesis by inducing angiogenesis, fibroblast proliferation and myofibroblast persistence [45–47]. It starts to increase its expression from a few days after injury, peaking after about seven days after injury and decreasing afterwards [48,49]. Many authors have reported that periostin is abnormally elevated in hypertrophic scars and keloids compared to normal tissues [45,46,50,51] and implicates periostin as a possible therapeutic target in the treatment of hypertrophic scars and keloid.

5.3. MicroRNAs

MicroRNAs (miRNAs) are a group of short noncoding RNAs that pair complementarily with target genes and silence that genes post-transcriptionally. It thereby regulates negatively the expression of their target genes. miRNAs are thought to be deregulated in many skin diseases such as malignant skin diseases and keloids [52–55]. Some researchers performed miRNA expression microarrays in keloids and normal tissues [55,56] and reported upregulated or downregulated miRNAs in keloid tissues compared to normal tissues. *miRNA-199a-5p* [57], *miRNA-21* [58–61], *miRNA-146a* [62], *miRNA-1224-5p* [56], *miRNA-31* [63], and so forth, were investigated and showed potential in the treatment of hypertrophic scars and keloids.

6. Preventions and Treatment Strategies for Hypertrophic Scars and Keloids

Because the processes are so complicated, the definitive processes that underlie excessive scar formation are yet to be elucidated. So far, preventions and treatment strategies mainly focus on reducing inflammation. Other therapies, targeting genes and molecules, require more study prior to being introduced in clinical practice. The current treatment strategies for hypertrophic scars and keloids are listed below and summarized in Table 1.

Table 1. Current treatment strategies for hypertrophic scars and keloids.

Categories	Modalities	Suggested Mechanisms	Use
Prophylaxis	Tension-free closure	-Reduce inflammation by reducing mechanotransduction	-Debridement of inviable tissues, adequate hemostasis -Rapid tension free primary closure
	Taping or silicone sheeting	-Reduce inflammation by reducing mechanotransduction: occlusion and hydration	-Start 2 weeks after primary wound treatment -12 h a day for at least 2 months
	Flavonoids	-Induction of MMPs -Inhibition of SMADs expression	-Start 2 weeks after primary wound treatment -Generally twice daily for 4 to 6 months
	Pressure therapy	-Occlusion of blood vessels -Inducing apoptosis	-Pressure of 15 to 40 mmHg -More than 23 h a day for at least 6 months
Treatment (current)	Corticosteroids	-Reducing inflammation and proliferation -Vasoconstriction	-Intralesional injection: triamcinolone 10 to 40 mg/mL -1 to 2 sessions a month (2 to 3 sessions, but can be extended) -Tapes/plasters, ointments are available -Combination is common
	Scar revision	-Direct reduction of scar volume	-At least 1 year after primary wound treatment -Combination is recommended
	Cryotherapy	-Scar tissue necrosis	-Deliver liquid nitrogen using spray, contact or intralesional needle cryoprobe -10 to 20 s freeze-thaw cycles -Combination is common
	Radiotherapy	-Anti-angiogenesis -Anti-inflammation	-Adjuvant after scar revision -24–48 h after scar revision surgery -Total of 40 Gray or less, over several divided sessions
	Laser therapy	-Vaporize blood vessel -Anti-inflammation	-585-nm pulsed dye laser: 6.0–7.5 J/cm^2 (7 mm spot) or 4.5–5.5 J/cm^2 (10 mm spot) -1064-nm Nd:YAG laser: 14 J/cm^2 (5 mm spot) -2 to 6 sessions, every 3–4 weeks
	5-Fluorouracil	-Anti-angiogenesis -Anti-inflammation	-Intralesional injection: 50 mg/mL -Weekly for 12 weeks -Combination is common
Treatment (Emerging)	MSC * therapy	-Modulation of proinflammatory cell activity -Anti-fibrosis -Promote normal angiogenetic activity	-Systemic injection -Local injection (at the wound) -Engineered MSC-seeded tissue scaffold
	Fat grafting	-Deliver adipose-tissue derived MSCs	-Fat injection or fat tissue grafting underneath or into the wound
	Interferon	-Downregulating TGF-β1 -Attenuates collagen synthesis and fibroblast proliferation	-Intralesional injection: 1.5 × 10^6 IU, twice daily over 4 days
	Human recombinant TGF-β3/TGF-β1 or 2 neutralizing antibody	-Adjust TGF-β3: TGF-β1 or 2 ratio	Not available currently
	Botulinum toxin type A	-Reduce muscle tension during wound healing -Arrest cell cycle in non-proliferative stage -Influence TGF-β1 expression	-Intralesional injection: 70~140 U, 1 or 3 months interval, 3 sessions
	Bleomycin	-Decreasing collagen synthesis -Reduce lysyl-oxidase levels -Induce apoptosis	-Intralesional injection: 1.5 IU/mL, 2 to 6 sessions at monthly interval

* MSC: mesenchymal stem cell; MMPs: matrix metalloproteinases; TGF: transforming growth factor.

6.1. Prevention

6.1.1. Tension-Free Primary Closure

Regardless of a patient's tendency to exhibit bad scars (or not), (1) debridement of inviable or severely contaminated tissues, (2) adequate hemostasis to prevent hematoma, seroma or abscess formation and (3) rapid primary closure using tension-free techniques are wound care basics and are very important for minimizing the effects of bad scars. Wound epithelialization that is delayed beyond 10–14 days increases the risk of hypertrophic scars, and quick primary closure to induce rapid epithelialization is necessary to achieve good scarring [64]. The importance of tension-free closure techniques cannot be overstated. Wounds that are subject to tension tend to develop into bad scars [65]. The exact molecular mechanisms that govern how our skin responds to physical tension remain uncertain; however, several pathways that convert mechanical forces into biochemical responses have been investigated and reported. This process is called mechanotransduction [66]. Gurtner et al. reported on the fibrotic effects of mechanical tension and described the preventive effect of offloading wound tension on scar formation [67].

6.1.2. Passive Mechanical Stabilization

To prevent wound stretching and consequential mechanotransduction, prolonged passive mechanical wound stabilization has been applied [68–71] using paper tapes or silicone sheets. Paper tapes help alleviate scar formation, and silicone sheeting is superior to paper tapes because it avoids repeated epidermal avulsion.

Other mechanisms of silicone sheets include occlusion and hydration of the scar surface. The inherent antifibrotic properties of silicone are not definite [72]. Silicone sheeting is recommended for use from two weeks after primary wound treatment for more than 12 h a day for at least two months. For body areas where silicone sheets do not easily fit, silicone gel can be applied.

6.1.3. Flavonoids

Flavonoids (or bioflavonoids) are naturally-derived substances from various plants and have been used for preventing severe scar formation. Several studies have reported the efficacy of flavonoid scar gels like Contractubex Gel (Merz Pharma, Frankfurt, Germany) or Mederma Skin Care Gel (Merz Pharmaceuticals, Greensboro, NC, USA). The efficacy of these gel products is controversial [73–77], but other flavonoids like quercetin exert antifibrotic actions. These actions may be mediated through induction of MMP-1 or inhibition of SMAD2, SMAD3 or SMAD4 expression [77,78]. The instructions of flavonoids, for instance, Contractubex Gel, is as follows: (1) start two weeks after primary wound treatment; and (2) twice daily for four to six months.

6.1.4. Pressure Therapy

Cutaneous wound compression has been used not only for prevention, but also for treatment of hypertrophic scars and keloids. Although pressure therapy reduces the subjective and objective signs and symptoms of hypertrophic scars and keloids, the scientific evidence supporting their use is weak, and their clinical efficacy is also controversial [79]. The suggested mechanisms underlying pressure therapy include occlusion of blood vessels and limiting the delivery of inflammatory cytokines, nutrients and oxygen from blood vessels to scar tissue [80–84]. Increasing apoptosis may be another mechanism of pressure therapy [85]. There are no comparative analyses of pressure amount, and the pressure amount that is used clinically relies on empirical reports. Currently, the recommended amount is 15–40 mm Hg for more than 23 h a day for at least six months [83,86].

6.2. *Current Treatment Strategies*

6.2.1. Corticosteroids

Intralesional steroid injection, steroid tapes/plasters and steroid ointments have been used to treat hypertrophic scars and keloids. Intralesional injection is the most popular method for steroid administration, although steroid tapes/plasters are gaining popularity [87]. The mechanism underlying this therapy is attributed to its anti-inflammatory effect [72]. In addition, steroid therapy seems to reduce collagen synthesis, glycosaminoglycan production, fibroblast proliferation and degeneration of collagen and fibroblasts [88,89]. Another suggested mechanism is induction of vasoconstriction mediated by binding of the topical steroid to classical glucocorticoid receptors [2]. Resolution rates for keloids treated with intralesional steroid injections are variable and range from

50% to 100% and recur in 9% to 50% [90]. Most previous studies used triamcinolone acetonide (TAC), injected alone or in combination with other treatment modalities such as 5-FU, verapamil, cryotherapy or surgery. The concentrations of injectable TAC vary from 10 to 40 mg/mL, but the recommended concentration of TAC in monotherapy is 40 mg/mL for keloid resolution [91]. The injection is performed 1–2 times a months until the scar has flattened. Intralesional steroid injections could cause side effects such as skin atrophy or telangiectasia.

6.2.2. Scar Revision Surgery

Surgical excision is a traditional treatment for hypertrophic scars and keloids. The remodeling phase of classical wound healing may last for more than one year; therefore, excision of hypertrophic scars or keloids should be considered after at least one year of primary wound treatment therapy. As time goes by, hypertrophic scars tend to regress naturally or with conservative treatment such as steroid injections. Therefore, in many cases, there is no need to perform scar revision surgery. For keloids, surgical excision alone frequently results in disappointing outcomes. To improve postoperative surgical outcomes, multimodal combination therapy such as postoperative steroid application or radiotherapy might be added. When surgeons perform scar revision surgery, they should establish tension-free wound closure in order to decrease tension-related inflammation and thereby reduce recurrence. Various techniques including three-layered sutures, subcutaneous/fascial tensile reduction sutures, Z-plastics or local flap reconstruction can be utilized on a case-by-case basis [92,93]. Recurrence rates of hypertrophic scars after scar revision surgery are low, but the recurrence rate of keloids after scar revision surgery is 45% to 100% [94–96].

6.2.3. Cryotherapy

Cryotherapy has been used to treat hypertrophic scars or keloids as a monotherapy or in conjunction with other therapies such as intralesional steroid injections [97]. Treatments that combine cryotherapy and intralesional triamcinolone injections significantly improve hypertrophic scars and keloids [98–100]. Delivery methods for cryotherapy are variable and include sprays, contact or the intralesional-needle cryoprobe method. The intralesional-needle cryoprobe method shows better results than the spray or contact method, producing rapid re-epithelialization [101]. The suggested mechanism underlying cryotherapy is tissue necrosis induced by vascular damage. It seems that necrotized tissues induced by frostbite (as opposed to burn injury) secrete unique inflammatory cytokines; therefore, the responses of fibroblasts may differ [2]. Cryotherapy success rates range from 32 to 74% after several sessions [102–104].

6.2.4. Radiotherapy

Several studies have shown the effectiveness of radiotherapy on keloid treatment. Both external beam therapy and brachytherapy (or internal radiation therapy) have been used and studied for treatment of keloids. Radiotherapy is generally conducted as an adjuvant treatment 24 to 48 h after scar revision surgery, and the recommend radiation dose is 40 Gray over several divided sessions to minimize adverse effects [105]. The suggested mechanism of radiotherapy for treating keloids is anti-angiogenesis and successive anti-fibroblast activity. Suppression of angiogenesis decreases delivery of inflammatory cytokines, and successive inhibition of fibroblast activity results in decreased collagen synthesis, thus suppressing keloid development [106,107]. Radiotherapy carries an inherent risk of carcinogenesis; therefore, even though the risk is low [108,109], radiation-vulnerable areas, including the thyroid and breast, should be treated after achieving informed consent and with abundant cautions. Shen et al. reported the recurrence rate of 9.59% [109]. Recently, radioactive skin patches have been used for localized skin diseases like skin cancers or keloids [110,111]. Radioactive skin patches use various kinds of radionuclides and have variable effectiveness for treating keloids. These patches are frequently used in combination with other available treatment.

6.2.5. Laser Therapy

Laser therapy was introduced for keloid treatment in the 1980s [112], and several kinds of lasers with various wavelengths were investigated and reported. Among these, the most popular laser used to treat hypertrophic scars and keloids is the 585-nm pulsed dye laser (PDL) [113]. The recommended energy is 6.0 to 7.5 J/cm^2 (7-mm spot) or 4.5 to 5.5 J/cm^2 (10-mm spot) [114], and two to six sessions of treatment may be needed [113]. The 1064-nm Nd:YAG laser is another popular laser for treating hypertrophic scars and keloids. For this laser, the recommended energy is 14 J/cm^2 (5-mm spot), with the procedure being repeated every three to four weeks [115,116]. These laser treatments vaporize blood vessels. By doing this, inflammatory cytokines are limited in their ability to reach hypertrophic scars and keloids, thereby suppressing the development of aberrant scars. Possible side effects of laser therapy include hyperpigmentation, hypopigmentation, blister formation and postoperative purpura [117–120].

6.2.6. 5-Fluorouracil

5-FU is a medication mainly used to treat cancer. By injecting it into a vein, it can be used for the treatment of esophageal, stomach, pancreatic, colon, breast and cervical cancers. It can also be used topically for actinic keratosis and basal cell carcinoma in a cream or solution formulation [87]. 5-FU has also been used to treat keloids [121]. The suggested mechanism is anti-angiogenesis, anti-fibroblast proliferation and anti-collagen Type I expression induced by TGF-β [122–124]. This therapy is used solely or in combination with another treatment, and intralesional injection is the preferred method of delivery. Nanda et al. reported scar size reduction in a majority of patients in whom 5-FU was injected intralesionally weekly for 12 weeks in a concentration of 50 mg/mL [122]. Possible side effects include pain and ulceration. A systematic review reported 45% to 96% of effectiveness [125].

6.3. Emerging Therapies

6.3.1. Mesenchymal Stem Cell Therapy

Mesenchymal stem cells (MSCs) have immunomodulatory and antifibrotic effects by secreting paracrine growth factors [126–129]. The antifibrotic effects of MSC on fibrotic diseases such as myocardial infarctions, renal fibrosis or liver cirrhosis have been investigated and reported [130–136]. MSCs are also used to prevent or attenuate excessive inflammatory processes that are characteristic of hypertrophic scars and keloids. MSC treatments have variable delivery methods and doses [137]. Delivery is conducted via systemic injections, local injections (at the wound, intradermal or subcutaneously) or via an engineered MSC-seeded tissue scaffold [138–141]. The possible mechanisms underlying MSC treatment include: (1) modulation and inhibition of proinflammatory cell activity; (2) antifibrotic activity via downregulation of myofibroblast differentiation and collagen type I and III production; and (3) promotion of normal angiogenetic activity that aids in normal wound healing [137,142]. Even though many researchers have reported anti-inflammatory and anti-fibrotic effects of MSC, there are reports of possible proinflammatory actions of MSC [143–145]. More investigations and long-term preclinical studies should be conducted to apply this method to clinical practice.

6.3.2. Fat Grafting

Autologous fat grafting or lipotransfer, underneath or into the wound, has been performed for patients with hypertrophic scars or keloids. Several studies have reported the effectiveness of fat grafting on severely-scarred lesions [146–148]. These reports showed beneficial effects on excessive scar lesions, and side effects were rarely reported. The mechanism underlying fat injections is believed to be that transferred fat tissues deliver adipose-tissue derived MSCs to the wound.

6.3.3. Interferon

Interferon (IFN) is comprised of cytokines that have anti-proliferative and anti-fibrotic effects. As mentioned earlier, IFNs attenuate collagen synthesis and fibroblast proliferation by downregulating TGF-β1. Although adverse effects including pain at the injection site and flu-like symptoms are relatively common in IFN treatment, some authors reported a good outcome of combination therapy of IFN α-2b with TAC injection [149,150].

6.3.4. Transforming Growth Factor-β

TGF-β isoforms (TGF-β1,2,3) had long been a target of anti-keloid therapy. Several studies showed that the ratio of TGF-β3 and TFG-β1 and 2 is important in scar progression or remission [151,152]. Many studies had been performed to investigate the effect of exogenous TGF-β1 and 2 neutralizing antibodies and exogenous TFG-β3 and had proven the effect of TGF-β isoforms; TGF-β1 and 2 increase fibrosis, and TGF-β3 attenuates fibrosis [153]. Recombinant human TGF-β3, avotermin (planned trade name Juvista) showed successful results in phase I/II clinical trials [154–156], but failed in phase III clinical trials.

6.3.5. Botulinum Toxin A

Botulinum toxin, which is derived from *Clostridium botulinum*, is a potent neurotoxin that blocks neuromuscular transmission. Some authors have reported that botulinum toxin type A can minimize scar formation by reducing muscle tension during wound healing, causing the fibroblast cell cycle to be paused in a non-proliferative state, G0 or G1, and influencing TGF-β1 expression [157–161]. Intralesional injection was the preferred delivery method, and 70–140 U of Type A botulinum toxin was delivered per sessions at one- or three-month intervals for three or nine months (three sessions) [160,162–164]. Treatment outcomes were generally favorable, and patient satisfaction was high. Improvement was also reported regarding pain, tenderness and itching sensation [160,162,163].

6.3.6. Bleomycin

Bleomycin is a cytotoxic, antineoplastic, antiviral and antibacterial agent [165], derived from *Streptomyces verticillus*, and has been used for dermatologic diseases such as warts. This agent has also been used for hypertrophic scars and keloids. Several studies have found that bleomycin-treated human dermal fibroblasts showed diminished collagen synthesis, even with the co-existence of TFG-β1, and a reduction in the levels of lysyl-oxidase, which is involved in the maturation of collagen. In addition, apoptosis was also induced by bleomycin treatment [166–169]. Intralesional injection is the preferred delivery method, and 1.5 IU/mL of bleomycin were injected two to six sessions at monthly intervals. Several studies reported that complete flattening was achieved in 54% to 73% of keloid patients [166,167] and other symptoms like itching and pain were also resolved. Possible side effects include injection site pain, ulceration, atrophy and hyperpigmentation, but systemic side effects were not observed [165,167].

7. Conclusions

Hypertrophic scars and keloids result from abnormal wound healing. Excessive ECM deposition is characteristic of these lesions. Increased inflammatory and proliferative processes and decreased remodeling processes cause excessive ECM deposition. Genetic and systemic factors are also related to these excessively scarring lesions. Although encouraging results of molecular- or cytokine-targeting therapies are being continuously reported, current prophylaxis and treatment strategies still mainly focus on decreasing inflammatory processes. Further understanding of the mechanisms underlying excessive scarring is needed to develop more effective prophylaxis and treatment strategies.

Acknowledgments: No funding was received for this study.

Author Contributions: Ho Jun Lee: conception of the work, acquisition and analysis of data, drafting the work, approved the submitted version, agreed to be personally accountable for the author's own contributions and for ensuring that questions related to the accuracy or integrity of any part of the work; Yong Ju Jang: conception of the work, substantively revised the draft, approved the submitted version, agreed to be personally accountable for the author's own contributions and for ensuring that questions related to the accuracy or integrity of any part of the work.

Conflicts of Interest: The authors declare no conflict of interest.

References

1. Sund, B. *New Development in Wound Care*; PJB Publications: London, UK, 2000; pp. 1–255.
2. Ogawa, R. Keloid and Hypertrophic Scars Are the Result of Chronic Inflammation in the Reticular Dermis. *Int. J. Mol. Sci.* **2017**, *18*, 606. [CrossRef] [PubMed]
3. Chiang, R.S.; Borovikova, A.A.; King, K.; Banyard, D.A.; Lalezari, S.; Toranto, J.D.; Paydar, K.Z.; Wirth, G.A.; Evans, G.R.; Widgerow, A.D. Current concepts related to hypertrophic scarring in burn injuries. *Wound Repair Regen.* **2016**, *24*, 466–477. [CrossRef] [PubMed]
4. Tredget, E.E.; Nedelec, B.; Scott, P.G.; Ghahary, A. Hypertrophic scars, keloids, and contractures. The cellular and molecular basis for therapy. *Surg. Clin. N. Am.* **1997**, *77*, 701–730. [CrossRef]
5. Imhof, B.A.; Jemelin, S.; Ballet, R.; Vesin, C.; Schapira, M.; Karaca, M.; Emre, Y. CCN1/CYR61-mediated meticulous patrolling by Ly6Clow monocytes fuels vascular inflammation. *Proc. Natl. Acad. Sci. USA* **2016**, *113*, E4847–E4856. [CrossRef] [PubMed]
6. Grose, R.; Werner, S. Wound-healing studies in transgenic and knockout mice. *Mol. Biotechnol.* **2004**, *28*, 147–166. [CrossRef]
7. Werner, S.; Krieg, T.; Smola, H. Keratinocyte-fibroblast interactions in wound healing. *J. Investig. Dermatol.* **2007**, *127*, 998–1008. [CrossRef] [PubMed]
8. Zhu, Z.; Ding, J.; Tredget, E.E. The molecular basis of hypertrophic scars. *Burns Trauma* **2016**, *4*, 2. [CrossRef] [PubMed]
9. Wang, P.; Jiang, L.Z.; Xue, B. Recombinant human endostatin reduces hypertrophic scar formation in rabbit ear model through down-regulation of VEGF and TIMP-1. *Afr. Health Sci.* **2016**, *16*, 542–553. [CrossRef] [PubMed]
10. Tuan, T.L.; Nichter, L.S. The molecular basis of keloid and hypertrophic scar formation. *Mol. Med. Today* **1998**, *4*, 19–24. [CrossRef]
11. Ishihara, H.; Yoshimoto, H.; Fujioka, M.; Murakami, R.; Hirano, A.; Fujii, T.; Ohtsuru, A.; Namba, H.; Yamashita, S. Keloid fibroblasts resist ceramide-induced apoptosis by overexpression of insulin-like growth factor I receptor. *J. Investig. Dermatol.* **2000**, *115*, 1065–1071. [CrossRef] [PubMed]
12. Butler, P.D.; Longaker, M.T.; Yang, G.P. Current progress in keloid research and treatment. *J. Am. Coll. Surg.* **2008**, *206*, 731–741. [CrossRef] [PubMed]
13. Ladak, A.; Tredget, E.E. Pathophysiology and management of the burn scar. *Clin. Plast. Surg.* **2009**, *36*, 661–674. [CrossRef] [PubMed]
14. Chen, X.; Thibeault, S.L. Role of tumor necrosis factor-α in wound repair in human vocal fold fibroblasts. *Laryngoscope* **2010**, *120*, 1819–1825. [CrossRef] [PubMed]
15. Mariani, T.J.; Sandefur, S.; Roby, J.D.; Pierce, R.A. Collagenase-3 induction in rat lung fibroblasts requires the combined effects of tumor necrosis factor-α and 12-lipoxygenase metabolites: A model of macrophage-induced, fibroblast driven extracellular matrix remodeling during inflammatory lung injury. *Mol. Biol. Cell* **1998**, *9*, 1411–1424. [CrossRef] [PubMed]
16. Yan, C.; Grimm, W.A.; Garner, W.L.; Qin, L.; Travis, T.; Tan, N.; Han, Y.P. Epithelial to mesenchymal transition in human skin wound healing is induced by tumor necrosis factor-α through bone morphogenic protein-2. *Am. J. Pathol.* **2010**, *176*, 2247–2258. [CrossRef] [PubMed]
17. Armour, A.; Scott, P.G.; Tredget, E.E. Cellular and molecular pathology of HTS: Basis for treatment. *Wound Repair Regen.* **2007**, *15* (Suppl. S1), S6–S17. [CrossRef] [PubMed]
18. Doucet, C.; Brouty-Boye, D.; Pottin-Clemenceau, C.; Canonica, G.W.; Jasmin, C.; Azzarone, B. Interleukin (IL) 4 and IL-13 act on human lung fibroblasts. Implication in asthma. *J. Clin. Investig.* **1998**, *101*, 2129–2139. [CrossRef] [PubMed]

19. Wynn, T.A. Fibrotic disease and the Th1/Th2 paradigm. *Nat. Rev. Immunol.* **2004**, *4*, 583–594. [CrossRef] [PubMed]

20. Van den Broek, L.J.; van der Veer, W.M.; de Jong, E.H.; Gibbs, S.; Niessen, F.B. Suppressed inflammatory gene expression during human hypertrophic scar compared to normotrophic scar formation. *Exp. Dermatol.* **2015**, *24*, 623–629. [CrossRef] [PubMed]

21. Namazi, M.R.; Fallahzadeh, M.K.; Schwartz, R.A. Strategies for prevention of scars: What can we learn from fetal skin? *Int. J. Dermatol.* **2011**, *50*, 85–93. [CrossRef] [PubMed]

22. Liechty, K.W.; Kim, H.B.; Adzick, N.S.; Crombleholme, T.M. Fetal wound repair results in scar formation in interleukin-10-deficient mice in a syngeneic murine model of scarless fetal wound repair. *J. Pediatr. Surg.* **2000**, *35*, 866–872. [CrossRef] [PubMed]

23. Bock, O.; Yu, H.; Zitron, S.; Bayat, A.; Ferguson, M.W.; Mrowietz, U. Studies of transforming growth factors β 1–3 and their receptors I and II in fibroblast of keloids and hypertrophic scars. *Acta Derm. Venereol.* **2005**, *85*, 216–220. [CrossRef] [PubMed]

24. Lee, T.Y.; Chin, G.S.; Kim, W.J.; Chau, D.; Gittes, G.K.; Longaker, M.T. Expression of transforming growth factor β 1, 2, and 3 proteins in keloids. *Ann. Plast. Surg.* **1999**, *43*, 179–184. [CrossRef] [PubMed]

25. Xia, W.; Phan, T.T.; Lim, I.J.; Longaker, M.T.; Yang, G.P. Complex epithelial-mesenchymal interactions modulate transforming growth factor-β expression in keloid-derived cells. *Wound Repair Regen.* **2004**, *12*, 546–556. [CrossRef] [PubMed]

26. Lu, L.; Saulis, A.S.; Liu, W.R.; Roy, N.K.; Chao, J.D.; Ledbetter, S.; Mustoe, T.A. The temporal effects of anti-TGF-β1, 2, and 3 monoclonal antibody on wound healing and hypertrophic scar formation. *J. Am. Coll. Surg.* **2005**, *201*, 391–397. [CrossRef] [PubMed]

27. Fujiwara, M.; Muragaki, Y.; Ooshima, A. Keloid-derived fibroblasts show increased secretion of factors involved in collagen turnover and depend on matrix metalloproteinase for migration. *Br. J. Dermatol.* **2005**, *153*, 295–300. [CrossRef] [PubMed]

28. Ghahary, A.; Ghaffari, A. Role of keratinocyte-fibroblast cross-talk in development of hypertrophic scar. *Wound Repair Regen.* **2007**, *15* (Suppl. S1), S46–S53. [CrossRef] [PubMed]

29. Mauviel, A. Cytokine regulation of metalloproteinase gene expression. *J. Cell. Biochem.* **1993**, *53*, 288–295. [CrossRef] [PubMed]

30. Zhang, Y.; McCluskey, K.; Fujii, K.; Wahl, L.M. Differential regulation of monocyte matrix metalloproteinase and TIMP-1 production by TNF-α, granulocyte-macrophage CSF, and IL-1 β through prostaglandin-dependent and -independent mechanisms. *J. Immunol.* **1998**, *161*, 3071–3076. [PubMed]

31. McQuibban, G.A.; Gong, J.H.; Tam, E.M.; McCulloch, C.A.; Clark-Lewis, I.; Overall, C.M. Inflammation dampened by gelatinase A cleavage of monocyte chemoattractant protein-3. *Science* **2000**, *289*, 1202–1206. [CrossRef] [PubMed]

32. McQuibban, G.A.; Gong, J.H.; Wong, J.P.; Wallace, J.L.; Clark-Lewis, I.; Overall, C.M. Matrix metalloproteinase processing of monocyte chemoattractant proteins generates CC chemokine receptor antagonists with anti-inflammatory properties in vivo. *Blood* **2002**, *100*, 1160–1167. [PubMed]

33. Rohani, M.G.; Parks, W.C. Matrix remodeling by MMPs during wound repair. *Matrix Biol.* **2015**, *44*, 113–121. [CrossRef] [PubMed]

34. Krumdieck, R.; Hook, M.; Rosenberg, L.C.; Volanakis, J.E. The proteoglycan decorin binds C1q and inhibits the activity of the C1 complex. *J. Immunol.* **1992**, *149*, 3695–3701. [PubMed]

35. Scott, P.G.; Dodd, C.M.; Tredget, E.E.; Ghahary, A.; Rahemtulla, F. Chemical characterization and quantification of proteoglycans in human post-burn hypertrophic and mature scars. *Clin. Sci.* **1996**, *90*, 417–425. [CrossRef] [PubMed]

36. Jarvelainen, H.; Sainio, A.; Wight, T.N. Pivotal role for decorin in angiogenesis. *Matrix Biol.* **2015**, *43*, 15–26. [CrossRef] [PubMed]

37. Zhang, Z.; Garron, T.M.; Li, X.J.; Liu, Y.; Zhang, X.; Li, Y.Y.; Xu, W.S. Recombinant human decorin inhibits TGF-β1-induced contraction of collagen lattice by hypertrophic scar fibroblasts. *Burns* **2009**, *35*, 527–537. [CrossRef] [PubMed]

38. Mukhopadhyay, A.; Wong, M.Y.; Chan, S.Y.; Do, D.V.; Khoo, A.; Ong, C.T.; Cheong, H.H.; Lim, I.J.; Phan, T.T. Syndecan-2 and decorin: Proteoglycans with a difference—Implications in keloid pathogenesis. *J. Trauma* **2010**, *68*, 999–1008. [CrossRef] [PubMed]

39. Sen, C.K.; Roy, S. Oxygenation state as a driver of myofibroblast differentiation and wound contraction: Hypoxia impairs wound closure. *J. Investig. Dermatol.* **2010**, *130*, 2701–2703. [CrossRef] [PubMed]

40. Nauta, T.D.; van Hinsbergh, V.W.; Koolwijk, P. Hypoxic signaling during tissue repair and regenerative medicine. *Int. J. Mol. Sci.* **2014**, *15*, 19791–19815. [CrossRef] [PubMed]

41. Ueda, K.; Yasuda, Y.; Furuya, E.; Oba, S. Inadequate blood supply persists in keloids. *Scand. J. Plast. Reconstr. Surg. Hand Surg.* **2004**, *38*, 267–271. [CrossRef] [PubMed]

42. Steinbrech, D.S.; Mehrara, B.J.; Chau, D.; Rowe, N.M.; Chin, G.; Lee, T.; Saadeh, P.B.; Gittes, G.K.; Longaker, M.T. Hypoxia upregulates VEGF production in keloid fibroblasts. *Ann. Plast. Surg.* **1999**, *42*, 514–519. [CrossRef] [PubMed]

43. Zhao, B.; Guan, H.; Liu, J.Q.; Zheng, Z.; Zhou, Q.; Zhang, J.; Su, L.L.; Hu, D.H. Hypoxia drives the transition of human dermal fibroblasts to a myofibroblast-like phenotype via the TGF-β1/Smad3 pathway. *Int. J. Mol. Med.* **2017**, *39*, 153–159. [CrossRef] [PubMed]

44. Horiuchi, K.; Amizuka, N.; Takeshita, S.; Takamatsu, H.; Katsuura, M.; Ozawa, H.; Toyama, Y.; Bonewald, L.F.; Kudo, A. Identification and characterization of a novel protein, periostin, with restricted expression to periosteum and periodontal ligament and increased expression by transforming growth factor β. *J. Bone Miner. Res.* **1999**, *14*, 1239–1249. [CrossRef] [PubMed]

45. Zhou, H.M.; Wang, J.; Elliott, C.; Wen, W.; Hamilton, D.W.; Conway, S.J. Spatiotemporal expression of periostin during skin development and incisional wound healing: Lessons for human fibrotic scar formation. *J. Cell Commun. Signal.* **2010**, *4*, 99–107. [CrossRef] [PubMed]

46. Crawford, J.; Nygard, K.; Gan, B.S.; O'Gorman, D.B. Periostin induces fibroblast proliferation and myofibroblast persistence in hypertrophic scarring. *Exp. Dermatol.* **2015**, *24*, 120–126. [CrossRef] [PubMed]

47. Elliott, C.G.; Wang, J.; Guo, X.; Xu, S.W.; Eastwood, M.; Guan, J.; Leask, A.; Conway, S.J.; Hamilton, D.W. Periostin modulates myofibroblast differentiation during full-thickness cutaneous wound repair. *J. Cell Sci.* **2012**, *125*, 121–132. [CrossRef] [PubMed]

48. Conway, S.J.; Izuhara, K.; Kudo, Y.; Litvin, J.; Markwald, R.; Ouyang, G.; Arron, J.R.; Holweg, C.T.; Kudo, A. The role of periostin in tissue remodeling across health and disease. *Cell. Mol. Life Sci.* **2014**, *71*, 1279–1288. [CrossRef] [PubMed]

49. Jackson-Boeters, L.; Wen, W.; Hamilton, D.W. Periostin localizes to cells in normal skin, but is associated with the extracellular matrix during wound repair. *J. Cell Commun. Signal.* **2009**, *3*, 125–133. [CrossRef] [PubMed]

50. Zhang, Z.; Nie, F.; Kang, C.; Chen, B.; Qin, Z.; Ma, J.; Ma, Y.; Zhao, X. Increased periostin expression affects the proliferation, collagen synthesis, migration and invasion of keloid fibroblasts under hypoxic conditions. *Int. J. Mol. Med.* **2014**, *34*, 253–261. [CrossRef] [PubMed]

51. Zhang, Z.; Nie, F.; Chen, X.; Qin, Z.; Kang, C.; Chen, B.; Ma, J.; Pan, B.; Ma, Y. Upregulated periostin promotes angiogenesis in keloids through activation of the ERK 1/2 and focal adhesion kinase pathways, as well as the upregulated expression of VEGF and angiopoietin1. *Mol. Med. Rep.* **2015**, *11*, 857–864. [CrossRef] [PubMed]

52. Kashiyama, K.; Mitsutake, N.; Matsuse, M.; Ogi, T.; Saenko, V.A.; Ujifuku, K.; Utani, A.; Hirano, A.; Yamashita, S. miR-196a downregulation increases the expression of type I and III collagens in keloid fibroblasts. *J. Investig. Dermatol.* **2012**, *132*, 1597–1604. [CrossRef] [PubMed]

53. Liu, Y.; Yang, D.; Xiao, Z.; Zhang, M. miRNA expression profiles in keloid tissue and corresponding normal skin tissue. *Aesthet. Plast. Surg.* **2012**, *36*, 193–201. [CrossRef] [PubMed]

54. Lovendorf, M.B.; Skov, L. miRNAs in inflammatory skin diseases and their clinical implications. *Expert Rev. Clin. Immunol.* **2015**, *11*, 467–477. [CrossRef] [PubMed]

55. Luan, Y.; Liu, Y.; Liu, C.; Lin, Q.; He, F.; Dong, X.; Xiao, Z. Serum miRNAs Signature Plays an Important Role in Keloid Disease. *Curr. Mol. Med.* **2016**, *16*, 504–514. [CrossRef] [PubMed]

56. Yao, X.; Cui, X.; Wu, X.; Xu, P.; Zhu, W.; Chen, X.; Zhao, T. Tumor suppressive role of miR-1224-5p in keloid proliferation, apoptosis and invasion via the TGF-β1/Smad3 signaling pathway. *Biochem. Biophys. Res. Commun.* **2017**, *495*, 713–720. [CrossRef] [PubMed]

57. Wu, Z.Y.; Lu, L.; Liang, J.; Guo, X.R.; Zhang, P.H.; Luo, S.J. Keloid microRNA expression analysis and the influence of miR-199a-5p on the proliferation of keloid fibroblasts. *Genet. Mol. Res.* **2014**, *13*, 2727–2738. [CrossRef] [PubMed]

58. Liu, Y.; Wang, X.; Yang, D.; Xiao, Z.; Chen, X. MicroRNA-21 affects proliferation and apoptosis by regulating expression of PTEN in human keloid fibroblasts. *Plast. Reconstr. Surg.* **2014**, *134*, 561e–573e. [CrossRef] [PubMed]
59. Li, Y.; Zhang, J.; Lei, Y.; Lyu, L.; Zuo, R.; Chen, T. MicroRNA-21 in Skin Fibrosis: Potential for Diagnosis and Treatment. *Mol. Diagn. Ther.* **2017**, *21*, 633–642. [CrossRef] [PubMed]
60. Zhou, R.; Wang, C.; Wen, C.; Wang, D. miR-21 promotes collagen production in keloid via Smad7. *Burns* **2017**, *43*, 555–561. [CrossRef] [PubMed]
61. Liu, Y.; Li, Y.; Li, N.; Teng, W.; Wang, M.; Zhang, Y.; Xiao, Z. TGF-β1 promotes scar fibroblasts proliferation and transdifferentiation via up-regulating microRNA-21. *Sci. Rep.* **2016**, *6*, 32231. [CrossRef] [PubMed]
62. Liu, Z.; Lu, C.L.; Cui, L.P.; Hu, Y.L.; Yu, Q.; Jiang, Y.; Ma, T.; Jiao, D.K.; Wang, D.; Jia, C.Y. MicroRNA-146a modulates TGF-β1-induced phenotypic differentiation in human dermal fibroblasts by targeting SMAD4. *Arch. Dermatol. Res.* **2012**, *304*, 195–202. [CrossRef] [PubMed]
63. Zhang, J.; Xu, D.; Li, N.; Li, Y.; He, Y.; Hu, X.; Lyu, L.; He, L. Downregulation of microRNA-31 inhibits proliferation and induces apoptosis by targeting HIF1AN in human keloid. *Oncotarget* **2017**, *8*, 74623–74634. [CrossRef] [PubMed]
64. Bond, J.S.; Duncan, J.A.; Mason, T.; Sattar, A.; Boanas, A.; O'Kane, S.; Ferguson, M.W. Scar redness in humans: How long does it persist after incisional and excisional wounding? *Plast. Reconstr. Surg.* **2008**, *121*, 487–496. [CrossRef] [PubMed]
65. Mutalik, S. Treatment of keloids and hypertrophic scars. *Indian J. Dermatol. Venereol. Leprol.* **2005**, *71*, 3–8. [CrossRef] [PubMed]
66. Wong, V.W.; Akaishi, S.; Longaker, M.T.; Gurtner, G.C. Pushing back: Wound mechanotransduction in repair and regeneration. *J. Investig. Dermatol.* **2011**, *131*, 2186–2196. [CrossRef] [PubMed]
67. Gurtner, G.C.; Dauskardt, R.H.; Wong, V.W.; Bhatt, K.A.; Wu, K.; Vial, I.N.; Padois, K.; Korman, J.M.; Longaker, M.T. Improving cutaneous scar formation by controlling the mechanical environment: Large animal and phase I studies. *Ann. Surg.* **2011**, *254*, 217–225. [CrossRef] [PubMed]
68. Atkinson, J.A.; McKenna, K.T.; Barnett, A.G.; McGrath, D.J.; Rudd, M. A randomized, controlled trial to determine the efficacy of paper tape in preventing hypertrophic scar formation in surgical incisions that traverse Langer's skin tension lines. *Plast. Reconstr. Surg.* **2005**, *116*, 1648–1656. [CrossRef] [PubMed]
69. Daya, M.; Nair, V. Traction-assisted dermatogenesis by serial intermittent skin tape application. *Plast. Reconstr. Surg.* **2008**, *122*, 1047–1054. [CrossRef] [PubMed]
70. Fulton, J.E., Jr. Silicone gel sheeting for the prevention and management of evolving hypertrophic and keloid scars. *Dermatol. Surg.* **1995**, *21*, 947–951. [CrossRef] [PubMed]
71. Sawada, Y.; Sone, K. Hydration and occlusion treatment for hypertrophic scars and keloids. *Br. J. Plast. Surg.* **1992**, *45*, 599–603. [CrossRef]
72. Reish, R.G.; Eriksson, E. Scar treatments: Preclinical and clinical studies. *J. Am. Coll. Surg.* **2008**, *206*, 719–730. [CrossRef] [PubMed]
73. Beuth, J.; Hunzelmann, N.; van Leendert, R.; Basten, R.; Noehle, M.; Schneider, B. Safety and efficacy of local administration of contractubex to hypertrophic scars in comparison to corticosteroid treatment. Results of a multicenter, comparative epidemiological cohort study in Germany. *In Vivo* **2006**, *20*, 277–283. [PubMed]
74. Chung, V.Q.; Kelley, L.; Marra, D.; Jiang, S.B. Onion extract gel versus petrolatum emollient on new surgical scars: Prospective double-blinded study. *Dermatol. Surg.* **2006**, *32*, 193–197. [CrossRef] [PubMed]
75. Ho, W.S.; Ying, S.Y.; Chan, P.C.; Chan, H.H. Use of onion extract, heparin, allantoin gel in prevention of scarring in chinese patients having laser removal of tattoos: A prospective randomized controlled trial. *Dermatol. Surg.* **2006**, *32*, 891–896. [CrossRef] [PubMed]
76. Jackson, B.A.; Shelton, A.J. Pilot study evaluating topical onion extract as treatment for postsurgical scars. *Dermatol. Surg.* **1999**, *25*, 267–269. [CrossRef] [PubMed]
77. Phan, T.T.; Lim, I.J.; Sun, L.; Chan, S.Y.; Bay, B.H.; Tan, E.K.; Lee, S.T. Quercetin inhibits fibronectin production by keloid-derived fibroblasts. Implication for the treatment of excessive scars. *J. Dermatol. Sci.* **2003**, *33*, 192–194. [CrossRef] [PubMed]
78. Cho, J.W.; Cho, S.Y.; Lee, S.R.; Lee, K.S. Onion extract and quercetin induce matrix metalloproteinase-1 in vitro and in vivo. *Int. J. Mol. Med.* **2010**, *25*, 347–352. [PubMed]
79. Atiyeh, B.S. Nonsurgical management of hypertrophic scars: Evidence-based therapies, standard practices, and emerging methods. *Aesthet. Plast. Surg.* **2007**, *31*, 468–492. [CrossRef] [PubMed]

80. Baur, P.S.; Larson, D.L.; Stacey, T.R.; Barratt, G.F.; Dobrkovsky, M. Ultrastructural analysis of pressure-treated human hypertrophic scars. *J. Trauma* **1976**, *16*, 958–967. [CrossRef] [PubMed]
81. Kelly, A.P. Medical and surgical therapies for keloids. *Dermatol. Ther.* **2004**, *17*, 212–218. [CrossRef] [PubMed]
82. Macintyre, L.; Baird, M. Pressure garments for use in the treatment of hypertrophic scars—An evaluation of current construction techniques in NHS hospitals. *Burns* **2005**, *31*, 11–14. [CrossRef] [PubMed]
83. Macintyre, L.; Baird, M. Pressure garments for use in the treatment of hypertrophic scars—A review of the problems associated with their use. *Burns* **2006**, *32*, 10–15. [CrossRef] [PubMed]
84. Macintyre, L.; Ferguson, R. Pressure garment design tool to monitor exerted pressures. *Burns* **2013**, *39*, 1073–1082. [CrossRef] [PubMed]
85. Reno, F.; Sabbatini, M.; Lombardi, F.; Stella, M.; Pezzuto, C.; Magliacani, G.; Cannas, M. In vitro mechanical compression induces apoptosis and regulates cytokines release in hypertrophic scars. *Wound Repair Regen.* **2003**, *11*, 331–336. [CrossRef] [PubMed]
86. Van den Kerckhove, E.; Stappaerts, K.; Fieuws, S.; Laperre, J.; Massage, P.; Flour, M.; Boeckx, W. The assessment of erythema and thickness on burn related scars during pressure garment therapy as a preventive measure for hypertrophic scarring. *Burns* **2005**, *31*, 696–702. [CrossRef] [PubMed]
87. Rauscher, G.E.; Kolmer, W.L. Treatment of recurrent earlobe keloids. *Cutis* **1986**, *37*, 67–68. [PubMed]
88. Boyadjiev, C.; Popchristova, E.; Mazgalova, J. Histomorphologic changes in keloids treated with Kenacort. *J. Trauma* **1995**, *38*, 299–302. [CrossRef] [PubMed]
89. Cruz, N.I.; Korchin, L. Inhibition of human keloid fibroblast growth by isotretinoin and triamcinolone acetonide in vitro. *Ann. Plast. Surg.* **1994**, *33*, 401–405. [CrossRef] [PubMed]
90. Robles, D.T.; Berg, D. Abnormal wound healing: Keloids. *Clin. Dermatol.* **2007**, *25*, 26–32. [CrossRef] [PubMed]
91. Wong, T.S.; Li, J.Z.; Chen, S.; Chan, J.Y.; Gao, W. The Efficacy of Triamcinolone Acetonide in Keloid Treatment: A Systematic Review and Meta-analysis. *Front. Med.* **2016**, *3*, 71. [CrossRef] [PubMed]
92. Ogawa, R.; Akaishi, S.; Huang, C.; Dohi, T.; Aoki, M.; Omori, Y.; Koike, S.; Kobe, K.; Akimoto, M.; Hyakusoku, H. Clinical applications of basic research that shows reducing skin tension could prevent and treat abnormal scarring: The importance of fascial/subcutaneous tensile reduction sutures and flap surgery for keloid and hypertrophic scar reconstruction. *J. Nippon Med. Sch.* **2011**, *78*, 68–76. [CrossRef] [PubMed]
93. Ogawa, R.; Akaishi, S.; Kuribayashi, S.; Miyashita, T. Keloids and Hypertrophic Scars Can Now Be Cured Completely: Recent Progress in Our Understanding of the Pathogenesis of Keloids and Hypertrophic Scars and the Most Promising Current Therapeutic Strategy. *J. Nippon Med. Sch.* **2016**, *83*, 46–53. [CrossRef] [PubMed]
94. Leventhal, D.; Furr, M.; Reiter, D. Treatment of keloids and hypertrophic scars: A meta-analysis and review of the literature. *Arch. Facial Plast. Surg.* **2006**, *8*, 362–368. [CrossRef] [PubMed]
95. Muir, I.F. On the nature of keloid and hypertrophic scars. *Br. J. Plast. Surg.* **1990**, *43*, 61–69. [CrossRef]
96. Mustoe, T.A.; Cooter, R.D.; Gold, M.H.; Hobbs, F.D.; Ramelet, A.A.; Shakespeare, P.G.; Stella, M.; Teot, L.; Wood, F.M.; Ziegler, U.E.; et al. International clinical recommendations on scar management. *Plast. Reconstr. Surg.* **2002**, *110*, 560–571. [CrossRef] [PubMed]
97. Har-Shai, Y.; Zouboulis, C.C. Intralesional Cryotherapy for the Treatment of Keloid Scars: A Prospective Study. *Plast. Reconstr. Surg.* **2015**, *136*, 397e–398e. [CrossRef] [PubMed]
98. Boutli-Kasapidou, F.; Tsakiri, A.; Anagnostou, E.; Mourellou, O. Hypertrophic and keloidal scars: An approach to polytherapy. *Int. J. Dermatol.* **2005**, *44*, 324–327. [CrossRef] [PubMed]
99. Jaros, E.; Priborsky, J.; Klein, L. Treatment of keloids and hypertrophic scars with cryotherapy. *Acta Med.* **1999**, *42*, 61–63.
100. Yosipovitch, G.; Widijanti Sugeng, M.; Goon, A.; Chan, Y.H.; Goh, C.L. A comparison of the combined effect of cryotherapy and corticosteroid injections versus corticosteroids and cryotherapy alone on keloids: A controlled study. *J. Dermatol. Treat.* **2001**, *12*, 87–90. [CrossRef] [PubMed]
101. Har-Shai, Y.; Amar, M.; Sabo, E. Intralesional cryotherapy for enhancing the involution of hypertrophic scars and keloids. *Plast. Reconstr. Surg.* **2003**, *111*, 1841–1852. [CrossRef] [PubMed]
102. Rusciani, L.; Paradisi, A.; Alfano, C.; Chiummariello, S.; Rusciani, A. Cryotherapy in the treatment of keloids. *J. Drugs Dermatol.* **2006**, *5*, 591–595. [CrossRef] [PubMed]

103. Rusciani, L.; Rossi, G.; Bono, R. Use of cryotherapy in the treatment of keloids. *J. Dermatol. Surg. Oncol.* **1993**, *19*, 529–534. [CrossRef] [PubMed]

104. Zouboulis, C.C.; Blume, U.; Buttner, P.; Orfanos, C.E. Outcomes of cryosurgery in keloids and hypertrophic scars. A prospective consecutive trial of case series. *Arch. Dermatol.* **1993**, *129*, 1146–1151. [CrossRef] [PubMed]

105. Ogawa, R.; Mitsuhashi, K.; Hyakusoku, H.; Miyashita, T. Postoperative electron-beam irradiation therapy for keloids and hypertrophic scars: Retrospective study of 147 cases followed for more than 18 months. *Plast. Reconstr. Surg.* **2003**, *111*, 547–553. [CrossRef] [PubMed]

106. Ji, J.; Tian, Y.; Zhu, Y.Q.; Zhang, L.Y.; Ji, S.J.; Huan, J.; Zhou, X.Z.; Cao, J.P. Ionizing irradiation inhibits keloid fibroblast cell proliferation and induces premature cellular senescence. *J. Dermatol.* **2015**, *42*, 56–63. [CrossRef] [PubMed]

107. Keeling, B.H.; Whitsitt, J.; Liu, A.; Dunnick, C.A. Keloid removal by shave excision with adjuvant external beam radiation therapy. *Dermatol. Surg.* **2015**, *41*, 989–992. [CrossRef] [PubMed]

108. McKeown, S.R.; Hatfield, P.; Prestwich, R.J.; Shaffer, R.E.; Taylor, R.E. Radiotherapy for benign disease; assessing the risk of radiation-induced cancer following exposure to intermediate dose radiation. *Br. J. Radiol.* **2015**, *88*, 20150405. [CrossRef] [PubMed]

109. Shen, J.; Lian, X.; Sun, Y.; Wang, X.; Hu, K.; Hou, X.; Sun, S.; Yan, J.; Yu, L.; Sun, X.; et al. Hypofractionated electron-beam radiation therapy for keloids: Retrospective study of 568 cases with 834 lesions. *J. Radiat. Res.* **2015**, *56*, 811–817. [CrossRef] [PubMed]

110. Vivante, H.; Salgueiro, M.J.; Ughetti, R.; Nicolini, J.; Zubillaga, M. 32P-patch contact brachyradiotherapy in the management of recalcitrant keloids and hypertrophic scars. *Indian J. Dermatol. Venereol. Leprol.* **2007**, *73*, 336–339. [PubMed]

111. Bhusari, P.; Shukla, J.; Kumar, M.; Vatsa, R.; Chhabra, A.; Palarwar, K.; Rathore, Y.; De, D.; Kumaran, S.; Handa, S.; et al. Noninvasive treatment of keloid using customized Re-188 skin patch. *Dermatol. Ther.* **2017**, *30*. [CrossRef] [PubMed]

112. Apfelberg, D.B.; Maser, M.R.; Lash, H.; White, D.; Weston, J. Preliminary results of argon and carbon dioxide laser treatment of keloid scars. *Lasers Surg. Med.* **1984**, *4*, 283–290. [CrossRef] [PubMed]

113. Alster, T.S.; Handrick, C. Laser treatment of hypertrophic scars, keloids, and striae. *Semin. Cutan. Med. Surg.* **2000**, *19*, 287–292. [CrossRef] [PubMed]

114. Tanzi, E.L.; Alster, T.S. Laser treatment of scars. *Skin Ther. Lett.* **2004**, *9*, 4–7.

115. Koike, S.; Akaishi, S.; Nagashima, Y.; Dohi, T.; Hyakusoku, H.; Ogawa, R. Nd:YAG Laser Treatment for Keloids and Hypertrophic Scars: An Analysis of 102 Cases. *Plast. Reconstr. Surg. Glob. Open* **2014**, *2*, e272. [CrossRef] [PubMed]

116. Akaishi, S.; Koike, S.; Dohi, T.; Kobe, K.; Hyakusoku, H.; Ogawa, R. Nd:YAG Laser Treatment of Keloids and Hypertrophic Scars. *Eplasty* **2012**, *12*, e1. [PubMed]

117. Alster, T. Laser scar revision: Comparison study of 585-nm pulsed dye laser with and without intralesional corticosteroids. *Dermatol. Surg.* **2003**, *29*, 25–29. [CrossRef] [PubMed]

118. Chan, H.H.; Wong, D.S.; Ho, W.S.; Lam, L.K.; Wei, W. The use of pulsed dye laser for the prevention and treatment of hypertrophic scars in chinese persons. *Dermatol. Surg.* **2004**, *30*, 987–994. [PubMed]

119. Fiskerstrand, E.J.; Svaasand, L.O.; Volden, G. Pigmentary changes after pulsed dye laser treatment in 125 northern European patients with port wine stains. *Br. J. Dermatol.* **1998**, *138*, 477–479. [CrossRef] [PubMed]

120. Hermanns, J.F.; Petit, L.; Hermanns-Le, T.; Pierard, G.E. Analytic quantification of phototype-related regional skin complexion. *Skin Res. Technol.* **2001**, *7*, 168–171. [CrossRef] [PubMed]

121. Fitzpatrick, R.E. Treatment of inflamed hypertrophic scars using intralesional 5-FU. *Dermatol. Surg.* **1999**, *25*, 224–232. [CrossRef] [PubMed]

122. Nanda, S.; Reddy, B.S. Intralesional 5-fluorouracil as a treatment modality of keloids. *Dermatol. Surg.* **2004**, *30*, 54–56. [PubMed]

123. Khan, M.A.; Bashir, M.M.; Khan, F.A. Intralesional triamcinolone alone and in combination with 5-fluorouracil for the treatment of Keloid and Hypertrophic scars. *J. Pak. Med. Assoc.* **2014**, *64*, 1003–1007. [PubMed]

124. Darougheh, A.; Asilian, A.; Shariati, F. Intralesional triamcinolone alone or in combination with 5-fluorouracil for the treatment of keloid and hypertrophic scars. *Clin. Exp. Dermatol.* **2009**, *34*, 219–223. [CrossRef] [PubMed]

125. Bijlard, E.; Steltenpool, S.; Niessen, F.B. Intralesional 5-fluorouracil in keloid treatment: A systematic review. *Acta Derm. Venereol.* **2015**, *95*, 778–782. [CrossRef] [PubMed]

126. Le Blanc, K. Immunomodulatory effects of fetal and adult mesenchymal stem cells. *Cytotherapy* **2003**, *5*, 485–489. [CrossRef] [PubMed]

127. Le Blanc, K.; Mougiakakos, D. Multipotent mesenchymal stromal cells and the innate immune system. *Nat. Rev. Immunol.* **2012**, *12*, 383–396. [CrossRef] [PubMed]

128. Ortiz, L.A.; Gambelli, F.; McBride, C.; Gaupp, D.; Baddoo, M.; Kaminski, N.; Phinney, D.G. Mesenchymal stem cell engraftment in lung is enhanced in response to bleomycin exposure and ameliorates its fibrotic effects. *Proc. Natl. Acad. Sci. USA* **2003**, *100*, 8407–8411. [CrossRef] [PubMed]

129. Zhang, J.; Guan, J.; Niu, X.; Hu, G.; Guo, S.; Li, Q.; Xie, Z.; Zhang, C.; Wang, Y. Exosomes released from human induced pluripotent stem cells-derived MSCs facilitate cutaneous wound healing by promoting collagen synthesis and angiogenesis. *J. Transl. Med.* **2015**, *13*, 49. [CrossRef] [PubMed]

130. Lee, R.H.; Pulin, A.A.; Seo, M.J.; Kota, D.J.; Ylostalo, J.; Larson, B.L.; Semprun-Prieto, L.; Delafontaine, P.; Prockop, D.J. Intravenous hMSCs improve myocardial infarction in mice because cells embolized in lung are activated to secrete the anti-inflammatory protein TSG-6. *Cell Stem Cell* **2009**, *5*, 54–63. [CrossRef] [PubMed]

131. Lin, J.S.; Zhou, L.; Sagayaraj, A.; Jumat, N.H.; Choolani, M.; Chan, J.K.; Biswas, A.; Wong, P.C.; Lim, S.G.; Dan, Y.Y. Hepatic differentiation of human amniotic epithelial cells and in vivo therapeutic effect on animal model of cirrhosis. *J. Gastroenterol. Hepatol.* **2015**, *30*, 1673–1682. [CrossRef] [PubMed]

132. Prockop, D.J. Concise review: Two negative feedback loops place mesenchymal stem/stromal cells at the center of early regulators of inflammation. *Stem Cells* **2013**, *31*, 2042–2046. [CrossRef] [PubMed]

133. Prockop, D.J. Inflammation, fibrosis, and modulation of the process by mesenchymal stem/stromal cells. *Matrix Boil.* **2016**, *51*, 7–13. [CrossRef] [PubMed]

134. Prockop, D.J.; Oh, J.Y. Mesenchymal stem/stromal cells (MSCs): Role as guardians of inflammation. *Mol. Ther.* **2012**, *20*, 14–20. [CrossRef] [PubMed]

135. Reinders, M.E.; de Fijter, J.W.; Rabelink, T.J. Mesenchymal stromal cells to prevent fibrosis in kidney transplantation. *Curr. Opin. Organ Transplant.* **2014**, *19*, 54–59. [CrossRef] [PubMed]

136. Shafiq, M.; Lee, S.H.; Jung, Y.; Kim, S.H. Strategies for recruitment of stem cells to treat myocardial infarction. *Curr. Pharm. Des.* **2015**, *21*, 1584–1597. [CrossRef] [PubMed]

137. Seo, B.F.; Jung, S.N. The Immunomodulatory Effects of Mesenchymal Stem Cells in Prevention or Treatment of Excessive Scars. *Stem Cells Int.* **2016**, *2016*, 6937976. [CrossRef] [PubMed]

138. Altman, A.M.; Matthias, N.; Yan, Y.; Song, Y.H.; Bai, X.; Chiu, E.S.; Slakey, D.P.; Alt, E.U. Dermal matrix as a carrier for in vivo delivery of human adipose-derived stem cells. *Biomaterials* **2008**, *29*, 1431–1442. [CrossRef] [PubMed]

139. Huang, S.P.; Hsu, C.C.; Chang, S.C.; Wang, C.H.; Deng, S.C.; Dai, N.T.; Chen, T.M.; Chan, J.Y.; Chen, S.G.; Huang, S.M. Adipose-derived stem cells seeded on acellular dermal matrix grafts enhance wound healing in a murine model of a full-thickness defect. *Ann. Plast. Surg.* **2012**, *69*, 656–662. [CrossRef] [PubMed]

140. Lam, M.T.; Nauta, A.; Meyer, N.P.; Wu, J.C.; Longaker, M.T. Effective delivery of stem cells using an extracellular matrix patch results in increased cell survival and proliferation and reduced scarring in skin wound healing. *Tissue Eng. A* **2013**, *19*, 738–747. [CrossRef] [PubMed]

141. Zonari, A.; Martins, T.M.; Paula, A.C.; Boeloni, J.N.; Novikoff, S.; Marques, A.P.; Correlo, V.M.; Reis, R.L.; Goes, A.M. Polyhydroxybutyrate-co-hydroxyvalerate structures loaded with adipose stem cells promote skin healing with reduced scarring. *Acta Biomater.* **2015**, *17*, 170–181. [CrossRef] [PubMed]

142. Kaigler, D.; Krebsbach, P.H.; Polverini, P.J.; Mooney, D.J. Role of vascular endothelial growth factor in bone marrow stromal cell modulation of endothelial cells. *Tissue Eng.* **2003**, *9*, 95–103. [CrossRef] [PubMed]

143. Liotta, F.; Angeli, R.; Cosmi, L.; Fili, L.; Manuelli, C.; Frosali, F.; Mazzinghi, B.; Maggi, L.; Pasini, A.; Lisi, V.; et al. Toll-like receptors 3 and 4 are expressed by human bone marrow-derived mesenchymal stem cells and can inhibit their T-cell modulatory activity by impairing Notch signaling. *Stem Cells* **2008**, *26*, 279–289. [CrossRef] [PubMed]

144. Waterman, R.S.; Tomchuck, S.L.; Henkle, S.L.; Bncourt, A.M. A new mesenchymal stem cell (MSC) paradigm: Polarization into a pro-inflammatory MSC1 or an Immunosuppressive MSC2 phenotype. *PLoS ONE* **2010**, *5*, e10088. [CrossRef] [PubMed]

145. Yan, H.; Wu, M.; Yuan, Y.; Wang, Z.Z.; Jiang, H.; Chen, T. Priming of Toll-like receptor 4 pathway in mesenchymal stem cells increases expression of B cell activating factor. *Biochem. Biophys. Res. Commun.* **2014**, *448*, 212–217. [CrossRef] [PubMed]

146. Bruno, A.; Delli Santi, G.; Fasciani, L.; Cempanari, M.; Palombo, M.; Palombo, P. Burn scar lipofilling: Immunohistochemical and clinical outcomes. *J. Craniofac. Surg.* **2013**, *24*, 1806–1814. [CrossRef] [PubMed]

147. Negenborn, V.L.; Groen, J.W.; Smit, J.M.; Niessen, F.B.; Mullender, M.G. The Use of Autologous Fat Grafting for Treatment of Scar Tissue and Scar-Related Conditions: A Systematic Review. *Plast. Reconstr. Surg.* **2016**, *137*, 31e–43e. [CrossRef] [PubMed]

148. Piccolo, N.S.; Piccolo, M.S.; Piccolo, M.T. Fat grafting for treatment of burns, burn scars, and other difficult wounds. *Clin. Plast. Surg.* **2015**, *42*, 263–283. [CrossRef] [PubMed]

149. Al-Khawajah, M.M. Failure of interferon-α 2b in the treatment of mature keloids. *Int. J. Dermatol.* **1996**, *35*, 515–517. [CrossRef] [PubMed]

150. Lee, J.H.; Kim, S.E.; Lee, A.Y. Effects of interferon-α2b on keloid treatment with triamcinolone acetonide intralesional injection. *Int. J. Dermatol.* **2008**, *47*, 183–186. [CrossRef] [PubMed]

151. Schrementi, M.E.; Ferreira, A.M.; Zender, C.; DiPietro, L.A. Site-specific production of TGF-β in oral mucosal and cutaneous wounds. *Wound Repair Regen.* **2008**, *16*, 80–86. [CrossRef] [PubMed]

152. O'Kane, S.; Ferguson, M.W. Transforming growth factor βs and wound healing. *Int. J. Biochem. Cell Biol.* **1997**, *29*, 63–78. [CrossRef]

153. Shah, M.; Foreman, D.M.; Ferguson, M.W. Neutralisation of TGF-β1 and TGF-β2 or exogenous addition of TGF-β3 to cutaneous rat wounds reduces scarring. *J. Cell Sci.* **1995**, *108 Pt 3*, 985–1002. [PubMed]

154. So, K.; McGrouther, D.A.; Bush, J.A.; Durani, P.; Taylor, L.; Skotny, G.; Mason, T.; Metcalfe, A.; O'Kane, S.; Ferguson, M.W. Avotermin for scar improvement following scar revision surgery: A randomized, double-blind, within-patient, placebo-controlled, phase II clinical trial. *Plast. Reconstr. Surg.* **2011**, *128*, 163–172. [CrossRef] [PubMed]

155. Occleston, N.L.; O'Kane, S.; Laverty, H.G.; Cooper, M.; Fairlamb, D.; Mason, T.; Bush, J.A.; Ferguson, M.W. Discovery and development of avotermin (recombinant human transforming growth factor β3): A new class of prophylactic therapeutic for the improvement of scarring. *Wound Repair Regen.* **2011**, *19* (Suppl. S1), s38–s48. [CrossRef] [PubMed]

156. Ferguson, M.W.; Duncan, J.; Bond, J.; Bush, J.; Durani, P.; So, K.; Taylor, L.; Chantrey, J.; Mason, T.; James, G.; et al. Prophylactic administration of avotermin for improvement of skin scarring: Three double-blind, placebo-controlled, phase I/II studies. *Lancet* **2009**, *373*, 1264–1274. [CrossRef]

157. Zhibo, X.; Miaobo, Z. Botulinum toxin type A affects cell cycle distribution of fibroblasts derived from hypertrophic scar. *J. Plast. Reconstr. Aesthet. Surg.* **2008**, *61*, 1128–1129. [CrossRef] [PubMed]

158. Xiao, Z.; Zhang, M.; Liu, Y.; Ren, L. Botulinum toxin type a inhibits connective tissue growth factor expression in fibroblasts derived from hypertrophic scar. *Aesthet. Plast. Surg.* **2011**, *35*, 802–807. [CrossRef] [PubMed]

159. Xiao, Z.; Zhang, F.; Lin, W.; Zhang, M.; Liu, Y. Effect of botulinum toxin type A on transforming growth factor β1 in fibroblasts derived from hypertrophic scar: A preliminary report. *Aesthet. Plast. Surg.* **2010**, *34*, 424–427. [CrossRef] [PubMed]

160. Xiao, Z.; Zhang, F.; Cui, Z. Treatment of hypertrophic scars with intralesional botulinum toxin type A injections: A preliminary report. *Aesthet. Plast. Surg.* **2009**, *33*, 409–412. [CrossRef] [PubMed]

161. Gassner, H.G.; Sherris, D.A.; Otley, C.C. Treatment of facial wounds with botulinum toxin A improves cosmetic outcome in primates. *Plast. Reconstr. Surg.* **2000**, *105*, 1948–1953. [CrossRef] [PubMed]

162. Elhefnawy, A.M. Assessment of intralesional injection of botulinum toxin type A injection for hypertrophic scars. *Indian J. Dermatol. Venereol. Leprol.* **2016**, *82*, 279–283. [CrossRef] [PubMed]

163. Shaarawy, E.; Hegazy, R.A.; Abdel Hay, R.M. Intralesional botulinum toxin type A equally effective and better tolerated than intralesional steroid in the treatment of keloids: A randomized controlled trial. *J. Cosmet. Dermatol.* **2015**, *14*, 161–166. [CrossRef] [PubMed]

164. Zhibo, X.; Miaobo, Z. Intralesional botulinum toxin type A injection as a new treatment measure for keloids. *Plast. Reconstr. Surg.* **2009**, *124*, 275e–277e. [CrossRef] [PubMed]

165. Jones, C.D.; Guiot, L.; Samy, M.; Gorman, M.; Tehrani, H. The Use of Chemotherapeutics for the Treatment of Keloid Scars. *Dermatol. Rep.* **2015**, *7*, 5880. [CrossRef] [PubMed]
166. Espana, A.; Solano, T.; Quintanilla, E. Bleomycin in the treatment of keloids and hypertrophic scars by multiple needle punctures. *Dermatol. Surg.* **2001**, *27*, 23–27. [PubMed]
167. Saray, Y.; Gulec, A.T. Treatment of keloids and hypertrophic scars with dermojet injections of bleomycin: A preliminary study. *Int. J. Dermatol.* **2005**, *44*, 777–784. [CrossRef] [PubMed]
168. Hendricks, T.; Martens, M.F.; Huyben, C.M.; Wobbes, T. Inhibition of basal and TGF β-induced fibroblast collagen synthesis by antineoplastic agents. Implications for wound healing. *Br. J. Cancer* **1993**, *67*, 545–550. [CrossRef] [PubMed]
169. Yeowell, H.N.; Marshall, M.K.; Walker, L.C.; Ha, V.; Pinnell, S.R. Regulation of lysyl oxidase mRNA in dermal fibroblasts from normal donors and patients with inherited connective tissue disorders. *Arch. Biochem. Biophys.* **1994**, *308*, 299–305. [CrossRef] [PubMed]

International Journal of
Molecular Sciences

MDPI

Review

Imaging Collagen in Scar Tissue: Developments in Second Harmonic Generation Microscopy for Biomedical Applications

Leila Mostaço-Guidolin [1,2], Nicole L. Rosin [1] and Tillie-Louise Hackett [1,2,*]

1   Centre for Heart Lung Innovation, University of British Columbia, Vancouver, BC V6Z 1Y6, Canada; Leila.Mostaco-Guidolin@hli.ubc.ca (L.M.-G.); Nicole.Rosin@hli.ubc.ca (N.L.R.)
2   Department of Anesthesiology, Pharmacology and Therapeutics, University of British Columbia, Vancouver, BC V6Z 1Y6, Canada
*   Correspondence: tillie.hackett@hli.ubc.ca; Tel.: +1-604-806-8346; Fax: +1-604-806-8351

Received: 26 June 2017; Accepted: 10 August 2017; Published: 15 August 2017

Abstract: The ability to respond to injury with tissue repair is a fundamental property of all multicellular organisms. The extracellular matrix (ECM), composed of fibrillar collagens as well as a number of other components is dis-regulated during repair in many organs. In many tissues, scaring results when the balance is lost between ECM synthesis and degradation. Investigating what disrupts this balance and what effect this can have on tissue function remains an active area of research. Recent advances in the imaging of fibrillar collagen using second harmonic generation (SHG) imaging have proven useful in enhancing our understanding of the supramolecular changes that occur during scar formation and disease progression. Here, we review the physical properties of SHG, and the current nonlinear optical microscopy imaging (NLOM) systems that are used for SHG imaging. We provide an extensive review of studies that have used SHG in skin, lung, cardiovascular, tendon and ligaments, and eye tissue to understand alterations in fibrillar collagens in scar tissue. Lastly, we review the current methods of image analysis that are used to extract important information about the role of fibrillar collagens in scar formation.

Keywords: second harmonic generation; nonlinear optical microscopy; scar tissue; fibrillar collagen; skin; lung; vessels; image analysis

1. Introduction

The ability to respond to injury with tissue repair is a fundamental property of all multicellular organisms. In the seconds after injury, various intracellular and intercellular pathways must be activated and synchronized to respond if tissue integrity and homeostasis are to be restored. In general, the wound repair process in almost all tissues involves activation of the cellular components of the immune system (neutrophils, monocytes, lymphocytes and dendritic cells), the blood coagulation cascade and the resultant activated inflammatory pathways. In addition to immune cells, multiple structural cell types are activated (endothelial cells, epithelium, keratinocytes, and fibroblasts), which undergo marked changes in phenotype and gene expression leading to cell proliferation, differentiation, migration and extracellular matrix production [1,2]. Once the wound repair response has successfully regained tissue homeostasis, it must be controlled and switched off. Interestingly, malignant transformation is an uncommon event in repairing wounds [3,4]. The human fetus and some adult eukaryotic organisms can respond to injury through regeneration leading to restoration of the original tissue architecture [5]. In adult humans, this ability is lost through unknown processes, and wound repair normally results in non-functional tissues formed by a patch of cells (mainly fibroblasts) and disorganized extracellular matrix (mainly fibrillar collagens) that is commonly referred to as scar tissue.

Demographically, the number of individuals suffering from chronic wounds and impaired healing conditions continues to rise leading to health and economic burdens to society [6–9]. As an example, defective wound repair and chronic wounds can result from myocardial scar tissue leading to congestive heart failure after myocardial infarction, cirrhosis of the liver and lung fibrosis in response to toxin-mediated injury, or hypertrophic scars to surgical wounds. Understanding the underlying molecular basis of tissue repair and its failure is therefore an important unmet clinical need.

Studying wound repair in various conditions and models will expand our understanding of the wound repair process in humans that will hopefully lead to the identification of novel pathways or molecular signals that can be therapeutically targeted to restore their lost regenerative capacity. Over the last 15 years, nonlinear optical microscopy (NLOM) has emerged as a powerful research tool [10–13] for visualizing the supramolecular assembly of collagen in tissues at an unprecedented level of detail [14–17]. Based on the physics of nonlinear light-matter interactions, and using these interactions as contrast mechanisms for cellular and tissue imaging investigations, modalities such as Second Harmonic Generation (SHG), Third Harmonic Generation (THG), Coherent anti-Stokes Raman (CARS), and two-photon excitation fluorescence (TPEF) have acquired a reputation as an excellent optical tool for answering multiple biological questions [13,18,19]. The development of NLOM imaging has enhanced basic biomedical research, and provided a new suite of quantitative metrics for the diagnosis of a wide range of diseases [20–38].

In this review, we describe the physical characteristic of SHG, and the current NLOM systems that are used by biologists for SHG imaging. We provide an extensive review of studies that have used SHG in wounding and repair of skin, lung, cardiovascular, tendon and ligament, and eye tissues to understand alterations in fibrillar collagens in scar tissue [39–42]. It is important to note that SHG imaging is used to study many other diseases, including cancer, but this review is limited to the tissues listed above. Lastly, we provide an overview of the current methods of image analysis that are used to extract important information about the role of fibrillar collagens in scar formation.

2. The Extracellular Matrix

The extracellular matrix (ECM) is essential for the normal development, function and homeostasis of all eukaryotic cells [43–46]. While providing the physical matrix that separates establishing and established tissues and organs, the ECM also actively participates in regulating the abundance of growth factors, receptors, level of hydration and pH of the local tissue environment. The exquisite tissue specific functions of the ECM are achieved through its complex biochemical composition (water, proteins and polysaccharides) and dynamic biophysical properties [43–46]. ECM molecules are generally formed from small, modular repeating subunits that form homo or heteropolymers that become supramolecule assemblies with highly specialized functions. In all cases, each class of ECM molecule has evolved the ability to interact with other classes of ECM molecules to produce unique structural and biochemical properties. Therefore ECM molecules can function as support for cells (complex adhesion surfaces, diffusion barriers) or act as active participants in cell signaling (binding domains for growth factors and chemokines) [47–50]. The ECM is primarily composed of two classes of macromolecules: glycoproteins (such as fibronectin, proteoglycans and laminin) and fibrous proteins (including collagens and elastin). This review focuses on the structural properties of fibrillar collagens and how the NLOM technique SHG can be applied to understand their biophysical properties in organs during health and disease.

3. Fibrillar Collagen

Collagens form a heterogeneous family of fibrous proteins of which there are 28 different types identified in vertebrates. In animals, the collagen family represents the most abundant protein [51] and collagen is the dominant protein within the ECM. Here, we focus on the fibrillar collagens, which are capable of withstanding tensile forces within the ECM and have a non-centrosymmetric structure that can be imaged using NLOM.

At the structural level, all collagen molecules are made up of three polypeptide α chains that form homo- or heterotrimers. The prototypical α chain in all fibrillar collagens consists of approximately 338 repeating Gly-X-Y- triplets called a triple-helical motif (where X and Y can be any amino acid but are frequently proline and hydroxyproline) that is flanked by two non-collagenous domains, the N- and C-propeptides [52–55]. Once transcribed and within the endoplasmic reticulum, the triple-helix structure results in the intertwining of the three α chains starting at the C-terminal propeptide, forming a right-handed superhelix that is further stabilized by hydroxylation of particular lysine and proline residues (O-linked glycosylation). This assembly results in a rod-like structure 300 nm in length and 1.5 nm in diameter, termed procollagen, which is packaged for export by the Golgi apparatus. Procollagen is then converted into mature collagen by the removal of N- and C-propeptides via collagen type-specific metalloproteinase enzymes, within plasma membrane extrusions (known as fibripositors) that project from the cell surface. Mature collagen molecules then have the ability to engage in self-association to form microfibrils at the cell surface that can merge and grow both longitudinally and axially. To form mature collagen fibres, lysyl oxidases covalently crosslink lysine residues within the supramolecular assembly, providing stability and mechanical properties [45,56–60]. Conversely, non-fibrillar collagens contain non-triple helix regions, which lead to kinks in the resulting macromolecular structure that straighten under small strains. This means that non-fibrillar collagens do not form centrosymmetric structures and cannot be visualized by SHG.

4. Collagen Imaging Using Second Harmonic Generation (SHG) Microscopy

SHG microscopy has emerged as a useful tool for studying key facets of collagen remodeling. SHG imaging is an attractive alternative to conventional or fluorescent-based histology for studying tissue composition and visualizing the molecular structure of collagen due to its label-free nature, high sensitivity and specificity [14,19,61–63]. The optical sectioning capability of SHG also provides a means of imaging bulk tissue in 3 dimensions (3D).

4.1. What Is SHG and How Does It Work?

In short, SHG is a process that occurs when two photons are combined in an optically nonlinear medium, lacking in centro-symmetry (such as collagen), creating a SHG photon with a wavelength exactly half of the excitation wavelength (or twice the frequency, ω), as illustrated in Figure 1.

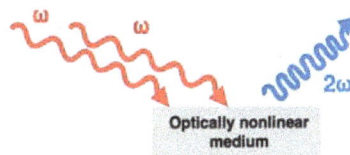

Figure 1. Second-harmonic generation (SHG) is a nonlinear optical process, in which two photons interacting within a nonlinear material are effectively "combined" to form a new photon with twice the energy (2ω), and therefore twice the frequency, or half the wavelength of the initial photons.

Before we begin explaining SHG, we need to first take a step back and define what light is. Light is part of the electromagnetic spectrum, and electromagnetic radiation waves are defined as fluctuations of electric and magnetic fields, which can transport energy between locations. Light, or electromagnetic radiation, can also be described as a stream of photons, mass-less particles that travel at the speed of light with wavelike properties.

The optical response of a material is expressed in terms of the induced polarization. Polarization describes the response of both the material and the applied electric field to one another. Polarization can be used to calculate the forces that result from these interactions [64–66]. For a linear material, the relationship between the polarization **P** and the electric field **E** of the incident radiation is linear:

$$\mathbf{P} = \varepsilon_0 \cdot \chi^{(1)} \, \mathbf{E} \tag{1}$$

where $\chi^{(1)}$ is the linear susceptibility (a dimensionless proportionality constant). The linear susceptibility due to an applied electric field indicates the degree of polarization of a dielectric material. A material with a higher linear susceptibility, has a greater ability to polarize in response to the field, thereby reduces the total electric field inside the material (and stores energy).

In nonlinear optics, such as SHG imaging the response of the material is often described as a polynomial expansion of the material polarization **P** in powers of the electric field **E**. The second term defines the SHG.

$$\mathbf{P} = \varepsilon_0 \cdot (\chi^{(1)} \, \mathbf{E} + \chi^{(2)} \, \mathbf{E} \cdot \mathbf{E} + \chi^{(3)} \, \mathbf{E} \cdot \mathbf{E} \cdot \mathbf{E} + \cdots\cdots) \tag{2}$$

The SHG signal originates from the nonlinear polarization. The incident wave generates dipoles inside the material. These dipoles radiate at twice the frequency of the incident wave. The relative phase of the induced dipoles is fixed because the incident beam has a well-defined amplitude and phase within the material at any given point and time. The SHG signal can be obtained only if the induced dipoles radiate in phase, as illustrated in Figure 2. This phase-matching ensures that the contributions add up constructively from all positions in the material.

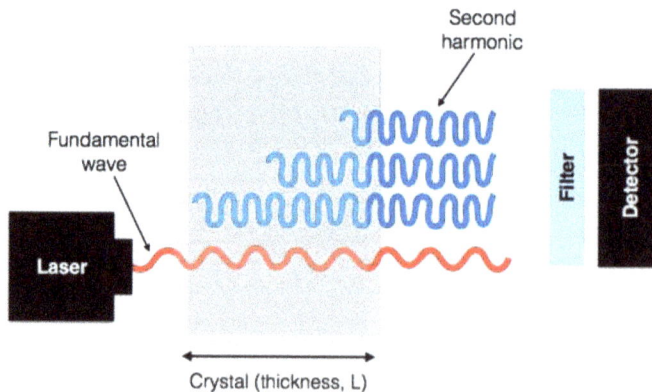

Figure 2. A sketch of the concept of phase matching. The fundamental wave at frequency ω has a well defined phase and amplitude everywhere in the crystal. The induced dipoles all radiate at a frequency 2ω with a phase dictated by the fundamental wave. The picture shows the case where all dipoles radiate in phase in the forward direction so that all contributions add up constructively.

There are several factors affecting the magnitude of SHG. It depends quadratically on the intensity of the excitation light, and is affected by the polarization and wavelength of the excitation light. SHG signal can also be dependent on the inherent properties of the material: the nonlinear susceptibility, the phase mismatch between the SHG and the excitation light, and the distribution and orientation of the SHG sources within the focal volume [67–70].

4.2. How Does Fibrillar Collagen Generate SHG Signal

Collagen fibres have a very suitable structure for generating SHG signal. Fibrillar collagen is highly anisotropic and the SHG signal generated is coherently amplified because of the tight alignment of repeating structures within the collagen triple helix and within fibrils. As SHG imaging is dependent on the signal remaining phase-matched within the material, a second harmonic wave generally co-propagates with the excitation beam, resulting in SHG signal in the same, forward direction as the excitation beam [19,69,71]. Typically, between 80% and 90% of the SHG signal from collagen in a tissue sample will propagate in the forward direction, depending on how much the sample scatters light.

A single, 40- to 300-nm collagen fibre would be expected to behave as a single dipole, which radiates in all directions except normal to the incident beam. The peptide bonds within the collagen chains generate a permanent dipole moment, which is a measure of the separation of positive and negative electrical charges within a system, allowing SHG to occur within collagen-rich samples, as shown in Figure 3a.

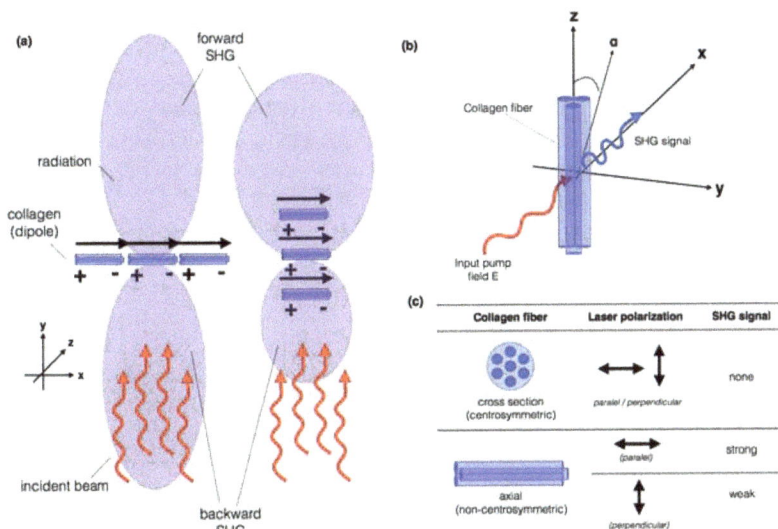

Figure 3. Illustration of (**a**) collagen fibres acting as dipoles, which radiates in all directions except normal to the incident beam; and (**b**) the geometric arrangement of a single collagen fibre relative to an applied electric field. The emitted SHG signal after spectral filtering is shown in blue; (**c**) if the light is polarized along the collagen fibre axis (z), the maximum SHG signal will be observed. On the other hand, if it is polarized perpendicular to the fibre axis (x), the weakest SHG signal will be observed.

Figure 3a illustrates how a planar collagen array, normal to the incident beam results in waves in the forward and backward direction being in phase, and therefore radiating strongly in both directions. Little lateral propagation occurs because the wave from one dipole and its neighbor will not be in phase laterally. An array orientated in the direction of the beam will propagate forward, because all dipoles will have the same phase relationship to the excitation beam in the forward direction, regardless of the spacing, but will randomly not be in phase in the backward direction [19,71].

An interesting characteristic of SHG microscopy is that the excited volume within the specimen is elliptical, with the long axis in the direction of the beam. In a collagenous tissue, the overall signal is predominantly propagated in the forward direction, compared to an isolated individual fibre that may radiate SHG signal both forward and backward. Any group of collagen fibres will have more excited dipoles in line with the beam than across it.

An important aspect to consider is that SHG is polarization sensitive. For many sources of SHG, the amount of signal produced is dependent on the polarization state of the incident laser light, relative to the scattering due to the molecular structure. To illustrate this, we shall focus on the polarization sensitivity of SHG from collagen fibrils, as it is the component most imaged by SHG microscopy in biomedical applications [11,13,72,73].

Most studies on the polarization sensitivity of collagen have been carried out on tendon which is composed of highly ordered parallel collagen-type I fibres. The intensity of the SHG signal produced from a collagen sample is dependent on the orientation of the polarization state of the laser excitation light with respect to the fibre axis [72–74], as illustrated in Figure 3b. For a linearly polarized laser beam incident on a collagen fibril, the amount of SHG signal produced for different polarization orientations is shown in Figure 3c.

For fibrils lying in the plane perpendicular to the direction of excitation laser light propagation, the amount of SHG produced depends on the angle (α) between the fibre axis (z) and the laser polarization. If the light is polarized along the fibre axis, the maximum SHG signal will be observed [74,75]. On the other hand, if it is polarized perpendicular to the fibre axis, the weakest SHG signal will be observed. This means that the polarization dependence of the SHG signal can be measured to study the orientation of the collagen fibrils within tissue. In the case of the fibre cross-section, due to the fibre structure's centrosymmetry, no SHG will ever be detected independent of the orientation of the laser polarization. The intensity of the SHG signal also depends on the angle between the collagen fibre and the imaging plane. The intensity of the SHG is maximized when the collagen fibre is in the imaging plane and very low when the fibre is perpendicular to the imaging plane.

The use of polarization analysis in SHG imaging permits the extraction of the tissue's structural information, including collagen fibre packing. However, for non-polarization resolved SHG imaging, circularly polarized light is preferred, as it will excite all fibre orientations equally. More recent systems use half or quarter wave plates to generate a circular polarized beam which can generate an uniform excitation in all fibre directions (both parallel or perpendicular collagen fibres), and therefore detect the signal generated from all fibres within the tissue, independent of their orientation.

In SHG microscopy the focal area in which the SHG signal is generated has sub-micron dimensions, which is much smaller than the coherence length in collagen. Therefore, the SHG signal should not be significantly reduced by destructive interference. SHG microscopy has proven to be an ideal tool for the analysis and quantification of the spatial arrangement of collagen fibres in tissue [11,13,72–74]. This information can be crucial when dealing with complex medical problems, such as atherosclerotic plaque development, fibrosis, airway wall remodeling, etc.

4.3. SHG Imaging Systems

There are a number of commercial multi-photon confocal microscopes now available on the market that can be used for SHG imaging (specifically collecting backscatter signal). In general, whether using in-house developed or ready-made equipment, a similar set up is used, as reviewed extensively previously [71] and shown in Figure 4. The basic requirements for a system to image SHG is a scanning confocal microscope equipped with a multi-photon femto-second pulse laser as an excitation source such as the Coherent Chameleon family of titanium:sapphire multi-photon tunable lasers (Coherent Inc., Santa Clara, CA, USA). Figure 4 illustrates the general schematic for the set up for collection of the SHG signal. When imaging the SHG of fibrillar collagen an excitation of 800 nm is used, the light path passes through a linear or circular polarizer, and an objective and reaches the specimen. The backscatter signal is not collected as in normal confocal imaging (such as with a pinhole). Instead, a non-descanned mode is used to collect the entire signal; the photons pass through a long wavelength dichroic mirror and the SHG filter (~410 nm). High gain photomultiplier tubes are then used to collect the SHG signals both in the forward and backward directions. As described above, the addition of polarization to a basic SHG imaging system (Figure 2) provides further information regarding the macrostructural information about the collagen fibre bundles. Therefore, most systems currently available additionally include a $\lambda/2$ or $\lambda/4$ plate to allow for uniform polarization of the tissue and maximum signal recovery. The maximum depth that can be imaged using such a SHG imaging system depends on the characteristics of the tissue. For example, the melanin in pigmented skin will burn at the laser power required to image more than ~50 µm into the superficial side, but if excised and flipped over the skin can be imaged ~300 µm into the dermal surface.

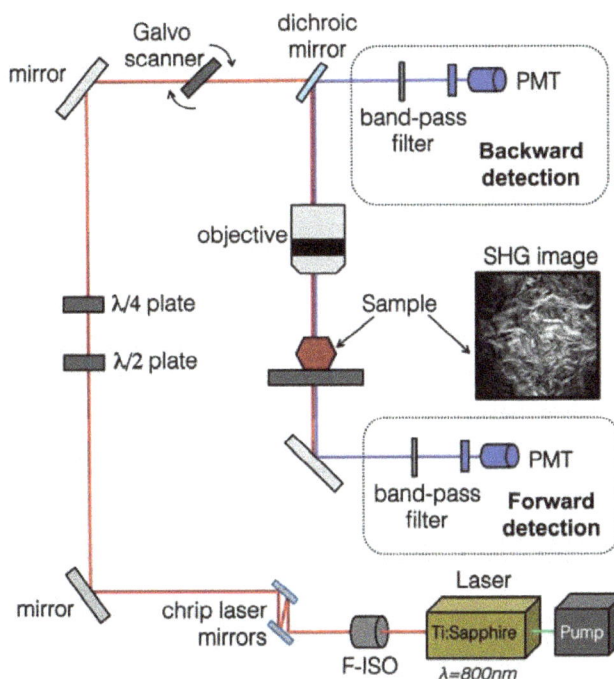

Figure 4. SHG imaging schematic and example image of mouse skin. The optical pathway schematic illustrates the general set-up for SHG imaging. The embedded mouse skin SHG image was taken using a Zeiss 710 confocal microscope was equipped with a Ti:Sa Chameleon multiphoton tunable laser (Coherent, Santa Clara, CA, USA) at 800 nm, a dichroic mirror, a custom filter set (BP:414/46, DC:495, BP:525/50), and a 20× water immersion objective. The resulting image was processed using Zen software (Zeiss Microscopy, Jena, Germany). PMT, photomultiplier tube; F-ISO, Faraday isolator.

4.4. Recent In Vivo Instrumentation Advances

Multimodal nonlinear optical (NLO) laser-scanning microscopes have been key to visualizing several specific biomolecules without the need of any specific tissue preparation and providing superior spatial and biochemical specificity. Coherent anti-Stokes Raman scattering (CARS), stimulated Raman scattering (SRS), two-photon-excitation fluorescence (TPEF), SHG and third harmonic generation (THR) have been extensively used in biomedical research. However, one of the biggest challenges faced by scientists are related to miniaturizing these laser-based microscopes to a point where these techniques could be easily used in clinic and in vivo.

Several groups have been intensively working towards this goal [76–78]. Crisafi et al. have presented the design of a multimodal NLO laser-scanning microscope on a compact fibre-format, integrating three NLO modalities (CARS, SRS, and TPEF) [79]. Their proposed system offers for the first time the possibility to develop a NLO microscope using off-the-shelf components, providing a cost-effective alternative to commercial systems. Song et al. were successful in adding SHG into the mix, developing a fully integrated multimodal microscopy that can provide photoacoustic (optical absorption), TPEF and SHG information from tissue in vivo [80]. The authors were able to visualize the cortex of a mouse and SHG was successfully used to reveal complementary tissue microstructures. Other groups, including Atsuta et al., were able to monitor collagen changes in human skin, by using fibre optic delivery of the pulse light, making a compact SHG microscope enclosed into a lens tube system [81]. This system provided the flexibility needed to perform measurements of several regions in the human skin.

Finally, another recent application of miniaturized SHG was proposed by Schnitzer et al. [82,83]. They have combined previous ideas of SHG imaging sarcomere organizations in human patients in vivo. They developed a miniaturized microscope which can be attached to the desired region (or limb), and therefore acquire information on sarcomere length coupled with electrical stimulation from normal patients, from stroke patients on both affected and unaffected limbs, and from patients recovering from injury. This application of SHG microscopy represents an important step towards the first true in vivo image system able to monitor muscle structure in patients.

5. Alterations in Fibrillar Collagens in the Disease State

SHG imaging has been used to study a number of tissues and assess the quality of the fibrillar collagen in both the normal and disease state. In Figure 5, we present some examples of label-free SHG images from different specimens of normal tissues. This imaging technology is capable of not only providing a snap-shot of the collagen structure, which is useful in itself, but also of helping to ascertain key biologically relevant information about disease pathology that will be key to furthering our understanding of the pathological process in many fibrotic diseases. Table 1 summarizes the breadth of studies that are possible in skin, lung, cardiovascular, tendon and ligaments, and eye tissues using SHG imaging in biomedical research (discussed in more detail below).

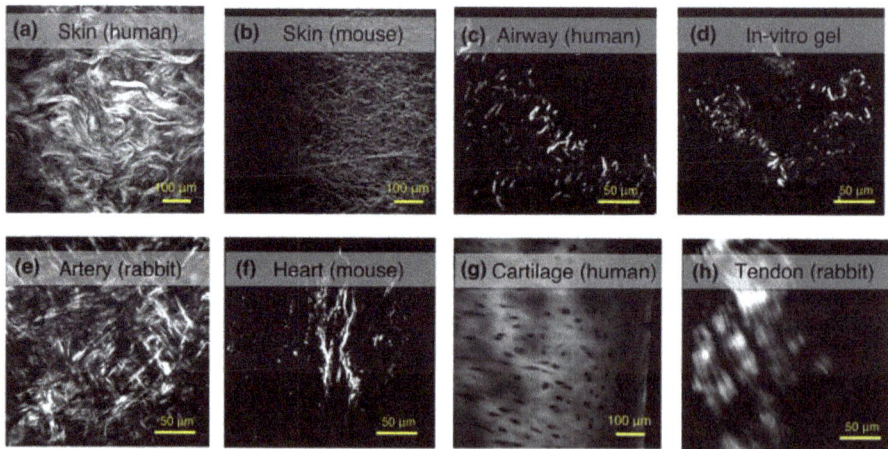

Figure 5. Examples of label-free SHG images from different tissue and specimens. The collagen network can be observed at the dermal layer of (**a**) human skin; or (**b**) mouse skin. The collagen deposition in human airways from a (**c**) human healthy donor; and (**d**) in vitro collagen gel model, showing fibrillar collagen synthesized by human fibroblasts from airways; (**e**) adventitia layer of a healthy aorta artery (rabbit); and (**f**) scar tissue formation in infarcted hearts from mouse; (**g**) human cartilage; and (**h**) tendons from rabbits can also be assessed using SHG imaging.

Table 1. Representative examples of Second Harmonic Generation (SHG) applications.

Organ/Tissue	Key Conclusions	Representative Reference
Skin	• Uniquely shows changes in collagen assembly upon thermal damage. • Highly valuable for diagnosing and screening early melanocytic lesions. • SHG to auto-fluorescence aging index of dermis (SAAID) can be a good indicator of the severity of photoaging; • Distinct morphological differences in melanoma compared with melanocytic nevi. • Used to distinguish between scar (keloid and hypertrophic) and normal tissue • Shows alteration in collagen structure in scar tissue after grafting	[23,30,84–86]
Tendon and Ligaments	• Good at determining the orientation of collagen fibrils in the fascicle and the ratio γ between the two independent elements of the second-order nonlinear susceptibility tensor; • Normal cartilage reveals a consistent pattern of variation in fibril orientation with depth. In lesions, the pattern is severely disrupted and there are changes in the pericellular matrix; • The differences in collagen fibre organization between normal and injured tendon indicate that the organization of collagen fibres is regularly oriented in normal tendons and randomly organized in injured tendons.	[75,87–89]
Cardiovascular	• SHG shows that collagen plaques intermingle with elastin; • Based on a measure of the collagen/elastin ratio, plaques were detected with a sensitivity of 65% and specificity of 81%. Furthermore, the technique gives detailed information on the structure of the collagen network in the fibrous cap; • Single parameter based on intensity changes derived from multi-channel nonlinear optical (NLO) images can classify plaque burden within the vessel; • Using the optical index for plaque burden (OIPB) it was possible to differentiate between healthy regions of the vessel and regions with plaque, as well as distinguish plaques relative to the age; • Texture analysis based on first-order statistics (FOS) and second-order statistics such as gray level co-occurrence matrix (GLCM) extracted SHG image features that are associated with the structural and biochemical changes of tissue collagen networks.	[28,40,90–93]
Lung	• Significant difference in collagen organization in airway tissue between chronic obstructive pulmonary disease (COPD) and non-diseased; • High resolution images can be generated without the need of staining and tissue damage; • Fibrillar collagen's subresolution structure is altered in usual interstitial pneumonia versus cryptogenic organizing pneumonia and healthy lung	[94–96]
Eye	• SHG can delineate the stroma from other corneal components. • SHG imaging revealed that corneal collagen fibrils are regularly packed as a polycrystalline lattice, accounting for the transparency of cornea. In contrast, scleral fibrils possess inhomogeneous, tubelike structures with thin hard shells, maintaining the high stiffness and elasticity of the sclera.	[24,97]
In-vitro models	• Self-assembled fibrillar gels can be imaged by SHG • SHG can characterize differential microscopic features of the collagen hydrogel that are strongly correlated with bulk mechanical properties • Depending on the collagen source, in vitro models yield homogeneous fibrillar texture with a quite narrow range of pore size variation, whereas all in vivo scaffolds comprise a range from low- to high-density fibrillar networks and heterogeneous pore sizes within the same tissue.	[34,98–100]

5.1. Skin

The skin is composed of three main layers: the epidermis, dermis and hypodermis. The epidermis is a thin layer of epidermal cells covered by the stratum corneum and provides the barrier function of the skin. The deeper dermal layer is primarily collagenous and provides structural support for cells and the dermal appendages (sabaceous glands, hair follicles, etc.), as well as the mechanical properties (strength, flexibility, distensibility, etc.) of the skin [101]. A large proportion of the cells within the dermis are mesenchymal in origin including interstitial fibroblasts, which maintain the ECM and the hair follicle dermal stem cells, which expand during injury to contribute to the interstitial fibroblast population [102]. The hypodermis is essentially highly vascularized adipose tissue. SHG imaging has proven useful in determining the directionality and organization of the fibrillar collagen in the skin, in both health and disease. Compared to cross-sections imaged after histochemical staining with Picro-Sirius Red stain or immunostaining for collagen, the 3D images that are created using SHG imaging of the whole tissue allow for determination of the arrangement of the collagen content in the dermis.

Recent studies have used SHG to distinguish between normal and pathologically scarred skin, suggesting potential diagnostic use for SHG imaging in the clinic [86,103]. Studies using SHG imaging have illustrated that the overall isotropic nature of dermal collagen provides the necessary distensability for mobility (to avoid puncture and for joint motion). Of particular note is that the proportion of collagen fibres that are aligned increases in keloid and surgical scar tissue [104–106]. The collagen fibre bundles in scarred skin after grafting have been shown to be straighter and thinner [85]. These factors appear to contribute to the decrease in the tensile strength of the scar compared with normal skin [85,107].

5.2. The Lung

The lung is formed from a dichotomous branching structure of airways that enables airflow that enters the nose and mouth to be directed to the 600 million terminal alveoli structures where gas exchange occurs. In terms of fibrillar collagens, the developing lung contains primarily fibrillar collagens I and III, which are deposited primarily by lung fibroblasts during the canalicular stage, in which the respiratory airways and alveolar ducts are formed, preceding the bulk of elastogenesis [108,109]. From the late stages of gestation to adult life, the collagen content in the lung increases five-fold. Specifically, in the rabbit lung, it has been shown during maturation that rapid growth after birth leads to a 15% increase in collagen synthesis relative to the rate of total protein synthesis which declines after two months [110]. In bronchopulmonary dysplasia (BPD) patients and animal models of BPD, ECM cross-linking enzymes are deregulated and aberrant late lung development blocks alveolarization, suggesting that perturbed ECM cross-linking may impact alveolarization [111]. Further, in many obstructive lung diseases that involve airway fibrosis such as asthma, chronic obstructive pulmonary disease (COPD) and idiopathic pulmonary fibrosis (IPF), alterations in fibrillar collagens due to deposition or turnover are the primary focus of research.

To date, SHG imaging of fibrillar collagen has primarily been used to confirm fibrotic regions in human IPF tissues [112] and mouse IPF models [113] identified by histological stains (Masson's trichrome). The technique has also been used to identify differences in the collagen organization in emphysematous COPD lung tissues compared to control donors [114,115]. In terms of quantification of SHG, signal intensity has previously been used to quantify the fibrosis-related increase in fibrillar collagen in a mouse model of pulmonary fibrosis [112], a human-mouse xenograft model of airway epithelial-induced fibrosis in asthma [116] and an in vitro collagen-gel contraction model using lung fibroblasts from COPD patients [117]. The SHG signal intensity has also been used to compare the ratio of collagen and elastin (measured by TPEF) in COPD lungs [115]. Using a standardized approach Tijn et al. reported on the potential to use forward SHG (F-SHG)/backward SHG (B-SHG) ratios as an approach to determine the amount of disorganized collagen in the airways of patients with COPD compared to controls [96]. Most recent studies have used the F-SHG/B-SHG ratio propagation SHG signal to demonstrate that the collagen microstructure is altered in regions of normal tissue and usual interstitial pneumonia, with increased mature fibrillar collagen compared to a reduced mature elastin fibre content. However, only preliminary data presented in abstract form has applied texture analysis to the human airways to demonstrate that collagen fibres are more disorganized in asthmatic airways compared to controls throughout the entire airway tree [118].

5.3. Cardiovascular System

Arteries are mainly composed of soft collagenous tissue; they serve as an elastic reservoir to transform the high pressure, pulsatile output into a flow with moderate fluctuations [119]. The vessel walls of large arteries have a number of structural features in common, although structural variations between the various arteries do exist. Fibrillar collagen is the ubiquitous load-bearing element which confers the macroscopic mechanical responses of the arterial wall, which are highly nonlinear, elastic and anisotropic [119]. The arrangement of collagen fibres in concentric anisotropic layers leads to the anisotropic mechanical behavior of arterial tissues. As the ECM is primarily composed of fibrillar collagen, in the arteries, it provides the majority of cellular attachment. Collagen tearing and defects result in disease and changes in the biomechanical behavior of the arterial tissue [14,42]. These alterations can result from the buildup of arterial plaques (or atherosclerosis), consisting of extracellular deposits of low-density lipids (LDLs), that gradually develop over a period of many years [120,121]. Current available clinical methods (ultrasound and magnetic resonance imaging) lack sufficient resolution to follow the structural changes in the ECM composition of the arterial lumen, as this requires imaging on the micron-scale [122,123]. SHG imaging may be suitable for identifying changes in arterial wall structure and composition as it could identify areas at risk of rupture earlier than currently available techniques, leading to the development of novel treatments specific for these alterations [14,124,125]. There are current limitations to SHG imaging in the cardiac system, as it has been shown ineffective at imaging arterial branch points, which are thought to be most susceptible to plaque formation due to local wall stress and heterogeneities of blood flow [90,91,93]. Studies using multimodal SHG microscopy have shown that mechanical considerations do not sufficiently convey the risk of rupture, but that local changes in biochemical composition must also be considered, which would not have been discerned by other clinical imaging modalities and standard histology alone [90–92].

In addition to arteries, SHG imaging of collagen has been used for the characterization of cardiac scar tissue formed due to myocardial infarction [40]. Mostaço-Guidolin et al. [40] showed significant reduction of collagen deposition in the adipose-derived stem cells (ASCs)-treated infracted heart. In the ASCs-treated infarcted myocardium, SHG imaging revealed highly-directional and organized collagen fibres compared to the un-treated infarcted myocardium, in which the collagen was significantly less organized.

5.4. Tendons and Ligaments

Tendons and ligaments facilitate a wide range of joint motion and considerable weight and energy savings associated with locomotor movement [87,126]. Tendons transmit forces between the muscle and bone, providing function, while ligaments provide bone to bone transmission of force, providing stability [87,126,127]. Although these two structures have different functions, both of them have similar characteristics. Specifically, both tendons and ligaments are predominantly composed of fibrillar collagen type I molecules arranged as fibrils, fibres, fibre bundles and fascicles [126,127]. The cellular components of tendons and ligaments consist of mature fibroblast and fibrocytes, although these are in low abundance and the majority of the tissue in both is composed of collagens and some glycosaminoglycans and elastic fibres [87,127–129]. A number of studies have now focused on the structural and molecular organization of the ECM in tendons and ligaments [87,126–128]. Tendons and ligament defects have largely been explained by abnormal collagen fibrillogenesis [130,131]. SHG imaging is capable of evaluating collagenase-induced tendon injury [89,94,132], with the ability to clearly differentiate normal and injured tendon collagen fibre organization. Injured tendons display less ordered collagen fibres in comparison to normal tendons. Biomechanically, abnormalities in crimp structure (which allow longitudinal elongation of tendons in response to loads) and collagen fibre organization result in a loss of performance of normal functions and thus degrade a subject's ability to move. SHG microscopy has been shown to be more sensitive to assessing tendon injury than standard clinical measurements using histology and light microscopy.

5.5. The Eye

The stroma of the eye is composed mainly of collagen type I, however common methods that are used for imaging the corneal epithelium (confocal microscopy) are ineffective for imaging this collagenous tissue [14,133]. Furthermore, due to the potential for permanent injury from resection of in vivo tissues, traditional histological techniques are not an option for diagnosing pathologies of the stroma. Because the collagenous composition of corneal stroma, SHG microscopy is an ideal option for the label-free imaging and diagnosis of pathological conditions [24]. For example, Tan et al. characterized collagen disruption in infectious keratitis using SHG imaging and simultaneously used two-photon fluorescence to identify infectious pathogens (bacterial, fungal and protozoan) ex vivo [97]. SHG imaging cannot compete with electron microscopy in terms of resolution, however under physiological conditions in vivo the structure of cornea can be investigated using SHG imaging [14,24,97].

5.6. In Vitro Models: Collagen Gels

Cellularized collagen gels are a common model used to understand several biological processes involving the interactions of cells with collagen [20,34,100,117,134,135]. Collagen gels have been successfully imaged by SHG microcopy, as they can provide information on structural rearrangements in the 3-D structure of the collagen matrix, which is modified by various cellular physiological processes [100]. Ajeti et al. [20] proposed the use of collagen gels, consisting of mixtures of collagen type I and type V isoforms to serve as a model of the ECM during cancer invasion in vivo. Using several metrics from SHG images, they found that SHG imaging was sensitive to the incorporation of Collagen V into Collagen I fibrils. They developed SHG microscopy as a tool to discriminate Collagen I/Collagen V composition in tissues to characterize and follow breast cancer invasion [20]. Another example of collagen gel characterization using intensity based analysis of SHG images was presented by Campbell et al., who concluded that defective collagen I remodeling and contraction was a feature of COPD parenchymal fibroblasts compared to fibroblasts derived from normal donors [117]. Finally, in tissue engineering, collagen gels and SHG images have played a vital role providing insights about the relationship between the microstructure and tissue bulk mechanical properties [134]. SHG has been proven to be an ideal non-invasive tool for examining collagen microstructure, cellularity and crosslink content in gels in both in vitro models and biological tissue.

6. Quantitative Image Analysis Methods

SHG microscopy has been a very powerful tool in biomedical research. However, most published SHG imaging work has described collagen organization without focusing on the quantitative measures which are possible to implement to characterize SHG images. In several studies, pathological conditions were described using empirical observations derived from collagen SHG images. While the ability to track associations between collagen SHG images and pathology is important, it is equally important to have the ability to track such correlations using quantifiable measures for objective comparison. Quantitative SHG imaging analysis methods have largely relied on image pixel-counting techniques, similar to those applied to histological tissue images. Several SHG collagen imaging studies have recently proposed novel methodologies for quantifying the features of SHG images.

Although commercially available laser scanning microscopes are supplied with software with some rudimentary data analysis capabilities, many groups have been working on developing methods to quantify the fibrillar structures detected in SHG images. Many external software packages have been successfully used for analysis of SHG data, including the measurement of intensities, the application of a threshold for measuring fibre lengths and division for F-SHG/B-SHG analysis. ImageJ or Fiji can run these analyses, or the can be automated using MATLAB or LabVIEW. Commercial packages including Improvision and Imaris are capable of rendering 3D images, as are some plugins especially written for Fiji. We present below the most commonly used analysis methods for SHG data; however, we would like to note that, while these processes are integral to measuring fibre length and distribution, they are not specific to SHG image analysis.

6.1. Intensity-Based Analysis

Intensity-based image analysis is the most straightforward form of quantification. The signal collected in each individual detector is associated with a specific pixel. Each pixel receives a value which can be associated to the signal intensity, and therefore, the amount of collagen deposited in that specific region. It is important to highlight that intensity-based analysis takes into account the intensities of individual pixels, and they are considered independently from their neighboring pixels [40]. Other than the absolute intensity value detected by each region of the sensor (which will later become the image's pixels), some first-order statistics can describe the gray levels (or intensities) of the histogram corresponding to an image. Among these parameters, the most commonly applied in SHG data analysis are the mean, standard deviation, integrated density, skewness, and kurtosis. The mean and integrated densities provide measures of the overall lightness/darkness of the image, while the standard deviation describes its overall contrast. These measures can be associated to the amount of collagen fibres, which is proportional to the detected SHG signal. The skewness quantitatively evaluates the asymmetry of the shape of the distribution of pixel intensities around the mean value of the histogram, while kurtosis measures the peakedness of the distribution relative to the length and size of the histogram tails [40]. Skewness and kurtosis can be useful when aiming to detect fibre features such as edges and how distinct they are from a certain background. In digital image processing, kurtosis values can be interpreted in combination with noise and resolution measurements. On the other hand, skewness tend to be positive in darker and glossier surfaces than lighter and matte surfaces, being therefore useful in making judgments regarding the fibre surfaces.

6.2. Forward–Backward SHG-Signal

To exploit the coherence of SHG and thus extract sub-resolution feature information, the SHG microscope could be set up with both F-SHG and B-SHG collection channels. The microscope's transmission pathway enables us to collect the F-SHG signal. This mode allows one to collect the majority of the SHG signal. To acquire B-SHG, the use of the confocal laser scanning head is necessary. It is important to highlight that to distinguish between TPEF and SHG signal it is necessary to perform a synchronous spectrum. In terms of quantitative analysis, the structural differences observed between the F-SHG and B-SHG images have not been well studied [136]. F-SHG/B-SHG ratio measurements are one of the most common quantitative measures presented in SHG image analysis. Some groups have claimed to be able to differentiate collagen types by calculating F-SHG/B-SHG ratio measurements. However, more rigorous experiments with standardized samples must be performed and correlated with traditional techniques, such as immunohistochemistry, before we can attribute this capability to SHG microscopy. F-SHG/B-SHG ratio measurements can be useful for assessing the fibre orientation content. Laterally oriented fibres appear primarily in the backward direction, whereas axial oriented fibres appear primarily in the forward direction. To successfully extract information from this measure, the relative collection efficiencies of the two detection pathways (transmission and confocal), including the detectors, need to be calibrated for each objective/condenser combination. This procedure must be done with isotropic emitters, such as KDP (monopotassium dihydrogen phosphate) crystal. One alternative to F-SHG/B-SHG images is the use of circularly polarized light; it can be obtained by placing a quarter-wave plate just before the objective. Circularly polarized nonlinear optics has the ability to quantify molecular symmetry after the acquisition of a minimal number of images using exclusively intensity information [137]. It can therefore be exploited to quantify the molecular alignment in arterial collagen with a single image during mechanical loading. The SHG intensity with circularly polarized light is not sensitive to the absolute orientation of the scattered signals, thus enabling characterization of their relative degree of order within a single image per tissue state.

6.3. Polarization

Polarization-resolved SHG is an alternative that can be used to extract information beyond simple visualization of fibre lengths or by pixel intensity. Polarization-resolved SHG is a form of measurement, which analyses the signal intensity as a function of laser polarization. Examples of this type of SHG measurement have been presented by some authors [75,138], where, by analyzing the signal anisotropy for constant linear polarization excitation, they were able to gather data on the protein helical pitch angle and the dipole alignment angle, respectively.

As the SHG signal generated by collagen fibres is highly dependent on their orientation, when working with polarization-resolved SHG, one can easily create a gamut of images showing the preferred fibre orientation within a certain region. As a general guideline, in the first measurement, the laser polarization is aligned with the long axis of a collagen fibre(s), and then rotated through 180°, where the intensity of these successive images is recorded. This collection of images can be used to verify changes occurring during collagen remodeling in scar formation, as the collagen fibres in this case tend not to present a preferable orientation.

6.4. Transform-Based Methods

Another way to quantitatively extract information from SHG images is by applying transform-based image analysis techniques. These techniques represent an image in a space whose coordinate system has an interpretation that is closely related to the characteristics of a texture, using the spatial frequency properties of the pixel intensity variations. Several methods are available and the success of these methods will strongly depend on the type of transform used to extract textural characteristics from the image and the final goal of the overall analysis. Methods based on Fast Fourier Transform (FFT) usually perform poorly in practice, due to its lack of spatial localization. However, Indhal and Næs [139] illustrated the use of spectra from 2-D FFT magnitude images for textural feature extraction, which can be used to determine the percentage of pixels and therefore collagen fibre bundles with the same alignment [85]. Gabor filters provide means for better spatial localization. Features derived from a set of Gabor filters have been widely used in texture analysis for image segmentation [140,141]. However, their usefulness is limited in practice because there is usually no single filter resolution at which one can localize a spatial structure in biological samples. Wavelet transform methods of feature extraction offer several advantages over FFT and Gabor-based methods. Wavelets have been used to characterize texture and to treat the problems of texture segmentation and classification [142,143]. FFT has been by far the most used method to characterize SHG images due to its simplicity and availability in several image analysis software packages. FFT has been able to show differences between different types of tissue and/or conditions, such as cancer-stage differentiation, scar formation and collagen remodeling in skin diseases. FFT analysis can be useful when combined with F-SHG/B-SHG or polarization-resolved SHG images, as it provides a quantitative measure of fibre orientation. However, further exploration of the capability of wavelets to aid in the interpretation of SHG images is still necessary.

6.5. Texture Analysis

Texture analysis has always been a powerful tool in the analysis of bio-medical imaging, remote sensing and industrial inspection, and its outputs are mainly classification, segmentation, and synthesis. Textures are very diverse and the approaches for analyzing them are likewise very diverse, and mainly differ from each other by the method used for extracting textural features. Texture analysis techniques,

Int. J. Mol. Sci. **2017**, *18*, 1772

also called higher order statistics, primarily describe characteristics of regions in an image through higher-order measures of their grayscale histograms. The most robust and frequently cited method for texture analysis is based on extracting various textural features from a gray level co-occurrence matrix (GLCM) [144]. The GLCM approach is based on the use of second-order statistics of the grayscale image histograms. Entropy, inverse difference moment (IDM), energy, inertia, entropy and correlation were reported as the most suitable textural parameter to evaluate collagen changes [40,117,118,144]. Entropy has been associated with the degree of fibre organization, and therefore can be very useful for the characterization of several tissue and conditions [144]. Alternatively, the run length matrix (RLM) texture analysis approach characterizes coarse textures as having many pixels in a constant gray level run and fine textures as having few [145]. For the purpose of collagen characterization, the choice between GLCM and RLM is clear: GLCM provides more flexibility and is able to provide quantitative measures of a varied of fibre features.

7. Limitations

There is no question that SHG microscopy is a powerful tool for examining biological tissue. However, as with any technology, it also presents limitations. As is the case with other NLOM modalities, SHG needs a laser as a light source. Most of the commercial femto-seconds lasers (needed to generate SHG signal) are based on Ti:Sapphire laser sources. The price of Ti:Sapphire lasers have decreased over the years, but are still quite an investment. One recent advance in this regard is the use of solid-state pump lasers, as they are less expensive, more compact and reliable sources for NLOM.

Another major factor that significantly affects making SHG microscopy applicable for broader use is the availability of compact systems small enough to fit into a catheter-sized probe that can be used to acquire in vivo images, as well as fully automated and maintenance-free systems. This next step for laser and NLOM development will allow these techniques to be used in routine, biological and clinical environments without the need for laser physicists.

The limited penetration depths (100–300 µm with laser excitation in the 800–1000 nm range) make SHG microscopy not suitable for certain applications, such as when the region of interested is located centimeters within the tissue or organ. Using conventional optical geometry, the spatial resolution of SHG microscopy is also limited by the diffraction limit, which is related to the wavelength (several hundred nm) of the incident waves. Super-resolution NLOM is starting to become available and this limitation should soon be overcome by the next generation of microscopes.

Lastly, for biological applications, the largest limitation to SHG microscopy is its ability to assess only a small number of structural proteins or harmonophores. In addition, to fibrillar collagens, this includes actin-myosin complexes, microtubules, centrosomes and mitotic spindles in cells (not discussed here). To date there are no methods available that can distinguish between fibrillar collagen types, which could greatly advance the field of wound repair and regeneration biology.

8. Conclusions

Understanding scar formation (fibrillar collagen formation) over time in humans has huge potential to deliver new therapeutics for correct wounds. Without the need for labeling, SHG imaging is particularly suitable for in vivo studies of collagenous connective tissue, partially due to strong contrast and large sensing depth. In future, the development of SHG endoscopes comprised of a pulsed laser source, a fibre-optic catheter, and a microelectromechanical systems (MEMS)-based scanning head, could offer large potential to clinical research. This would enable achievable penetration depths of 100–200 micrometers to analyze scar tissue non-invasively in human subjects using current clinical endoscopy techniques. Such a device could aid in assessment of scar tissue and enable new protocols to non-invasively monitor the efficiency of therapeutics and their effects on tissue function in vivo.

Acknowledgments: This study was funded by the Canadian Institutes of Health Research (CIHR) and British Columbia Lung Association operating grants. Leila Mostaço-Guidolin is supported by Michael Smith Health Research Foundation (MSHRF), Canadian Thoracic Society (CTS), and AllerGen National Centre of Excellence Inc fellowship. Tillie-Louise Hackett is supported by CIHR, MSFHR and Parker B. Francis new investigator awards. Nicole L. Rosin was supported by the University of Calgary Eye's High fellowship.

Author Contributions: Leila Mostaço-Guidolin, Nicole L. Rosin, and Tillie-Louise Hackett co-wrote and edited the review article.

Conflicts of Interest: The authors have no conflicts of interest.

Abbreviations

SHG	Second harmonic generation
THG	Third harmonic generation
B-SHG	Backward second harmonic generation
F-SHG	Forward second harmonic generation
Col	Collagen
COPD	Chronic obstructive pulmonary disease
CARS	Coherent anti-Stokes raman spectroscopy
MEMS	Microelectromechanical system
GLCM	Gray level co-occurrence matrix
IDM	Inverse difference moment
RLM	Run length matrix
AMT	Angle Measure Technique
FFT	Fast Fourier Transform
ECM	Extracellular matrix
3D	3-dimension
LDL	Low density lipids
ASC	Apdipose stem cell
NLOM	Non-linear optical microscopy
TPEF	Two photon electron fluorescence
BPD	Bronchopulmonary dysplasia

References

1. Konigova, R.; Rychterova, V. Marjolin's ulcer. *Acta Chir. Plast.* **2000**, *42*, 91–94. [PubMed]
2. Trent, J.T.; Kirsner, R.S. Wounds and malignancy. *Adv. Skin Wound Care* **2003**, *16*, 31–34. [CrossRef] [PubMed]
3. Aarabi, S.; Longaker, M.T.; Gurtner, G.C. Hypertrophic scar formation following burns and trauma: New approaches to treatment. *PLoS Med.* **2007**, *4*, e234. [CrossRef] [PubMed]
4. Singer, A.J.; Clark, R.A. Cutaneous wound healing. *N. Engl. J. Med.* **1999**, *341*, 738–746. [CrossRef] [PubMed]
5. Colwell, A.S.; Longaker, M.T.; Lorenz, H.P. Fetal wound healing. *Front. Biosci.* **2003**, *8*, s1240–s1248. [PubMed]
6. Brem, H.; Tomic-Canic, M. Cellular and molecular basis of wound healing in diabetes. *J. Clin. Investig.* **2007**, *117*, 1219–1222. [CrossRef] [PubMed]
7. Sen, C.K.; Gordillo, G.M.; Roy, S.; Kirsner, R.; Lambert, L.; Hunt, T.K.; Gottrup, F.; Gurtner, G.C.; Longaker, M.T. Human skin wounds: A major and snowballing threat to public health and the economy. *Wound Repair Regen.* **2009**, *17*, 763–771. [CrossRef] [PubMed]
8. Reish, R.G.; Eriksson, E. Scars: A review of emerging and currently available therapies. *Plast. Reconstr. Surg.* **2008**, *122*, 1068–1078. [CrossRef] [PubMed]
9. Richmond, N.A.; Lamel, S.A.; Davidson, J.M.; Martins-Green, M.; Sen, C.K.; Tomic-Canic, M.; Vivas, A.C.; Braun, L.R.; Kirsner, R.S. US-National Institutes of Health-funded research for cutaneous wounds in 2012. *Wound Repair Regen.* **2013**, *21*, 789–792. [CrossRef] [PubMed]
10. Mertz, J. Nonlinear microscopy: New techniques and applications. *Curr. Opin. Neurobiol.* **2004**, *14*, 610–616. [CrossRef] [PubMed]
11. Williams, R.M.; Zipfel, W.R.; Webb, W.W. Multiphoton microscopy in biological research. *Curr. Opin. Chem. Biol.* **2001**, *5*, 603–608. [CrossRef]
12. Yue, S.; Slipchenko, M.N.; Cheng, J.X. Multimodal Nonlinear Optical Microscopy. *Laser Photon. Rev.* **2011**, *5*, 10. [CrossRef] [PubMed]

13. Zipfel, W.R.; Williams, R.M.; Webb, W.W. Nonlinear magic: Multiphoton microscopy in the biosciences. *Nat. Biotechnol.* **2003**, *21*, 1369–1377. [CrossRef] [PubMed]

14. Campagnola, P. Second harmonic generation imaging microscopy: Applications to diseases diagnostics. *Anal. Chem.* **2011**, *83*, 3224–3231. [CrossRef] [PubMed]

15. Campagnola, P.J.; Loew, L.M. Second-harmonic imaging microscopy for visualizing biomolecular arrays in cells, tissues and organisms. *Nat. Biotechnol.* **2003**, *21*, 1356–1360. [CrossRef] [PubMed]

16. Campagnola, P.J.; Millard, A.C.; Terasaki, M.; Hoppe, P.E.; Malone, C.J.; Mohler, W.A. Three-dimensional high-resolution second-harmonic generation imaging of endogenous structural proteins in biological tissues. *Biophys. J.* **2002**, *82*, 493–508. [CrossRef]

17. Zipfel, W.R.; Williams, R.M.; Christie, R.; Nikitin, A.Y.; Hyman, B.T.; Webb, W.W. Live tissue intrinsic emission microscopy using multiphoton-excited native fluorescence and second harmonic generation. *Proc. Natl. Acad. Sci. USA* **2003**, *100*, 7075–7080. [CrossRef] [PubMed]

18. Masihzadeh, O.; Schlup, P.; Bartels, R.A. Label-free second harmonic generation holographic microscopy of biological specimens. *Opt. Express* **2010**, *18*, 9840–9851. [CrossRef] [PubMed]

19. Pfeffer, C.P.; Olsen, B.R.; Ganikhanov, F.; Legare, F. Multimodal nonlinear optical imaging of collagen arrays. *J. Struct. Biol.* **2008**, *164*, 140–145. [CrossRef] [PubMed]

20. Ajeti, V.; Nadiarnykh, O.; Ponik, S.M.; Keely, P.J.; Eliceiri, K.W.; Campagnola, P.J. Structural changes in mixed Col I/Col V collagen gels probed by SHG microscopy: Implications for probing stromal alterations in human breast cancer. *Biomed. Opt. Express* **2011**, *2*, 2307–2316. [CrossRef] [PubMed]

21. Cicchi, R.; Massi, D.; Sestini, S.; Carli, P.; De Giorgi, V.; Lotti, T.; Pavone, F.S. Multidimensional non-linear laser imaging of Basal Cell Carcinoma. *Opt. Express* **2007**, *15*, 10135–10148. [CrossRef] [PubMed]

22. Conklin, M.W.; Eickhoff, J.C.; Riching, K.M.; Pehlke, C.A.; Eliceiri, K.W.; Provenzano, P.P.; Friedl, A.; Keely, P.J. Aligned collagen is a prognostic signature for survival in human breast carcinoma. *Am. J. Pathol.* **2011**, *178*, 1221–1232. [CrossRef] [PubMed]

23. Dimitrow, E.; Ziemer, M.; Koehler, M.J.; Norgauer, J.; Konig, K.; Elsner, P.; Kaatz, M. Sensitivity and specificity of multiphoton laser tomography for in vivo and ex vivo diagnosis of malignant melanoma. *J. Investig. Dermatol.* **2009**, *129*, 1752–1758. [CrossRef] [PubMed]

24. Han, M.; Giese, G.; Bille, J. Second harmonic generation imaging of collagen fibrils in cornea and sclera. *Opt. Express* **2005**, *13*, 5791–5797. [CrossRef] [PubMed]

25. Kirkpatrick, N.D.; Brewer, M.A.; Utzinger, U. Endogenous optical biomarkers of ovarian cancer evaluated with multiphoton microscopy. *Cancer Epidemiol. Biomark. Prev.* **2007**, *16*, 2048–2057. [CrossRef] [PubMed]

26. Kwon, G.P.; Schroeder, J.L.; Amar, M.J.; Remaley, A.T.; Balaban, R.S. Contribution of macromolecular structure to the retention of low-density lipoprotein at arterial branch points. *Circulation* **2008**, *117*, 2919–2927. [CrossRef] [PubMed]

27. Lacomb, R.; Nadiarnykh, O.; Campagnola, P.J. Quantitative second harmonic generation imaging of the diseased state osteogenesis imperfecta: Experiment and simulation. *Biophys. J.* **2008**, *94*, 4504–4514. [CrossRef] [PubMed]

28. Le, T.T.; Langohr, I.M.; Locker, M.J.; Sturek, M.; Cheng, J.X. Label-free molecular imaging of atherosclerotic lesions using multimodal nonlinear optical microscopy. *J. Biomed. Opt.* **2007**, *12*, 054007. [CrossRef] [PubMed]

29. Lin, S.J.; Jee, S.H.; Kuo, C.J.; Wu, R.J.; Lin, W.C.; Chen, J.S.; Liao, Y.H.; Hsu, C.J.; Tsai, T.F.; Chen, Y.F.; et al. Discrimination of basal cell carcinoma from normal dermal stroma by quantitative multiphoton imaging. *Opt. Lett.* **2006**, *31*, 2756–2758. [CrossRef] [PubMed]

30. Lin, S.J.; Wu, R., Jr.; Tan, H.Y.; Lo, W.; Lin, W.C.; Young, T.H.; Hsu, C.J.; Chen, J.S.; Jee, S.H.; Dong, C.Y. Evaluating cutaneous photoaging by use of multiphoton fluorescence and second-harmonic generation microscopy. *Opt. Lett.* **2005**, *30*, 2275–2277. [CrossRef] [PubMed]

31. Lo, W.; Teng, S.W.; Tan, H.Y.; Kim, K.H.; Chen, H.C.; Lee, H.S.; Chen, Y.F.; So, P.T.; Dong, C.Y. Intact corneal stroma visualization of GFP mouse revealed by multiphoton imaging. *Microsc. Res. Tech.* **2006**, *69*, 973–975. [CrossRef] [PubMed]

32. Nadiarnykh, O.; LaComb, R.B.; Brewer, M.A.; Campagnola, P.J. Alterations of the extracellular matrix in ovarian cancer studied by Second Harmonic Generation imaging microscopy. *BMC Cancer* **2010**, *10*, 94. [CrossRef] [PubMed]

33. Provenzano, P.P.; Eliceiri, K.W.; Campbell, J.M.; Inman, D.R.; White, J.G.; Keely, P.J. Collagen reorganization at the tumor-stromal interface facilitates local invasion. *BMC Med.* **2006**, *4*, 38. [CrossRef] [PubMed]

34. Raub, C.B.; Suresh, V.; Krasieva, T.; Lyubovitsky, J.; Mih, J.D.; Putnam, A.J.; Tromberg, B.J.; George, S.C. Noninvasive assessment of collagen gel microstructure and mechanics using multiphoton microscopy. *Biophys. J.* **2007**, *92*, 2212–2222. [CrossRef] [PubMed]
35. Sahai, E.; Wyckoff, J.; Philippar, U.; Segall, J.E.; Gertler, F.; Condeelis, J. Simultaneous imaging of GFP, CFP and collagen in tumors in vivo using multiphoton microscopy. *BMC Biotechnol.* **2005**, *5*, 14. [CrossRef] [PubMed]
36. Schenke-Layland, K.; Xie, J.; Angelis, E.; Starcher, B.; Wu, K.; Riemann, I.; MacLellan, W.R.; Hamm-Alvarez, S.F. Increased degradation of extracellular matrix structures of lacrimal glands implicated in the pathogenesis of Sjogren's syndrome. *Matrix Biol.* **2008**, *27*, 53–66. [CrossRef] [PubMed]
37. Strupler, M.; Pena, A.M.; Hernest, M.; Tharaux, P.L.; Martin, J.L.; Beaurepaire, E.; Schanne-Klein, M.C. Second harmonic imaging and scoring of collagen in fibrotic tissues. *Opt. Express* **2007**, *15*, 4054–4065. [CrossRef] [PubMed]
38. Sun, W.; Chang, S.; Tai, D.C.; Tan, N.; Xiao, G.; Tang, H.; Yu, H. Nonlinear optical microscopy: Use of second harmonic generation and two-photon microscopy for automated quantitative liver fibrosis studies. *J. Biomed. Opt.* **2008**, *13*, 064010. [CrossRef] [PubMed]
39. Ko, A.C.; Ridsdale, A.; Mostaço-Guidolin, L.B.; Major, A.; Stolow, A.; Sowa, M.G. Nonlinear optical microscopy in decoding arterial diseases. *Biophys. Rev.* **2012**, *4*, 323–334. [CrossRef] [PubMed]
40. Mostaço-Guidolin, L.B.; Ko, A.C.; Wang, F.; Xiang, B.; Hewko, M.; Tian, G.; Major, A.; Shiomi, M.; Sowa, M.G. Collagen morphology and texture analysis: From statistics to classification. *Sci. Rep.* **2013**, *3*, 2190. [CrossRef] [PubMed]
41. Wang, H.W.; Langohr, I.M.; Sturek, M.; Cheng, J.X. Imaging and quantitative analysis of atherosclerotic lesions by CARS-based multimodal nonlinear optical microscopy. *Arterioscler. Thromb. Vasc. Biol.* **2009**, *29*, 1342–1348. [CrossRef] [PubMed]
42. Zoumi, A.; Yeh, A.; Tromberg, B.J. Imaging cells and extracellular matrix in vivo by using second-harmonic generation and two-photon excited fluorescence. *Proc. Natl. Acad. Sci. USA* **2002**, *99*, 11014–11019. [CrossRef] [PubMed]
43. Lu, P.; Takai, K.; Weaver, V.M.; Werb, Z. Extracellular matrix degradation and remodeling in development and disease. *Cold Spring Harb. Perspect. Biol.* **2011**, *3*, a005058. [CrossRef] [PubMed]
44. Mecham, R.P. Overview of extracellular matrix. *Curr. Protoc. Cell Biol.* **2012**. [CrossRef]
45. Mouw, J.K.; Ou, G.; Weaver, V.M. Extracellular matrix assembly: A multiscale deconstruction. *Nat. Rev. Mol. Cell Biol.* **2014**, *15*, 771–785. [CrossRef] [PubMed]
46. Rozario, T.; DeSimone, D.W. The extracellular matrix in development and morphogenesis: A dynamic view. *Dev. Biol.* **2010**, *341*, 126–140. [CrossRef] [PubMed]
47. Barczyk, M.M.; Olsen, L.H.; da Franca, P.; Loos, B.G.; Mustafa, K.; Gullberg, D.; Bolstad, A.I. A role for α11β1 integrin in the human periodontal ligament. *J. Dent. Res.* **2009**, *88*, 621–626. [CrossRef] [PubMed]
48. Dzamba, B.J.; Keene, D.R.; Isogai, Z.; Charbonneau, N.L.; Karaman-Jurukovska, N.; Simon, M.; Sakai, L.Y. Assembly of epithelial cell fibrillins. *J. Investig. Dermatol.* **2001**, *117*, 1612–1620. [PubMed]
49. Ramirez, F.; Sakai, L.Y. Biogenesis and function of fibrillin assemblies. *Cell Tissue Res.* **2010**, *339*, 71–82. [CrossRef] [PubMed]
50. Durbeej, M. Laminins. *Cell Tissue Res.* **2010**, *339*, 259. [CrossRef] [PubMed]
51. Gordon, M.K.; Hahn, R.A. Collagens. *Cell Tissue Res.* **2010**, *339*, 247–257. [CrossRef] [PubMed]
52. Brodsky, B.; Persikov, A.V. Molecular structure of the collagen triple helix. *Adv. Protein Chem.* **2005**, *70*, 301–339. [PubMed]
53. Celerin, M.; Ray, J.M.; Schisler, N.J.; Day, A.W.; Stetler-Stevenson, W.G.; Laudenbach, D.E. Fungal fimbriae are composed of collagen. *EMBO J.* **1996**, *15*, 4445–4453. [PubMed]
54. King, N.; Westbrook, M.J.; Young, S.L.; Kuo, A.; Abedin, M.; Chapman, J.; Fairclough, S.; Hellsten, U.; Isogai, Y.; Letunic, I.; et al. The genome of the choanoflagellate Monosiga brevicollis and the origin of metazoans. *Nature* **2008**, *451*, 783–788. [CrossRef] [PubMed]
55. Rasmussen, M.; Jacobsson, M.; Bjorck, L. Genome-based identification and analysis of collagen-related structural motifs in bacterial and viral proteins. *J. Biol. Chem.* **2003**, *278*, 32313–32316. [CrossRef] [PubMed]
56. Fratzl, P.; Misof, K.; Zizak, I.; Rapp, G.; Amenitsch, H.; Bernstorff, S. Fibrillar structure and mechanical properties of collagen. *J. Struct. Biol.* **1998**, *122*, 119–122. [CrossRef] [PubMed]

57. Hulmes, D.J. Building collagen molecules, fibrils, and suprafibrillar structures. *J. Struct. Biol.* **2002**, *137*, 2–10. [CrossRef] [PubMed]

58. Hulmes, D.J. *Collagen Structure and Mechanics*; Springer: Berlin, Germany, 2008.

59. Myllyharju, J.; Kivirikko, K.I. Collagens, modifying enzymes and their mutations in humans, flies and worms. *Trends Genet.* **2004**, *20*, 33–43. [CrossRef] [PubMed]

60. Ricard-Blum, S.; Ruggiero, F. The collagen superfamily: From the extracellular matrix to the cell membrane. *Pathol. Biol.* **2005**, *53*, 430–442. [CrossRef] [PubMed]

61. Cicchi, R.; Vogler, N.; Kapsokalyvas, D.; Dietzek, B.; Popp, J.; Pavone, F.S. From molecular structure to tissue architecture: Collagen organization probed by SHG microscopy. *J. Biophotonics* **2013**, *6*, 129–142. [CrossRef] [PubMed]

62. Dunn, K.W.; Young, P.A. Principles of multiphoton microscopy. *Nephron Exp. Nephrol.* **2006**, *103*, e33–e40. [CrossRef] [PubMed]

63. Xu, C.; Zipfel, W.; Shear, J.B.; Williams, R.M.; Webb, W.W. Multiphoton fluorescence excitation: New spectral windows for biological nonlinear microscopy. *Proc. Natl. Acad. Sci. USA* **1996**, *93*, 10763–10768. [CrossRef] [PubMed]

64. Bass, M.; van Strvland, E.W.; Williams, D.R.; Wolfe, W.L. *Handbook of Optics*; McGraw-Hill: New York, NY, USA, 2001; Volume 2.

65. Sutherland, R.L. *Handbook of Nonlinear Optics*; CRC Press: Boca Raton, FL, USA, 2003.

66. Svirko, Y.P.; Zheludev, N.I. *Polarization of Light in Nonlineear Optics*; Wiley-VCH: Weinheim, Germany, 2000.

67. Boyd, G.D.; Kleinman, D.A. Parametric interation of focused Gaussian light beams. *J. Appl. Phys.* **1968**, *39*, 3597. [CrossRef]

68. Brjorkholm, J.E. Opitcal second-harmonic generation using a focused Gaussian laser beam. *Phys. Rev.* **1966**, *142*, 126–136. [CrossRef]

69. Dmitriev, V.G.; Gurzadyan, G.G.; Nikogosyan, D.N.; Lotsch, H.K.V. *Handbook of Nonlinear Optical Crystals*; Springer: Berlin, Germany, 1999.

70. Sapaev, U.K.; Kulagin, I.A.; Usmanov, T. Theory of second-harmonic generation for limited laser beams in nonlinear crystals. *J. Opt. B Quantum Semiclass. Opt.* **2003**, *5*, 355–356. [CrossRef]

71. Chen, X.; Nadiarynkh, O.; Plotnikov, S.; Campagnola, P.J. Second harmonic generation microscopy for quantitative analysis of collagen fibrillar structure. *Nat. Protoc.* **2012**, *7*, 654–669. [CrossRef] [PubMed]

72. Masters, B.R.; So, P. *Handbook of Biomedical Nonlinear Optical Microscopy*; Oxford University Press: Oxford, UK, 2008.

73. Roth, S.; Freund, I. Second harmonic generation in collagen. *J. Chem. Phys.* **1979**, *70*, 1637–1643. [CrossRef]

74. Theodossiou, T.A.; Thrasivoulou, C.; Ekwobi, C.; Becker, D.L. Second harmonic generation confocal microscopy of collagen type I from rat tendon cryosections. *Biophys. J.* **2006**, *91*, 4665–4677. [CrossRef] [PubMed]

75. Stoller, P.; Kim, B.M.; Rubenchik, A.M.; Reiser, K.M.; Da Silva, L.B. Polarization-dependent optical second-harmonic imaging of a rat-tail tendon. *J. Biomed. Opt.* **2002**, *7*, 205–214. [CrossRef] [PubMed]

76. Helmchen, F.; Fee, M.S.; Tank, D.W.; Denk, W. A miniature head-mounted two-photon microscope: High-resolution brain imaging in freely moving animals. *Neuron* **2001**, *31*, 903–912. [CrossRef]

77. Llewellyn, M.E.; Barretto, R.P.; Delp, S.L.; Schnitzer, M.J. Minimally invasive high-speed imaging of sarcomere contractile dynamics in mice and humans. *Nature* **2008**, *454*, 784–788. [CrossRef] [PubMed]

78. Wu, Y.; Leng, Y.; Xi, J.; Li, X. Scanning all-fibre-optic endomicroscopy system for 3D nonlinear optical imaging of biological tissues. *Opt. Express* **2009**, *17*, 7907–7915. [CrossRef] [PubMed]

79. Crisafi, F.; Kumar, V.; Perri, A.; Marangoni, M.; Cerullo, G.; Polli, D. Multimodal nonlinear microscope based on a compact fibre-format laser source. *Spectrochim. Acta A Mol. Biomol. Spectrosc.* **2017**, *188*, 135–140. [CrossRef] [PubMed]

80. Song, W.; Xu, Q.; Zhang, Y.; Zhan, Y.; Zheng, W.; Song, L. Fully integrated reflection-mode photoacoustic, two-photon, and second harmonic generation microscopy in vivo. *Sci. Rep.* **2016**, *6*, 32240. [CrossRef] [PubMed]

81. Atsuta, K.; Ogura, Y.; Hase, E.; Minamikawa, T.; Yasui, T. In situ monitoring of collagen fibres in human skin using a photonic-crystal-fibre-coupled, hand-held, second-harmonic-generation microscope. *Proc. SPIE* **2017**. [CrossRef]

82. Sanchez, G.N.; Sinha, S.; Liske, H.; Chen, X.; Nguyen, V.; Delp, S.L.; Schnitzer, M.J. In Vivo Imaging of Human Sarcomere Twitch Dynamics in Individual Motor Units. *Neuron* **2015**, *88*, 1109–1120. [CrossRef] [PubMed]

83. Williams, J.C.; Campagnola, P.J. Wearable Second Harmonic Generation Imaging: The Sarcomeric Bridge to the Clinic. *Neuron* **2015**, *88*, 1067–1069. [CrossRef] [PubMed]

84. Chen, G.; Chen, J.; Zhuo, S.; Xiong, S.; Zeng, H.; Jiang, X.; Chen, R.; Xie, S. Nonlinear spectral imaging of human hypertrophic scar based on two-photon excited fluorescence and second-harmonic generation. *Br. J. Dermatol.* **2009**, *161*, 48–55. [CrossRef] [PubMed]

85. Rosin, N.L.; Agabalyan, N.; Olsen, K.; Martufi, G.; Gabriel, V.; Biernaskie, J.; Di Martino, E.S. Collagen structural alterations contribute to stiffening of tissue after split-thickness skin grafting. *Wound Repair Regen.* **2016**, *24*, 263–274. [CrossRef] [PubMed]

86. Su, P.J.; Chen, W.L.; Hong, J.B.; Li, T.H.; Wu, R.J.; Chou, C.K.; Chen, S.J.; Hu, C.; Lin, S.J.; Dong, C.Y. Discrimination of collagen in normal and pathological skin dermis through second-order susceptibility microscopy. *Opt. Express* **2009**, *17*, 11161–11671. [CrossRef] [PubMed]

87. Liu, S.H.; Yang, R.S.; al-Shaikh, R.; Lane, J.M. Collagen in tendon, ligament, and bone healing. A current review. *Clin. Orthop. Relat. Res.* **1995**, *318*, 265–278.

88. Mansfield, J.C.; Winlove, C.P.; Moger, J.; Matcher, S.J. Collagen fibre arrangement in normal and diseased cartilage studied by polarization sensitive nonlinear microscopy. *J. Biomed. Opt.* **2008**, *13*, 044020. [CrossRef] [PubMed]

89. Sivaguru, M.; Durgam, S.; Ambekar, R.; Luedtke, D.; Fried, G.; Stewart, A.; Toussaint, K.C., Jr. Quantitative analysis of collagen fibre organization in injured tendons using Fourier transform-second harmonic generation imaging. *Opt. Express* **2010**, *18*, 24983–24993. [CrossRef] [PubMed]

90. Doras, C.; Taupier, G.; Barsella, A.; Mager, L.; Boeglin, A.; Bulou, H.; Bousquet, P.; Dorkenoo, K.D. Polarization state studies in second harmonic generation signals to trace atherosclerosis lesions. *Opt. Express* **2011**, *19*, 15062–15068. [CrossRef] [PubMed]

91. Lilledahl, M.B.; Haugen, O.A.; de Lange Davies, C.; Svaasand, L.O. Characterization of vulnerable plaques by multiphoton microscopy. *J. Biomed. Opt.* **2007**, *12*, 044005. [CrossRef] [PubMed]

92. Megens, R.T.; oude Egbrink, M.G.; Merkx, M.; Slaaf, D.W.; van Zandvoort, M.A. Two-photon microscopy on vital carotid arteries: Imaging the relationship between collagen and inflammatory cells in atherosclerotic plaques. *J. Biomed. Opt.* **2008**, *13*, 044022. [CrossRef] [PubMed]

93. Mostaço-Guidolin, L.B.; Sowa, M.G.; Ridsdale, A.; Pegoraro, A.F.; Smith, M.S.; Hewko, M.D.; Kohlenberg, E.K.; Schattka, B.; Shiomi, M.; Stolow, A.; et al. Differentiating atherosclerotic plaque burden in arterial tissues using femtosecond CARS-based multimodal nonlinear optical imaging. *Biomed. Opt. Express* **2010**, *1*, 59–73. [CrossRef] [PubMed]

94. Abraham, T.; Fong, G.; Scott, A. Second harmonic generation analysis of early Achilles tendinosis in response to in vivo mechanical loading. *BMC Musculoskelet. Disord.* **2011**, *12*, 26. [CrossRef] [PubMed]

95. Kottmann, R.M.; Sharp, J.; Owens, K.; Salzman, P.; Xiao, G.Q.; Phipps, R.P.; Sime, P.J.; Brown, E.B.; Perry, S.W. Second harmonic generation microscopy reveals altered collagen microstructure in usual interstitial pneumonia versus healthy lung. *Respir. Res.* **2015**, *16*, 61. [CrossRef] [PubMed]

96. Tjin, G.; Xu, P.; Kable, S.H.; Kable, E.P.; Burgess, J.K. Quantification of collagen I in airway tissues using second harmonic generation. *J. Biomed. Opt.* **2014**, *19*, 36005. [CrossRef] [PubMed]

97. Tan, H.Y.; Sun, Y.; Lo, W.; Teng, S.W.; Wu, R.J.; Jee, S.H.; Lin, W.C.; Hsiao, C.H.; Lin, H.C.; Chen, Y.F.; et al. Multiphoton fluorescence and second harmonic generation microscopy for imaging infectious keratitis. *J. Biomed. Opt.* **2007**, *12*, 024013. [CrossRef] [PubMed]

98. Lutz, V.; Sattler, M.; Gallinat, S.; Wenck, H.; Poertner, R.; Fischer, F. Impact of collagen crosslinking on the second harmonic generation signal and the fluorescence lifetime of collagen autofluorescence. *Skin Res. Technol.* **2012**, *18*, 168–179. [CrossRef] [PubMed]

99. Pena, A.M.; Fagot, D.; Olive, C.; Michelet, J.F.; Galey, J.B.; Leroy, F.; Beaurepaire, E.; Martin, J.L.; Colonna, A.; Schanne-Klein, M.C. Multiphoton microscopy of engineered dermal substitutes: Assessment of 3-D collagen matrix remodeling induced by fibroblast contraction. *J. Biomed. Opt.* **2010**, *15*, 056018. [CrossRef] [PubMed]

100. Wolf, K.A.S.; Schacht, V.; Coussens, L.M.; von Andrian, U.H.; van Rheenen, J.; Deryugina, E.; Friedl, P. Collagen-based cell migration models in vitro and in vivo. *Semin. Cell Dev. Biol.* **2009**, *20*, 931–941. [CrossRef] [PubMed]

101. Gurtner, G.C.; Werner, S.; Barrandon, Y.; Longaker, M.T. Wound repair and regeneration. *Nature* **2008**, *453*, 314–321. [CrossRef] [PubMed]

102. Rahmani, W.; Abbasi, S.; Hagner, A.; Raharjo, E.; Kumar, R.; Hotta, A.; Magness, S.; Metzger, D.; Biernaskie, J. Hair follicle dermal stem cells regenerate the dermal sheath, repopulate the dermal papilla, and modulate hair type. *Dev. Cell* **2014**, *31*, 543–558. [CrossRef] [PubMed]

103. Liu, Y.; Zhu, X.; Huang, Z.; Cai, J.; Chen, R.; Xiong, S.; Chen, G.; Zeng, H. Texture analysis of collagen second-harmonic generation images based on local difference local binary pattern and wavelets differentiates human skin abnormal scars from normal scars. *J. Biomed. Opt.* **2015**, *20*, 016021. [CrossRef] [PubMed]

104. Junker, J.P.; Philip, J.; Kiwanuka, E.; Hackl, F.; Caterson, E.J.; Eriksson, E. Assessing quality of healing in skin: Review of available methods and devices. *Wound Repair Regen.* **2014**, *22*, 2–10. [CrossRef] [PubMed]

105. Kumar, N.; Kumar, P.; Nayak Badagabettu, S.; Prasad, K.; Kudva, R.; Vasudevarao, R.C. Surgical implications of asymmetric distribution of dermal collagen and elastic fibres in two orientations of skin samples from extremities. *Plast. Surg. Int.* **2014**, *2014*, 364573. [CrossRef] [PubMed]

106. Verhaegen, P.D.; van Zuijlen, P.P.; Pennings, N.M.; van Marle, J.; Niessen, F.B.; van der Horst, C.M.; Middelkoop, E. Differences in collagen architecture between keloid, hypertrophic scar, normotrophic scar, and normal skin: An objective histopathological analysis. *Wound Repair Regen.* **2009**, *17*, 649–656. [CrossRef] [PubMed]

107. Hollinsky, C.; Sandberg, S. Measurement of the tensile strength of the ventral abdominal wall in comparison with scar tissue. *Clin. Biomech.* **2007**, *22*, 88–92. [CrossRef] [PubMed]

108. Bateman, E.D.; Turner-Warwick, M.; Adelmann-Grill, B.C. Immunohistochemical study of collagen types in human foetal lung and fibrotic lung disease. *Thorax* **1981**, *36*, 645–653. [CrossRef] [PubMed]

109. Thibeault, D.W.; Mabry, S.M.; Ekekezie, I.I.; Zhang, X.; Truog, W.E. Collagen scaffolding during development and its deformation with chronic lung disease. *Pediatrics* **2003**, *111*, 766–776. [CrossRef] [PubMed]

110. Bradley, K.H.; McConnell, S.D.; Crystal, R.G. Lung collagen composition and synthesis. Characterization and changes with age. *J. Biol. Chem.* **1974**, *249*, 2674–2683. [PubMed]

111. Mizikova, I.; Ruiz-Camp, J.; Steenbock, H.; Madurga, A.; Vadasz, I.; Herold, S.; Mayer, K.; Seeger, W.; Brinckmann, J.; Morty, R.E. Collagen and elastin cross-linking is altered during aberrant late lung development associated with hyperoxia. *Am. J. Physiol. Lung Cell. Mol. Physiol.* **2015**, *308*, L1145–L1158. [CrossRef] [PubMed]

112. Pena, A.M.; Fabre, A.; Debarre, D.; Marchal-Somme, J.; Crestani, B.; Martin, J.L.; Beaurepaire, E.; Schanne-Klein, M.C. Three-dimensional investigation and scoring of extracellular matrix remodeling during lung fibrosis using multiphoton microscopy. *Microsc. Res. Tech.* **2007**, *70*, 162–170. [CrossRef] [PubMed]

113. Raub, C.B.; Mahon, S.; Narula, N.; Tromberg, B.J.; Brenner, M.; George, S.C. Linking optics and mechanics in an in vivo model of airway fibrosis and epithelial injury. *J. Biomed. Opt.* **2010**, *15*, 015004. [CrossRef] [PubMed]

114. Abraham, T.; Hirota, J.A.; Wadsworth, S.; Knight, D.A. Minimally invasive multiphoton and harmonic generation imaging of extracellular matrix structures in lung airway and related diseases. *Pulm. Pharmacol. Ther.* **2011**, *24*, 487–496. [CrossRef] [PubMed]

115. Abraham, T.; Hogg, J. Extracellular matrix remodeling of lung alveolar walls in three dimensional space identified using second harmonic generation and multiphoton excitation fluorescence. *J. Struct. Biol.* **2010**, *171*, 189–196. [CrossRef] [PubMed]

116. Hackett, T.L.; Ferrante, S.C.; Hoptay, C.E.; Engelhardt, J.F.; Ingram, J.L.; Zhang, Y.; Alcala, S.E.; Shaheen, F.; Matz, E.; Pillai, D.K.; et al. A Heterotopic Xenograft Model of Human Airways for Investigating Fibrosis in Asthma. *Am. J. Respir. Cell Mol. Biol.* **2017**, *56*, 291–299. [CrossRef] [PubMed]

117. Campbell, J.D.; McDonough, J.E.; Zeskind, J.E.; Hackett, T.L.; Pechkovsky, D.V.; Brandsma, C.A.; Suzuki, M.; Gosselink, J.V.; Liu, G.; Alekseyev, Y.O.; et al. A gene expression signature of emphysema-related lung destruction and its reversal by the tripeptide GHK. *Genome Med.* **2012**, *4*, 67. [CrossRef] [PubMed]

118. Mostaço-Guidolin, L.B.; Osei, E.T.; Hajimohammadi, S.; Ullah, J.; Hackett, T.-L. Novel non-linear optical imaging to understand the composition of firbrilar collagen and elastin in remodeled asthmatic airways. *Am. J. Respir. Crit. Care Med.* **2016**, *193*, A6173.

119. Khan, M.G. *Encyclopedia of Heart Disease*; Academic Press: Cambridge, MA, USA, 2005.

120. Furchgott, R.F. Role of endothelium in responses of vascular smooth muscle. *Circ. Res.* **1983**, *53*, 557–573. [CrossRef] [PubMed]

121. Ross, R. Atherosclerosis is an inflammatory disease. *Am. Heart J.* **1999**, *138*, S419–S420. [CrossRef]
122. Chen, X.; Huang, Z.; Xi, G.; Chen, Y.; Lin, D.; Wang, J.; Li, Z.; Sun, L.; Chen, J.; Chen, R. Quantitative analysis of collagen change between normal and cancerous thyroid tissues based on SHG method. *Proc. SPIE* **2011**. [CrossRef]
123. Han, X.; Burke, R.M.; Zettel, M.L.; Tang, P.; Brown, E.B. Second harmonic properties of tumor collagen: Determining the structural relationship between reactive stroma and healthy stroma. *Opt. Express* **2008**, *16*, 1846–1859. [CrossRef] [PubMed]
124. Boulesteix, T.; Pena, A.M.; Pages, N.; Godeau, G.; Sauviat, M.P.; Beaurepaire, E.; Schanne-Klein, M.C. Micrometer scale ex vivo multiphoton imaging of unstained arterial wall structure. *Cytometry A* **2006**, *69*, 20–26. [CrossRef] [PubMed]
125. Van Zandvoort, M.; Engels, W.; Douma, K.; Beckers, L.; Oude Egbrink, M.; Daemen, M.; Slaaf, D.W. Two-photon microscopy for imaging of the (atherosclerotic) vascular wall: A proof of concept study. *J. Vasc. Res.* **2004**, *41*, 54–63. [CrossRef] [PubMed]
126. Fratzl, P. *Collagen: Structure and Mechanics*; Springer Science & Business Media: Berlin, Germany, 2008.
127. Ottani, V.; Raspanti, M.; Ruggeri, A. Collagen structure and functional implications. *Micron* **2001**, *32*, 251–260. [CrossRef]
128. Heybeli, N.K.B.; Yilmaz, B.; Guler, O. *Musculoskeletal Research and Basic Science—"Tendons and ligaments"*; Springer: Berlin, Germany, 2016; Volume 3.
129. Kannus, P. Structure of the tendon connective tissue. *Scand. J. Med. Sci. Sports* **2000**, *10*, 312–320. [CrossRef] [PubMed]
130. Keene, D.R.; Sakai, L.Y.; Bachinger, H.P.; Burgeson, R.E. Type III collagen can be present on banded collagen fibrils regardless of fibril diameter. *J. Cell Biol.* **1987**, *105*, 2393–2402. [CrossRef] [PubMed]
131. Shimomura, T.; Jia, F.; Niyibizi, C.; Woo, S.L. Antisense oligonucleotides reduce synthesis of procollagen α1 (V) chain in human patellar tendon fibroblasts: Potential application in healing ligaments and tendons. *Connect. Tissue Res.* **2003**, *44*, 167–172. [CrossRef] [PubMed]
132. Sivaguru, M.D.S.; Ambekar, R.; Luedtke, D.; Fried, G.; Stewart, A.; Toussaint, K.C. Quantitative analysis of diseased horse tendons using Fourier-tranform-second-harmonic generation imaging. *Proc. SPIE* **2011**. [CrossRef]
133. Malik, R.A.; Kallinikos, P.; Abbott, C.A.; van Schie, C.H.; Morgan, P.; Efron, N.; Boulton, A.J. Corneal confocal microscopy: A non-invasive surrogate of nerve fibre damage and repair in diabetic patients. *Diabetologia* **2003**, *46*, 683–688. [CrossRef] [PubMed]
134. Raub, C.B.; Putnam, A.J.; Tromberg, B.J.; George, S.C. Predicting bulk mechanical properties of cellularized collagen gels using multiphoton microscopy. *Acta Biomater.* **2010**, *6*, 4657–4665. [CrossRef] [PubMed]
135. Kirkpatrick, N.D.; Hoying, J.B.; Botting, S.K.; Weiss, J.A.; Utzinger, U. In vitro model for endogenous optical signatures of collagen. *J. Biomed. Opt.* **2006**, *11*, 054021. [CrossRef] [PubMed]
136. Turcotte, R.; Mattson, J.M.; Wu, J.W.; Zhang, Y.; Lin, C.P. Molecular Order of Arterial Collagen Using Circular Polarization Second-Harmonic Generation Imaging. *Biophys. J.* **2016**, *110*, 530–533. [CrossRef] [PubMed]
137. Rao, R.A.; Mehta, M.R.; Leithem, S.; Toussaint, K.C., Jr. Quantitative analysis of forward and backward second-harmonic images of collagen fibres using Fourier transform second-harmonic-generation microscopy. *Opt. Lett.* **2009**, *34*, 3779–3781. [CrossRef] [PubMed]
138. Williams, R.M.; Zipfel, W.R.; Webb, W.W. Interpreting second-harmonic generation images of collagen I fibrils. *Biophys. J.* **2005**, *88*, 1377–1386. [CrossRef] [PubMed]
139. Indahl, U.; Næs, T. Evaluation of alternative spectral feature extraction methods of textural images for multivariate modeling. *J. Chemom.* **1998**, *12*, 261–278. [CrossRef]
140. Chang, T.; Kuo, C.J. Texture analysis and classification with tree-structured wavelet transform. *IEEE Trans Image Process* **1993**, *2*, 429–441. [CrossRef] [PubMed]
141. Tuceryan, M.J.; Jain, A.K. Texture Analysis. In *Handbook of Pattern Recognition and Comuter Vision*; Chen, C.H., Ed.; Word Scientific: Singapore, 1993; Volume 2, pp. 235–276.
142. Unser, M. Texture classification and segmentation using wavelet frames. *IEEE Trans. Image Process.* **1995**, *4*, 1549–1560. [CrossRef] [PubMed]
143. Arivazhagan, S.G.L. Texture classfication using wavelet transform. *Pattern Recognit. Lett.* **2003**, *24*, 1513–1521. [CrossRef]

144. Mostaço-Guidolin, L.B.; Ko, A.C.; Popescu, D.P.; Smith, M.S.; Kohlenberg, E.K.; Shiomi, M.; Major, A.; Sowa, M.G. Evaluation of texture parameters for the quantitative description of multimodal nonlinear optical images from atherosclerotic rabbit arteries. *Phys. Med. Biol.* **2011**, *56*, 5319–5334. [CrossRef] [PubMed]
145. Bharati, M.H.; Liu, J.J.; MacGregor, J.F. Image texture analysis: Methods and comparisons. *Chemom. Intell. Lab. Syst.* **2004**, *72*, 57–71. [CrossRef]

International Journal of
Molecular Sciences

MDPI

Article

Burn Eschar Stimulates Fibroblast and Adipose Mesenchymal Stromal Cell Proliferation and Migration but Inhibits Endothelial Cell Sprouting

Hanneke N. Monsuur [1], Lenie J. van den Broek [1], Renushka L. Jhingoerie [1],
Adrianus F. P. M. Vloemans [2] and Susan Gibbs [1,3,*]

[1]   Department of Dermatology, VU University Medical Center, Amsterdam Movement Sciences,
       1081 HZ Amsterdam, The Netherlands; h.monsuur@vumc.nl (H.N.M.);
       l.vandenbroek@vumc.nl (L.J.v.d.B.); renushkajhingoerie@gmail.com (R.L.J.)
[2]   Burn Center, Red Cross Hospital, 1942 LE Beverwijk, The Netherlands; jvloemans@rkz.nl
[3]   Department of Oral Cell Biology, Academic Center for Dentistry Amsterdam (ACTA),
       University of Amsterdam and VU University Amsterdam, Amsterdam Movement Sciences,
       1081 HZ Amsterdam, The Netherlands
*   Correspondence: s.gibbs@vumc.nl; Tel.: +31-204-442-815

Received: 10 July 2017; Accepted: 12 August 2017; Published: 18 August 2017

Abstract: The majority of full-thickness burn wounds heal with hypertrophic scar formation. Burn eschar most probably influences early burn wound healing, since granulation tissue only forms after escharotomy. In order to investigate the effect of burn eschar on delayed granulation tissue formation, burn wound extract (BWE) was isolated from the interface between non-viable eschar and viable tissue. The influence of BWE on the activity of endothelial cells derived from dermis and adipose tissue, dermal fibroblasts and adipose tissue-derived mesenchymal stromal cells (ASC) was determined. It was found that BWE stimulated endothelial cell inflammatory cytokine (CXCL8, IL-6 and CCL2) secretion and migration. However, BWE had no effect on endothelial cell proliferation or angiogenic sprouting. Indeed, BWE inhibited basic Fibroblast Growth Factor (bFGF) induced endothelial cell proliferation and sprouting. In contrast, BWE stimulated fibroblast and ASC proliferation and migration. No difference was observed between cells isolated from dermis or adipose tissue. The inhibitory effect of BWE on bFGF-induced endothelial proliferation and sprouting would explain why excessive granulation tissue formation is prevented in full-thickness burn wounds as long as the eschar is still present. Identifying the eschar factors responsible for this might give indications for therapeutic targets aimed at reducing hypertrophic scar formation which is initiated by excessive granulation tissue formation once eschar is removed.

Keywords: burn; wound healing; wound extract; granulation tissue; endothelial cell; fibroblast; ASC; skin

---

1. Introduction

One of the most frequent causes of full-thickness burn wounds is exposure to hot water or (flash) fire. A full-thickness burn wound results in loss of viable epidermis and dermis. Currently, small full-thickness burns (<15% Total Body Surface Area; TBSA) are treated conservatively for 10–14 days, followed by debridement of eschar and application of a split skin autograft to deeper, non healing regions. To prevent a Severe Systemic Inflammation Syndrom (SIRS), burns larger than 15% TBSA require earlier excision followed by debridement and application of a split skin autograft [1,2]. The majority of full-thickness burn wounds result in the formation of hypertrophic scars, independent of the treatment strategy [3]. In order to develop improved treatment strategies to prevent hypertrophic scar formation, a better understanding of the early stages of wound healing and the influence of eschar on the early healing process is required.

Wound healing of full-thickness burns differs from normal wound healing in several aspects, notably alterations in haemostasis, inflammation and granulation tissue formation [4]. Burn injury coagulates the superficial blood vessels hindering fibrin clot formation, and when the wound is debrided excessive bleeding is triggered, leading to haemostasis, followed by the formation of granulation tissue within a few days [4]. Clinical observations show that the presence of eschar on the wound bed prohibits granulation tissue formation. Granulation tissue is characterized by a high density of fibroblasts, granulocytes, macrophages and microcapillaries. In full-thickness burn wounds the cells required for wound healing have to migrate from the wound edges, from the subcutaneous adipose tissue or other origins. In addition to dermal fibroblasts and dermal-endothelial cells (dermal-EC) migrating from the wound edges also adipose tissue-derived mesenchymal stromal cells (ASC) and adipose-endothelial cells (adipose-EC) are likely to be involved and may contribute to less favorable wound healing and hypertrophic scar formation. For example, the persistent myofibroblasts in hypertrophic scars may originate from the adipose tissue, since a high percentage of ASC express the myofibroblast marker α-smooth muscle actin [5–7]. Also a possible contribution of endothelial cells to hypertrophic scar formation has been suggested as hypertrophic scars contain more microcapillaries than normal scars [8,9]. The alterations in haemostasis, granulation tissue formation and the contribution to wound healing by cells from alternative origins, such as adipose tissue can contribute to the increased risk of hypertrophic scar formation as seen in many full-thickness burn wounds [4].

Since granulation tissue only forms after the burn wound eschar has been removed, burn eschar is most likely to strongly influence early healing of the burn wound. To investigate this further, burn wound extract (BWE) can be isolated from the interface between non-viable eschar and viable tissue and used to represent the burn wound environment, allowing us to study the cellular and molecular components involved in burn wound healing [10,11]. Previously we have shown that this BWE is highly bioactive containing abundant levels of many cytokines, chemokines and growth factors such as CCL2, CCL5, CCL18, CCL20, CCL27, IL-1α, IL-6, CXCL1, CXCL8, basic fibroblast growth Factor (bFGF), hepatocyte growth factor and transforming growth factor-β. BWE could further stimulate ASC and fibroblasts to secrete more mediators related to inflammation, angiogenesis and granulation tissue formation resulting in an amplified inflammatory response [11].

In this study, to further investigate the effect of eschar-derived BWE on delayed granulation tissue formation we focused on endothelial cells derived from the dermis and adipose tissue. The influence of BWE on endothelial cell inflammation, migration, proliferation and angiogenic sprouting was determined. Our results indicate that BWE from full-thickness burn wounds stimulates the secretion of inflammatory proteins and endothelial cell migration, but inhibits endothelial cell proliferation and vessel sprouting. In contrast to the findings with dermal- and adipose-EC, stimulation of both proliferation and migration was seen with fibroblasts and ASC in the presence of BWE.

2. Results

2.1. Eschar

Eschar was removed from the patient by (tangential) excision (Figure 1). Eschar at the interface between non-viable and viable tissue was used to obtain an acellular BWE for the experiments described in this study. Characteristics of the eschar and BWE are shown in Table 1. The histology of this eschar showed absence of an epidermis and a tissue containing many small, rounded cells in the lower eschar layers (Figure 1c). It has previously been reported that eschar contains viable cells resembling ASC [12].

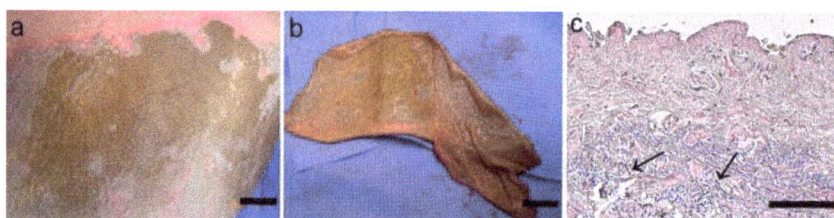

Figure 1. Macroscopic pictures and morphology of human eschar tissue. (**a**) Eschar tissue on patient; (**b**) Eschar after tangential excision; (**c**) Eschar at interface of non-viable and viable tissue from which acellular Burn Wound Extract (BWE) is derived (see Materials and Methods); hematoxylin and eosin staining showing the absence of epidermis; small, rounded cells are present in the lower levels of the dermis (areas with blue nuclei; see arrows). Scale bar = 1 cm (**a,b**) or 200 μm (**c**).

Table 1. Characteristics of burn wounds, properties of the burn wound extract (BWE) and experiments where each BWE donor was used.

#	Gender	Age	Cause of Burn	TBSA	Time after Injury (Days)	Protein Concentration BWE (μg/mL)	BWE Used in Figure
1	female	49	hot water	9%	21	2850	F2
2	female	62	hot object	2%	13	2420	F2
3	female	30	flame	60%	6	1120	F2
4	female	73	flame	2%	17	3260	F4
5	male	49	chemicals	48%	6	720	F3,4,5
6	male	64	hot object	7%	12	1780	F3,4,5
7	male	45	chemicals	0.5%	14	830	F3,4,5
8	female	46	hot object	0.5%	10	1620	F3,4,5
9	male	52	hot water	2.5%	10	2970	F3,4,5

2.2. Burn Wound Extract Inhibits Endothelial Cell Proliferation and Sprouting

Previously we have shown that eschar BWE contains a large reservoir of bioactive cytokines and chemokines [11]. In order to determine the effect of BWE on cells underneath the eschar, both dermal-EC and adipose-EC were exposed to BWE. The influence of BWE on (i) inflammatory cytokine secretion was determined by ELISA; (ii) cell migration was determined using the wound healing scratch assay and (iii) proliferation by 3H incorporation.

BWE exposure increased CXCL8, IL-6 and CCL2 secretion by dermal- and adipose-EC in a dose dependent manner in line with our previous findings for fibroblasts and ASC (Figure 2). CXCL8 secretion was induced by 8.2 and 8.9-fold, IL-6 by 37.3 and 28.1-fold and CCL2 by 4.4 and 4.7-fold (dermal- and adipose-EC respectively; 100 μg/mL). Migration of adipose-EC was also stimulated during a time period of 16 h in the scratch wound-healing assay. For dermal-EC a relative increase of 1.19-fold (100 μg/mL) was observed compared to the bFGF positive control, which showed 1.29-fold increase (not significant) (Figure 3a). For adipose-EC a relative increase of 1.33-fold (100 μg/mL) was observed compared to the bFGF positive control, which showed 1.50-fold increase ($p < 0.01$) (Figure 3a). The morphology of the cells was not affected by the addition of BWE (Supplementary Figure S1a). In contrast to cell migration, BWE did not influence the basal level of proliferation of endothelial cells (Figure 3b,c). Endothelial cell proliferation was stimulated by the addition of bFGF, for dermal-EC a relative increase of 4.63-fold was achieved by 10 ng/mL bFGF and for adipose-EC a relative increase of 3.35-fold (Figure 3b,c). Notably, when BWE was added in combination with bFGF, the bFGF stimulated increase in proliferation was inhibited in a dose-dependent manner (Figure 3b,c). The inhibitory effect was more pronounced for dermal-EC than for adipose-EC. The relative proliferation for dermal-EC was reduced by 49% and for adipose-EC by 37% when 10 ng/mL bFGF was combined with 100 μg/mL BWE.

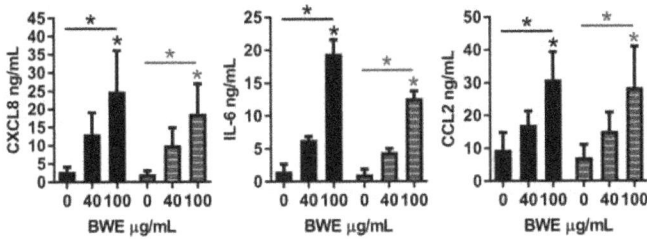

Figure 2. Secretion of inflammation factors by dermal- and adipose-endothelial cells. Secretion of CXCL8, IL-6 and CCL2 after a 24 h exposure to 0, 40 or 100 μg/mL BWE. Basal amounts of protein in culture medium containing 100 μg/mL BWE without cells: CXCL8: <2 ng/mL; IL-6: <1 ng/mL; CCL2: <0.5 ng/mL. Significance of the dose response curve was calculated using a one-way ANOVA followed by a Dunn's multiple comparison test and significance of basic Fibroblast Growth Factor (bFGF) induction was tested with a *t*-test; * $p < 0.05$. Data is shown for 3 independent experiments as mean ± SEM. Each experiment represents a different cell donor and a different BWE donor (see Table 1). Black, solid bars represent dermal-endothelial cells (dermal-EC) and grey, striped bars adipose-endothelial cells (adipose-EC).

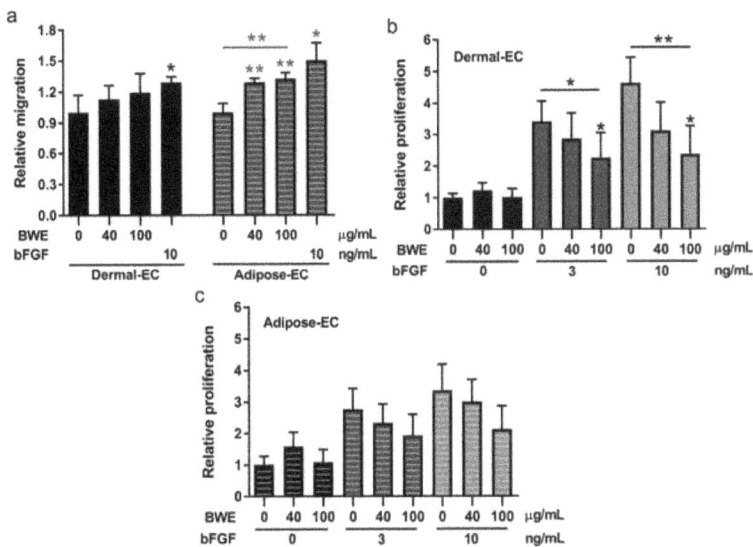

Figure 3. Migration scratch assay and proliferation assay using dermal- and adipose-endothelial cells. (a) Relative migration values of dermal- and adipose-EC cultured in the presence of 0, 40 or 100 μg/mL BWE or 10 ng/mL bFGF. Relative migration is calculated from the scratch area closed compared to unexposed endothelial cells; (b) Proliferation (3H incorporation) values, relative to unexposed endothelial cells, of dermal-EC cultured in the presence of 0, 3 or 10 ng/mL bFGF in combination with BWE; (c) Proliferation values, relative to unexposed endothelial cells, of adipose-EC cultured in the presence of 0, 3 or 10 ng/mL bFGF combined with BWE. Significance of the dose response curve was calculated using a one-way ANOVA followed by a Dunn's multiple comparison test and significance of bFGF induction was tested with a *t*-test; * $p < 0.05$, ** $p < 0.01$. Data is shown for 4 independent experiments as mean ± SEM. Each experiment represents a different cell donor and a different BWE donor (see Table 1). Solid bars represent dermal-EC and striped bars adipose-EC.

Since angiogenesis involves a combination of cell proliferation, migration and matrix degradation we then determined the influence of BWE in a vessel sprouting assay. Sprout formation, as a measure for angiogenic response, was investigated using a 3D fibrin matrix. Endothelial cells seeded on top of this matrix will form sprouts into the matrix when an angiogenic stimulus is added to the medium [13]. Dermal- and adipose-EC did not form sprouts when exposed to BWE alone. When dermal- and adipose-EC were exposed to the angiogenic stimulus bFGF (10 ng/mL) induction of sprouting was clearly observed (Figure 4). Notably, this bFGF mediated increase in sprouting was inhibited by BWE in a dose-dependent manner. Dermal-EC showed 72% inhibition and adipose-EC showed 82% inhibition when 10 ng/mL bFGF was combined with 100 µg/mL BWE (Figure 4).

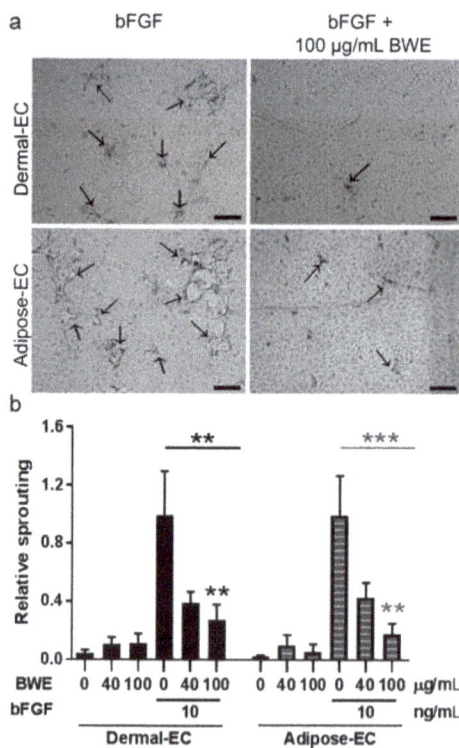

Figure 4. Sprouting assay using dermal- and adipose-endothelial cells. (**a**) Representative pictures of sprout formation of dermal- and adipose-EC into 3D fibrin matrices when exposed to 10 ng/mL bFGF or 10 ng/mL bFGF with 100 µg/mL BWE. Arrows indicate the sprouts; (**b**) Relative sprouting values of dermal- and adipose-EC in the presence of 0 or 10 ng/mL bFGF combined with 0, 40 or 100 µg/mL BWE compared to 10 ng/mL bFGF stimulated cultures. Significance of the dose response curve was calculated using a one-way ANOVA followed by a Dunn's multiple comparison test; ** $p < 0.01$, *** $p < 0.001$. Data is shown for 6 independent experiments as mean $\pm$ SEM. Each experiment represents a different cell donor and a different BWE donor (see Table 1). Black, solid bars represent dermal-EC and grey, striped bars adipose-EC. Scale bars represent 50 µm.

2.3. Burn Wound Extract Stimulates Both Migration and Proliferation of Fibroblasts and ASC

Next the influence of BWE on fibroblast and ASC proliferation and migration was investigated. In contrast to endothelial cells, fibroblasts and ASC both showed a significant increase in migration in the scratch assay in the same order of magnitude as the epidermal growth factor (EGF) positive control

(Figure 5a). For fibroblasts the highest relative increase of 2.31-fold was observed using 100 µg/mL BWE whereas EGF only showed 1.73-fold increase. For ASC the highest relative increase of 2.31-fold was observed using 40 µg/mL BWE whereas EGF only showed 1.94-fold increase. The morphology of the cells was not affected by the addition of BWE (Supplementary Figure S1b). BWE stimulated proliferation of fibroblasts and ASC, to the same extent as EGF (5 ng/mL) (Figure 5b). For fibroblasts the highest relative increase of 1.50-fold using 100 µg/mL BWE was observed whereas EGF showed 1.42-fold increase. For ASC the highest relative increase of 1.88 fold using 40 µg/mL BWE was observed whereas EGF showed 1.68 fold increase.

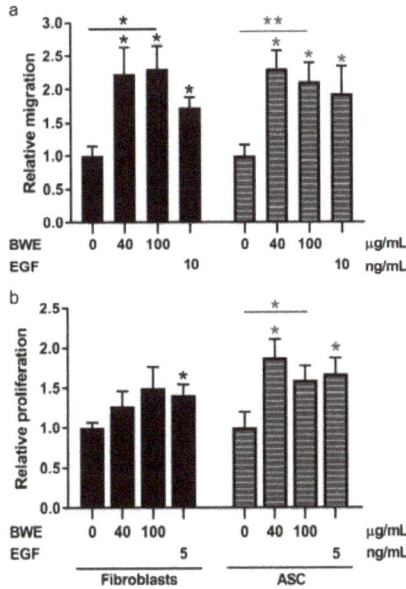

Figure 5. Migration scratch assay and proliferation assay using dermal fibroblasts and adipose tissue-derived mesenchymal stromal cells. (**a**) Relative migration values of fibroblasts and Adipose tissue-derived mesenchymal Stromal Cells (ASC) cultured in the presence of BWE or epidermal growth factor (EGF); (**b**) Relative proliferation values of fibroblasts and ASC cultured in the presence of BWE or EGF. The 5 and 10 ng/mL concentrations of EGF are optimal concentrations to serve as a positive control for the proliferation and migration experiments respectively. Significance of the dose response curve was calculated using a one-way ANOVA followed by a Dunn's multiple comparison test and significance of EGF induction was tested with a *t*-test; * $p < 0.05$, ** $p < 0.01$. Data is shown for 4 independent experiments as mean ± SEM. Each experiment represents a different cell donor and a different BWE donor (see Table 1). Black, solid bars represent fibroblasts and grey, striped bars ASC.

3. Discussion

The BWE derived from full-thickness burn wounds contains a very potent cocktail of bioactive cytokines, chemokines and growth factors representative of burn wound eschar [11]. In this study our focus was on the effect of BWE on endothelial cells from dermis and adipose tissue. BWE stimulated endothelial cells to secrete inflammatory proteins and enhanced endothelial cell migration. However, BWE had no effect on endothelial cell proliferation and angiogenic sprouting, and actually inhibited bFGF-mediated proliferation and sprouting. In contrast BWE stimulated both migration and proliferation of fibroblasts and ASC.

Our observation that BWE stimulated endothelial cells to secrete IL-6, CXCL8 and CCL2 was in agreement with our previous findings in which we showed that BWE stimulated inflammatory protein secretion by fibroblasts and ASC (but not keratinocytes) [11]. In the BWE there are many proteins present that can elicit an inflammatory response in endothelial cells, such as CCL2, IL-6, IL-1α, CXCL1 and CXCL8 [11]. No differences were found between dermal-EC and adipose-EC, however ASC were found to secrete more CCL2, IL-6 and CXCL8 in response to BWE than fibroblasts. The increased migration of endothelial cells, fibroblasts and ASC in the wound healing scratch assay may be attributed to the highly bioactive composition of the BWE that contains many chemotactic proteins such as bFGF and CCL5 [14,15]. Notably, we found that BWE inhibited endothelial cell proliferation and vessel sprouting. Vessel sprouting requires a combination of proliferation, migration and matrix breakdown [16,17]. Other studies investigating the effect of burn blister fluid on endothelial cells showed conflicting results, as endothelial cell proliferation, chemotaxis and angiogenesis were either stimulated or not affected [18–20]. However, blister fluid cannot be compared to our BWE, since blister fluid is obtained very early after injury (within 6–72 h) from (deep) partial-thickness burn wounds (compared to BWE which is isolated from full-thickness burn wounds between day 6 and 21). However, our results for fibroblasts and ASC were in line with results from blister fluid, which had a clear stimulatory effect on fibroblast proliferation and contraction [21,22].

In full-thickness burn wounds the eschar is often left on the burn wound for 10–14 days before a decision is made to remove the eschar by (tangential) excision. During this time period not only an alteration in wound healing is seen with regards to hemostasis, but also in granulation tissue formation. The removal of eschar causes excessive bleeding followed by hemostasis and (excessive) granulation tissue forms a few days after escharotomy [4]. An explanation for the inhibitory effect of BWE on bFGF-induced proliferation and sprouting may be the presence of inhibitory factors in the BWE, e.g., plasminogen activator inhibitor-1 or angiopoietin-2. The non-viable burned tissue might also release collagen-4 derived angiogenesis inhibitors into the BWE, for example arresten, canstatin or tumstatin [19]. Our findings give an explanation as to why excessive granulation tissue formation is prevented in full-thickness wounds as long as the eschar is still present. Further research is required to identify the factors present at the interface of non-viable eschar and viable tissue as this can give indications for therapeutic targets aimed at reducing hypertrophic scar formation which is initiated by excessive granulation tissue formation.

4. Materials and Methods

4.1. Human Tissue

Human adult skin with underlying adipose tissue was obtained from healthy individuals undergoing abdominal dermolipectomy. The discarded skin was collected anonymously if patients had not objected to use of their rest material (opt-out system). Eschar was obtained from patients with full-thickness burn wounds undergoing escharotomy. Anonymous tissue collection procedures were performed in compliance with the "Code for Proper Secondary Use of Human Tissue" as formulated by the Dutch Federation of Medical Scientific Societies (www.federa.org) and following procedures approved by the institutional review board of the VU University medical center.

4.2. Burn Wound Extract

Eschar was removed 6–21 days post burn from 9 patients with full-thickness burn wounds. Characteristics of the burn wounds and properties of the BWE are shown in Table 1. The upper layers of the eschar were removed and discarded until just above the viable layer. Then the eschar at the interface between non-viable and viable tissue was collected, cut into 0.4 cm^2 pieces and placed in either 1 mL PBS or 1 mL PBS containing protease inhibitor cocktail (1:100; PIC; Sigma-Aldrich, St. Louis, MO, USA). After two hours gentle shaking at 4 °C the remaining tissue was removed and the solution was centrifuged to pellet any remaining tissue. The supernatant was then filtered using a sterile 0.4 μm pore size filter (Merck Millipore, Amsterdam, the Netherlands) to ensure that all cell and tissue debris was removed as well as any bacteria. This acellular supernatant extracted from the tissue interface between non-viable and viable tissue was collected and stored at −80 °C, and further referred to as BWE. The total protein concentration in the BWE was determined using the Bradford Bio-Rad Protein Assay (BioRad Laboratories, Hercules, CA, USA) as described by the supplier. The BWE contained varying protein concentrations, 2200 ± 1000 μg/mL. For each independent experiment a different cell donor and a BWE isolated from a different donor was used. In the experiments BWE was diluted in the culture medium to 40 and 100 μg/mL to standardize the protein content within each experiment.

4.3. Cell Culture

Adipose tissue was carefully dissected from the skin. The remaining skin was then treated with dispase to remove the epidermis from the dermis. The adipose stromal vascular cell fraction and dermal stromal vascular cell fraction were then isolated using collagenase type II/dispase II adipose tissue or dermis as previously described [14].

Dermal fibroblasts (fibroblasts) and ASC were cultured in DMEM (Lonza, Verviers, Belgium), 1% UltroSerG (UG) (BioSepra SA, Cergy-Saint-Christophe, France) and 1% penicillin/streptomycin (P/S) (Invitrogen, Carlsbad, CA, USA).

Endothelial cells were purified from the dermal stromal vascular cell fraction (dermal-EC) and from the adipose stromal vascular cell fraction (adipose-EC) using a MidiMACS separator with microbeads against CD31 as previously described [13]. A >99% pure population (CD31+/CD90−) was obtained at passage 3. The endothelial cells were further cultured on 1% gelatin (Sigma-Aldrich) coated flasks in endothelial cell medium (EC medium): M199 medium (Lonza), 1% P/S, 2 mM L-glutamin (Invitrogen), 10% heat-inactivated New Born Calf Serum (Invitrogen), 10% heat-inactivated Human Serum (Invitrogen), 5 U/mL heparin (Pharmacy VUmc, Amsterdam, The Netherlands) and 3.7 μg/mL endothelial cell growth factor (ECGF; crude extract from bovine brain) (Physiology department VUmc, Amsterdam, The Netherlands).

The cells were stored in the vapor phase of liquid nitrogen until required. For experiments dermal-EC and adipose-EC between passage 5 and 7 were used and fibroblasts and ASC between passage 1 and 3. In all experiments donor-matched cells were used.

4.4. Exposure of Endothelial Cells to BWE

Dermal-EC and adipose-EC were seeded in an equal density of 1×10^4 cells/cm^2 on gelatin-coated culture plates in EC medium. After 16 h the wells were washed twice with HBSS/0.5 mM EDTA before replacing the medium with M199 medium, 10% HS, 10% NBCS, 1% P/S, 2 mM L-glutamin. Monolayers of endothelial cells were exposed to 0, 40 or 100 μg/mL BWE in PIC in HMEC medium for 24 h. Culture supernatants were collected for ELISA.

4.5. Cell Migration Assay

Migration of dermal- and adipose-EC was studied as previously described [13]. In short: Dermal- and adipose-EC were seeded in an equal density of 2×10^4 cells/cm^2 on gelatin-coated culture plates in EC medium. The EC medium was replaced when the cells reached confluency by M199 medium with 10% HS, 10% NBCS, 2 mM L-glutamin and 1% P/S for 24 h before the start of the experiment. A scratch was drawn in a confluent monolayer of dermal- and adipose-EC with a plastic disposable pipette tip. After washing, the cells were exposed to M199 medium with 10% HS, 10% NBCS, 2 mM L-glutamin and 1% P/S supplemented with different concentrations of BWE (0, 40 or 100 μg/mL) or 10 ng/mL bFGF. Phase contrast pictures were taken directly after drawing the scratch and after 16 h of exposure.

Fibroblasts and ASC were seeded in a density of 3.5×10^4 cells/cm^2 on culture plates in DMEM medium with 1% UltroSerG and 1% P/S [14]. The medium was replaced when the cells reached confluency by DMEM medium with 0.1% Bovine serum albumin and 1% P/S for four days. Then the scratch was drawn through the confluent monolayer with a plastic disposable pipette tip. After washing, the cells were exposed to DMEM medium with 0.1% BSA and 1% P/S supplemented with different concentrations of BWE (0, 40 or 100 µg/mL) or 5 ng/mL EGF. Phase contrast pictures were taken directly after drawing the scratch and 72 h of exposure. Data were analyzed using an image processing algorithm [23]. The closed area was determined by subtracting the open area at time point $t = 16$ h or $t = 72$ h from $t = 0$ h.

4.6. Proliferation Assay

Proliferation of dermal- and adipose-EC in response to BWE was determined using ^{3}H-thymidine incorporation, adapted from Monsuur et al. [13]. Endothelial cell proliferation was studied in triplicate in low nutrient medium in order to determine their response to BWE. The endothelial cells were seeded on gelatin-coated culture plates in a density of 8×10^3 cells/cm^2 in M199 medium with 5% HS, 10% NBCS, 2 mM L-glutamin and 1% P/S. Cells were left to adhere to the culture plates for 16 h, followed by 72 h stimulation with either BWE (0, 40, 100 µg/mL in PBS) or bFGF (0, 3 or 10 ng/mL; ReliaTech GmbH, Wolfenbuttel, Germany) or combinations between growth factor and BWE (0, 40, 100 µg/mL). During the last 16 h of growth, 1 µCi ^{3}H-thymidine (Perkin Elmer, Belgium) was added to quantify the amount of DNA replication as a measure for proliferation. The β-emission was measured with Ultima Gold scintillation fluid on a Tri-Carb 2800TR Liquid Scintillation Analyzer (PerkinElmer, Zaventem, Belgium).

Proliferation of fibroblasts and ASC in response to BWE was determined in triplicate by manual cell counting. Fibroblasts were seeded on uncoated culture plates in a density of 5×10^3 cells/cm^2 in DMEM medium with 0.1% Bovine serum albumin and 1% P/S. Cells were left to adhere to the culture plates for 16 h, followed by 56 h stimulation with either BWE (0, 40, 100 µg/mL in PBS) or EGF (0, 10 ng/mL; Sigma-Aldrich). Phase contrast pictures were taken directly after exposure to BWE and after 56 h of exposure. Manual cell counting was performed to determine relative proliferation compared to control.

4.7. In Vitro Sprouting Assay

In vitro tube formation was studied using 3D fibrin matrices and dermal- or adipose-EC, as previously described [13]. Briefly, fibrin matrices were prepared by addition of thrombin (0.5 U/mL) (MSD, Haarlem, the Netherlands) to a 3 mg/mL fibrinogen (Enzyme Research Laboratories, Leiden, the Netherlands) solution in M199 medium and 100 µL was added to the wells of a 96-well plate. After polymerization, thrombin was inactivated by incubating the matrices with M199 medium with 10% HS, 10% NBCS, 2 mM L-glutamin and 1% P/S. Dermal- or adipose-EC were seeded to reach a confluent density of 6×10^4 cells/cm^2. After 16 h, the adipose- and dermal-EC were stimulated with M199 medium with 10% HS, 10% NBCS, 2 mM L-glutamin and 1% P/S and 2 ng/mL TNF-α (ReliaTech GmbH) supplemented with BWE (0, 40, 100 µg/mL in PBS) or bFGF (0, 3 or 10 ng/mL) or BWE and bFGF combined. The sprouts formed by dermal- or adipose-EC into the fibrin matrices were photographed and analyzed using a Nikon Eclipse 80i microscope (Nikon, Tokyo, Japan) and NIS-elements AR software 3.2 (Nikon). The amount of sprouting was measured as percentage surface area of the sprouts of the total surface of the picture.

4.8. Histological Staining

Paraffin embedded sections of eschar of 5 µm were stained for morphological analysis (hematoxylin and eosin; HE). The sections were photographed using a Nikon Eclipse 80i microscope (Nikon).

4.9. Secretion of Cytokines and Chemokines

ELISAs were performed using commercially available ELISA antibodies. All reagents were used in accordance to the manufacturer's specifications. IL-6 and CCL2 (both R&D Systems, Abingdon, UK) and CXCL8 (Sanquin, Amsterdam, The Netherlands). ELISA results are expressed in ng/mL.

4.10. Statistical Analysis

Statistical analyses were performed using *t*-tests or one-way ANOVA followed by a Dunn's multiple comparison test. All data was obtained from three to six independent experiments using different cell donors and duplicate wells. The cells in each experiment were donor-matched. Each cell donor was combined with a different BWE donor. Differences were considered significant when * $p < 0.05$, ** $p < 0.01$, *** $p < 0.001$. Results are shown as mean $\pm$ SEM.

Supplementary Materials: Supplementary materials can be found at www.mdpi.com/1422-0067/18/8/1790/s1.

Acknowledgments: This study was financed by the Dutch Burns Foundation grant number 13.101, the Netherlands. The authors wish to acknowledge the patients and their surgeons that made this study possible. The authors would like to thank M. Thon for practical assistance and P. van Zuijlen for sharing clinical observations regarding burn wound healing.

Author Contributions: Hanneke N. Monsuur, Lenie J. van den Broek and Susan Gibbs conceived and designed the experiments, Hanneke N. Monsuur and Renushka Jhingoerie performed the experiments, Hanneke N. Monsuur analyzed the data, Adrianus F. P. M. Vloemans provided eschar, Lenie J. van den Broek, Susan Gibbs supervised the project, Hanneke N. Monsuur wrote the manuscript together with Lenie J. van den Broek, Adrianus F. P. M. Vloemans and Susan Gibbs.

Conflicts of Interest: We have the following interests: Susan Gibbs is co-founder of A-Skin BV, which is a VU university medical center startup company (SME).

Abbreviations

ASC	Adipose tissue-derived mesenchymal stromal cell
bFGF	Basic fibroblast growth factor
BWE	Burn wound extract
EC	Endothelial cell
EGF	Epidermal growth factor
HE	Hematoxylin and eosin
PIC	Protease inhibitor cocktail
TBSA	Total body surface area

References

1. Van der Wal, M.B.; Vloemans, J.F.; Tuinebreijer, W.E.; van de Ven, P.; van Unen, E.; van Zuijlen, P.P.; Middelkoop, E. Outcome after burns: An observational study on burn scar maturation and predictors for severe scarring. *Wound Repair Regen.* **2012**, *20*, 676–687. [CrossRef] [PubMed]
2. DeSanti, L. Pathophysiology and current management of burn injury. *Adv. Skin Wound Care* **2005**, *18*, 323–332. [CrossRef] [PubMed]
3. Bloemen, M.C.; van der Veer, W.M.; Ulrich, M.M.; van Zuijlen, P.P.; Niessen, F.B.; Middelkoop, E. Prevention and curative management of hypertrophic scar formation. *Burns* **2009**, *35*, 463–475. [CrossRef] [PubMed]
4. Van der Veer, W.M.; Bloemen, M.C.; Ulrich, M.M.; Molema, G.; van Zuijlen, P.P.; Middelkoop, E.; Niessen, F.B. Potential cellular and molecular causes of hypertrophic scar formation. *Burns* **2009**, *35*, 15–29. [CrossRef] [PubMed]
5. Van den Bogaerdt, A.J.; van der Veen, V.C.; van Zuijlen, P.P.; Reijnen, L.; Verkerk, M.; Bank, R.A.; Middelkoop, E.; Ulrich, M.M. Collagen cross-linking by adipose-derived mesenchymal stromal cells and scar-derived mesenchymal cells: Are mesenchymal stromal cells involved in scar formation? *Wound Repair Regen.* **2009**, *17*, 548–558. [CrossRef] [PubMed]
6. Van den Broek, L.J.; Niessen, F.B.; Scheper, R.J.; Gibbs, S. Development, validation and testing of a human tissue engineered hypertrophic scar model. *ALTEX* **2012**, *29*, 389–402. [CrossRef] [PubMed]
7. Matsumura, H.; Engrav, L.H.; Gibran, N.S.; Yang, T.M.; Grant, J.H.; Yunusov, M.Y.; Fang, P.; Reichenbach, D.D.; Heimbach, D.M.; Isik, F.F. Cones of skin occur where hypertrophic scar occurs. *Wound Repair Regen.* **2001**, *9*, 269–277. [CrossRef] [PubMed]

8. Van der Veer, W.M.; Niessen, F.B.; Ferreira, J.A.; Zwiers, P.J.; de Jong, E.H.; Middelkoop, E.; Molema, G. Time course of the angiogenic response during normotrophic and hypertrophic scar formation in humans. *Wound Repair Regen.* **2011**, *19*, 292–301. [CrossRef] [PubMed]

9. Amadeu, T.; Braune, A.; Mandarim-de-Lacerda, C.; Porto, L.C.; Desmouliere, A.; Costa, A. Vascularization pattern in hypertrophic scars and keloids: A stereological analysis. *Pathol. Res. Pract.* **2003**, *199*, 469–473. [CrossRef] [PubMed]

10. Widgerow, A.D.; King, K.; Tocco-Tussardi, I.; Banyard, D.A.; Chiang, R.; Awad, A.; Afzel, H.; Bhatnager, S.; Melkumyan, S.; Wirth, G.; et al. The burn wound exudate-an under-utilized resource. *Burns* **2015**, *41*, 11–17. [CrossRef] [PubMed]

11. Van den Broek, L.J.; Kroeze, K.L.; Waaijman, T.; Breetveld, M.; Sampat-Sardjoepersad, S.C.; Niessen, F.B.; Middelkoop, E.; Scheper, R.J.; Gibbs, S. Differential response of human adipose tissue-derived mesenchymal stem cells, dermal fibroblasts, and keratinocytes to burn wound exudates: Potential role of skin-specific chemokine ccl27. *Tissue Eng. Part A* **2014**, *20*, 197–209. [CrossRef] [PubMed]

12. Van den Bogaerdt, A.J.; van Zuijlen, P.P.; van Galen, M.; Lamme, E.N.; Middelkoop, E. The suitability of cells from different tissues for use in tissue-engineered skin substitutes. *Arch. Dermatol. Res.* **2002**, *294*, 135–142. [CrossRef] [PubMed]

13. Monsuur, H.N.; Weijers, E.M.; Niessen, F.B.; Gefen, A.; Koolwijk, P.; Gibbs, S.; van den Broek, L.J. Extensive characterization and comparison of endothelial cells derived from dermis and adipose tissue: Potential use in tissue engineering. *PLoS ONE* **2016**, *11*, e0167056. [CrossRef] [PubMed]

14. Kroeze, K.L.; Jurgens, W.J.; Doulabi, B.Z.; van Milligen, F.J.; Scheper, R.J.; Gibbs, S. Chemokine-mediated migration of skin-derived stem cells: Predominant role for ccl5/rantes. *J. Investig. Dermatol.* **2009**, *129*, 1569–1581. [CrossRef] [PubMed]

15. Suffee, N.; Hlawaty, H.; Meddahi-Pelle, A.; Maillard, L.; Louedec, L.; Haddad, O.; Martin, L.; Laguillier, C.; Richard, B.; Oudar, O.; et al. Rantes/ccl5-induced pro-angiogenic effects depend on ccr1, ccr5 and glycosaminoglycans. *Angiogenesis* **2012**, *15*, 727–744. [CrossRef] [PubMed]

16. De Smet, F.; Segura, I.; de Bock, K.; Hohensinner, P.J.; Carmeliet, P. Mechanisms of vessel branching: Filopodia on endothelial tip cells lead the way. *Arterioscler. Thromb. Vasc. Biol.* **2009**, *29*, 639–649. [CrossRef] [PubMed]

17. Carmeliet, P. Angiogenesis in health and disease. *Nat. Med.* **2003**, *9*, 653–660. [CrossRef] [PubMed]

18. Pan, S.C.; Wu, L.W.; Chen, C.L.; Shieh, S.J.; Chiu, H.Y. Deep partial thickness burn blister fluid promotes neovascularization in the early stage of burn wound healing. *Wound Repair Regen.* **2010**, *18*, 311–318. [CrossRef] [PubMed]

19. Pan, S.C.; Wu, L.W.; Chen, C.L.; Shieh, S.J.; Chiu, H.Y. Angiogenin expression in burn blister fluid: Implications for its role in burn wound neovascularization. *Wound Repair Regen.* **2012**, *20*, 731–739. [CrossRef] [PubMed]

20. Nissen, N.N.; Gamelli, R.L.; Polverini, P.J.; DiPietro, L.A. Differential angiogenic and proliferative activity of surgical and burn wound fluids. *J. Trauma Acute Care Surg.* **2003**, *54*, 1205–1210. [CrossRef] [PubMed]

21. Wilson, A.M.; McGrouther, D.A.; Eastwood, M.; Brown, R.A. The effect of burn blister fluid on fibroblast contraction. *Burns* **1997**, *23*, 306–312. [CrossRef]

22. Inoue, M.; Zhou, L.J.; Gunji, H.; Ono, I.; Kaneko, F. Effects of cytokines in burn blister fluids on fibroblast proliferation and their inhibition with the use of neutralizing antibodies. *Wound Repair Regen.* **1996**, *4*, 426–432. [CrossRef] [PubMed]

23. Topman, G.; Sharabani-Yosef, O.; Gefen, A. A standardized objective method for continuously measuring the kinematics of cultures covering a mechanically damaged site. *Med. Eng. Phys.* **2012**, *34*, 225–232. [CrossRef] [PubMed]

International Journal of
Molecular Sciences

MDPI

Article

The Role of Focal Adhesion Kinase in Keratinocyte Fibrogenic Gene Expression

Michael Januszyk [1,2], Sun Hyung Kwon [1], Victor W. Wong [1], Jagannath Padmanabhan [1], Zeshaan N. Maan [1], Alexander J. Whittam [1], Melanie R. Major [1] and Geoffrey C. Gurtner [1,*]

[1] Hagey Laboratory, Division of Plastic Surgery, Department of Surgery,
Stanford University School of Medicine, Stanford, CA 94305-5148, USA; mjanuszyk@mednet.ucla.edu (M.J.);
kwonsunh@stanford.edu (S.H.K.); vicw.wong@gmail.com (V.W.W.); jaganpa@stanford.edu (J.P.);
zmaan@stanford.edu (Z.N.M.); alexander.whittam@gmail.com (A.J.W.); melaniermajor@gmail.com (M.R.M.)
[2] Program in Biomedical Informatics, Stanford University School of Medicine, Stanford,
CA 94305-5148, USA
* Correspondence: ggurtner@stanford.edu; Tel.: +1-650-736-2776; Fax: +1-650-724-9501

Received: 28 June 2017; Accepted: 1 September 2017; Published: 7 September 2017

Abstract: Abnormal skin scarring causes functional impairment, psychological stress, and high socioeconomic cost. Evidence shows that altered mechanotransduction pathways have been linked to both inflammation and fibrosis, and that focal adhesion kinase (FAK) is a key mediator of these processes. We investigated the importance of keratinocyte FAK at the single cell level in key fibrogenic pathways critical for scar formation. Keratinocytes were isolated from wildtype and keratinocyte-specific FAK-deleted mice, cultured, and sorted into single cells. Keratinocytes were evaluated using a microfluidic-based platform for high-resolution transcriptional analysis. Partitive clustering, gene enrichment analysis, and network modeling were applied to characterize the significance of FAK on regulating keratinocyte subpopulations and fibrogenic pathways important for scar formation. Considerable transcriptional heterogeneity was observed within the keratinocyte populations. FAK-deleted keratinocytes demonstrated increased expression of genes integral to mechanotransduction and extracellular matrix production, including Igtbl, Mmpla, and Col4a1. Transcriptional activities upon FAK deletion were not identical across all single keratinocytes, resulting in higher frequency of a minor subpopulation characterized by a matrix-remodeling profile compared to wildtype keratinocyte population. The importance of keratinocyte FAK signaling gene expression was revealed. A minor subpopulation of keratinocytes characterized by a matrix-modulating profile may be a keratinocyte subset important for mechanotransduction and scar formation.

Keywords: focal adhesion kinase; keratinocyte; mechanotransduction; extracellular matrix; single-cell transcriptional analysis; skin fibrosis; hypertrophic scar; transcriptomics

1. Introduction

Tissue repair is among the most complex biological processes and occurs through a highly regulated cascade of overlapping biochemical and cellular events [1]. The underlying orchestrators of the wound healing cascade have yet to be fully elucidated [2]. A fundamental paradigm of cutaneous healing in the adult human is that every injury provokes a fibroproliferative response resulting in the formation of a scar [3]. Scar tissue, formed through normal wound healing to re-establish continuity of the integument, represents the "midpoint" in a spectrum of wound healing responses [4]. Aberrations in the process, such as hypertrophic scarring and keloid formation, represent an "over-healing" response, whereas some patients may suffer from "under-healing" in chronic and/or delayed wounds [5]. Both extremes of wound healing can lead to significant functional impairment, psychosocial morbidity, and constitute a significant socioeconomic burden [6].

Although the pathogenesis of over- and under-healing is not completely understood, recent studies have provided significant insight into the pathophysiologic basis of dysfunctional wound healing. Mechanical cues have been identified to play a major role in both chronic wound development as well as hypertrophic scar (HTS) formation [7]. Mechanotransduction pathways influence the homeostasis in virtually all tissues and organ systems [8–10]. Mechanical homeostasis in the skin is also a critical regulator of skin biology. For example, the human skin has been shown to have static lines of maximal stress called the Langer's lines, which surgeons use to orient incisions to minimize scar formation [11]. In addition, joint movement, muscle activity, and gravity can cause dynamic stress to the skin. Mechanical strain introduced by pregnancy, weight gain, or subcutaneously implanted devices can induce an increase in mass, volume, and area of the skin [12]. Disruption of skin homeostasis by wounds, genetic alterations, or diseases can lead to aberrant mechanobiology of the skin [10]. Improved understanding of skin mechanotransduction pathways at baseline levels, under homeostasis and under mechanical stress will clarify skin mechanobiology and its role in normal and pathologic skin disorders.

Recent evidence has linked a central mechanotransduction pathway, i.e., the non-receptor protein tyrosine kinase, focal adhesion kinase (FAK), to both mechanical homeostasis and pro-fibrotic mechanotransduction and other wound healing aberrations [8,9]. Our laboratory was the first to develop and publish the murine model of HTS based on mechanical loading and showed that sustained mechanical forces were able to induce FAK-mediated pro-survival signaling and modify inflammation. In addition, using microarray analysis we have shown in a porcine model and human studies that elevated wound tension caused a pro-fibrotic phenotype and HTS formation. In addition, we recently reported that in diabetic delayed wound healing, FAK degradation by calpain might decelerate wound repair. Integrin-FAK mechanotransduction cascades are involved in fibro-proliferative states such as hepatic fibrosis [13], cardiac hypertrophy [14], vascular smooth muscle atherosclerosis [15], and pulmonary fibrosis [16]. FAK can activate numerous downstream components involved in fibrogenic events such as PI3K/Akt and mitogen-activated protein kinases (MAPK) [17–23]. How FAK modulates skin cell behavior, especially on other mechanotransduction and wound healing-associated proteins, is not well understood.

Our laboratory has previously demonstrated that FAK is important for mechanotransduction in cutaneous fibroblasts and that a fibroblast-specific deletion of FAK results in reduced fibrosis after injury in a mouse model of scar formation [24]. In contrast, the loss of FAK specifically in keratinocytes leads to significantly delayed wound healing and pathologic dermal proteolysis in mice [25]. Keratinocyte FAK-deleted mice were also found to have decreased dermal thickness and collagen density, findings linked to over-activation of the matrix-remodeling enzyme matrix metalloproteinase 9 (MMP9) [25]. However, a paradoxical upregulation of collagen I and III, the predominant collagen subtypes in cutaneous healing, in these wounds suggests that FAK signaling has a complex effect on extracellular matrix (ECM) repair.

Keratinocytes have been shown to express high levels of collagen based on transcriptome-wide microarray studies [26,27], which suggest a role for these cells in ECM deposition. While global gene expression analyses are sufficient for population wide transcriptional screening, transcriptional profiling of higher resolution is needed to uncover aberrations in cellular signaling only affecting a subset of cells [28]. Since rare, but important subpopulations of cells can significantly alter the biological process of wound healing and mechanical homeostasis, our laboratory has developed a microfluidics-based approach to gene expression analysis that enables the discovery of altered gene expression patterns on a single-cell level [29]. Here we employ this method to evaluate how dysregulation of FAK signaling affects mechanically unstimulated keratinocyte gene transcription at baseline and whether the loss of this mechanical mediator leads to perturbations in cellular subpopulations that may explain the impairments observed at a physiological level.

2. Results

2.1. Knockout of Keratinocyte FAK (Focal Adhesion Kinase) Alters the Expression of Numerous Genes Integral to Tissue Repair

Keratinocytes have been shown to influence cutaneous fibrosis and repair [25,30] and are highly mechanoresponsive cells [31–33]. Mechanoresponsive cellular components, in particular through activation of FAK, have recently been identified as key mediators in the development of hypertrophic scars, as well as physiological wound healing [24,25]. In an attempt to elucidate how FAK affects gene expression in epithelial cells during cutaneous repair, we compared wildtype (WT) and FAK knockout (KO) keratinocytes (Figure 1) utilizing microfluidic-based single-cell transcriptomics [29]. Single-cell analysis of freshly isolated keratinocytes from WT and keratinocyte-specific FAK-deleted mice demonstrated significant transcriptional heterogeneity both within and across these two groups (Figure 1A,B), with clear lack of any FAK expression in the KO group. Numerous genes were differentially expressed between these cells as a result of FAK deletion (Figure 1B). These include ECM genes such as Collagen type IV (*Col4*) subunits and Keratin 6 (*Krt6*) as well genes that regulate cell-ECM adhesion and ECM-mediated mechanotransduction such as *Fak*, *Cd44*, *Pax*, integrins such as *Itgav* and *Itgb1*, *Itgb4*, *Itgb6*, and *Itgb8*. Factors involved in tissue repair and matrix remodeling such as MMPs and tissue inhibitors of MMPs (TIMPs) were altered in KO cells. For example, expression of *Cd44*, *Itgav*, and *Itgb1* were differentially up-regulated with FAK deletion, suggesting that these signaling regulators are closely associated with FAK and FAK-mediated mechanotransduction network. Interestingly, genes involved in cancer progression including tyrosine protein kinase (*Src*) and breast cancer anti-estrogen resistance protein (*Bcar*) were also altered in FAK KO cells. These data indicate that keratinocyte FAK expression is closely linked to mechanoregulatory factors, as well as mediators implicated in tissue repair and remodeling, underscoring the variety of molecular pathways affected by mechanotransduction component FAK.

2.2. FAK-Deleted Keratinocytes Demonstrate Alterations in Key Mechanotransduction and Collagen Signaling Pathways

To elucidate the effects of FAK deletion on keratinocyte intracellular signaling, we next identified canonical pathways whose expression was significantly altered using Ingenuity Pathway Analysis (IPA). Analyzing known canonical pathways based on genes up- and down-regulated in FAK-deleted keratinocytes, we found that integrin signaling, FAK signaling, and ERK/MAPK signaling were most highly affected by the loss of keratinocyte FAK. We further utilized IPA to generate transcriptional networks based on over- (Figure S2A) or under- (Figure S2B) expressed genes in the KO cells compared to WT keratinocytes. These included numerous collagen and integrin genes, as well as major upstream regulators such as *Fak*, *Akt*, and *Erk*. When we merged these pathways (Figure 2) we obtained a comprehensive signaling network centered on the FAK-AKT-ERK axis, again demonstrating the critical role of FAK in the regulation of collagen/integrin expression. Notably, an upstream analysis of these differentially expressed genes implicated two key molecular regulators of the transcriptional changes in KO cells. The merged transcriptome network shown in Figure 2 represents features of activation of fibrogenic regulator *Tgfbr2* and suppression of the proto-oncogene *Mycn* that produced the majority of transcriptional changes observed in KO cells.

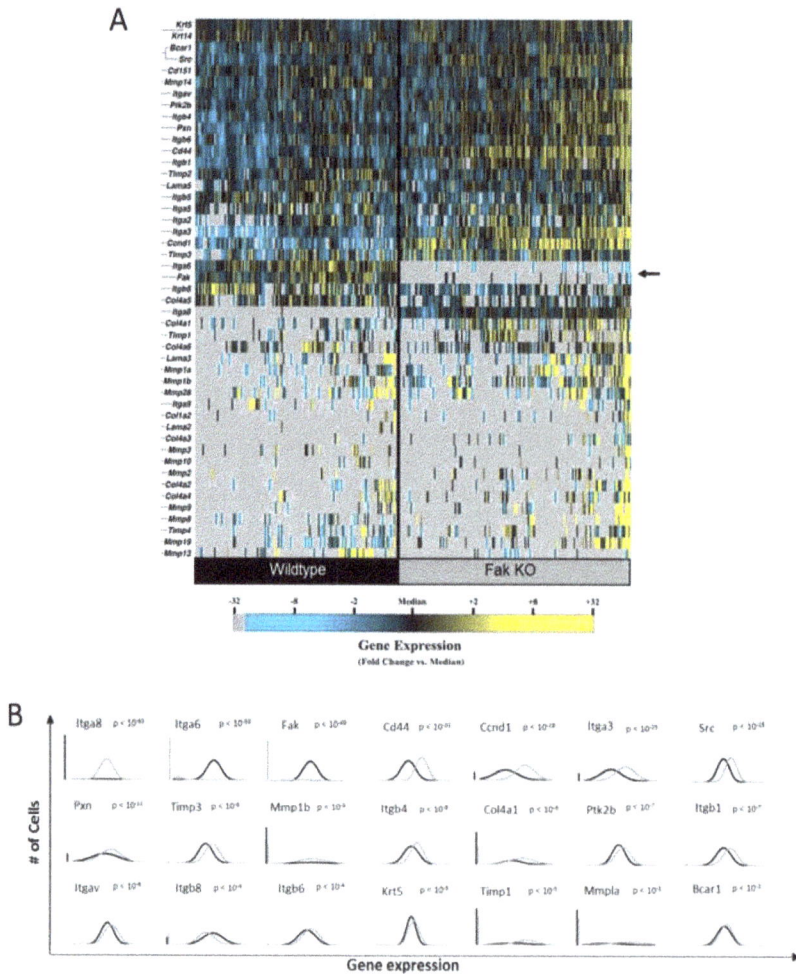

Figure 1. Keratinocyte FAK deletion alters the expression of numerous genes integral to tissue repair. (**A**) Hierarchical clustering of simultaneous gene expression for single cells from WT (left) and FAK KO (right) mice. Gene expression is presented as fold change from median on a color scale from yellow (high expression, 32-fold above median) to blue (low expression, 32-fold below median). Cell/gene qPCR reactions failing to amplify after 40 cycles are designated as non-expressers and represented in gray; (**B**) Differential gene expression between WT and FAK KO cells identified using nonparametric two-sample Kolmogorov-Smirnov testing. Twenty-one genes exhibit significantly different ($p < 0.01$ following Bonferroni correction for multiple comparisons) distributions of single cell expression between populations, illustrated here using median-centered Gaussian curve fits. Black and gray color histogtams denote WT and FAK KO expression, respectively. The left bar for each panel represents the fraction of qPCR reactions that failed to amplify in each group.

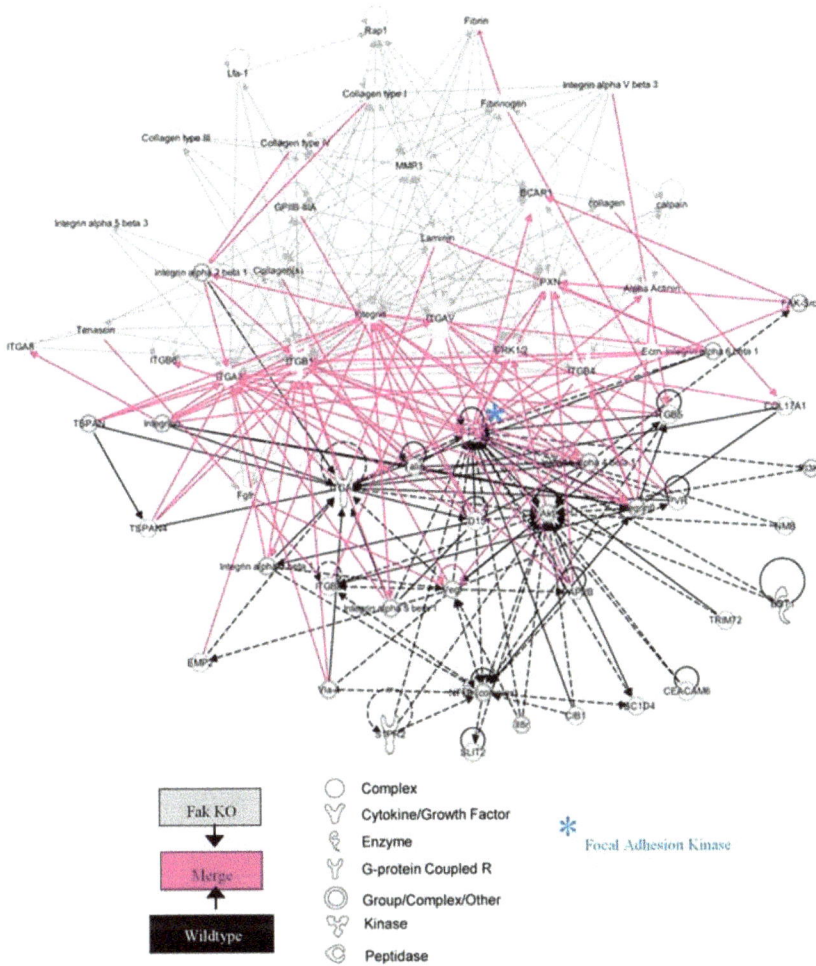

Figure 2. FAK-deleted keratinocytes demonstrate alterations in key mechanotransduction and collagen signaling pathways. The top scoring Ingenuity Pathway Analysis (IPA)-constructed transcriptome network generated from genes significantly up-regulated (grey) and down-regulated (black) in FAK-deleted keratinocytes compared to WT cells were merged using IPA's Ingenuity Knowledge Base, creating a super-network centered on the FAK-AKT-ERK axis. Direct relationships are indicated by solid lines, and dashed lines represent indirect relationships. Known relationships among molecules across the original two networks are represented in magenta. * denotes FAK (PTK2).

2.3. FAK Deletion Affects Keratinocyte Gene Expression Asymmetrically and Induces a Transcriptionally Activated Subpopulation

Given the considerable transcriptional heterogeneity observed at the single-cell level and earlier description of keratinocytes as a cell pool composed of distinct subsets [34,35], we applied partitional clustering to identify transcriptionally distinct (and potentially functionally distinct) subpopulations (Figure 3A). We found that keratinocytes could be grouped into three discrete subgroups based on their transcriptional signatures, designated here as clusters 1, 2, and 3. Interestingly, cluster 1 cells were almost exclusively found in WT mice. In contrast, cluster 2 cells were predominantly found in the FAK KO mice. These subgroups were defined by differing expression patterns similar to those of aggregate WT vs KO cells, and the added granularity of this analysis identified additional differentially expressed genes including multiple collagen and MMP targets (Figure 3B). Furthermore, we identified an additional population of cells (cluster 3) that appear to be activated keratinocytes defined by significant overexpression of collagen and MMP transcripts, which were predominantly found in the KO mice. Higher frequency of cluster 3 cells in the KO mice (approximately 80%) suggests that suppression of FAK may trigger activation and proliferation of these cells, thus resulting in the paradoxical over-expression of collagen observed in the wounds of keratinocyte FAK KO mice [25]. As described previously, keratinocyte FAK KO mice demonstrate dermal proteolysis and clinical features of wound chronicity but also over-express certain subtypes of collagen [25]. Identification of cluster 3 keratinocytes in our current analysis provides an explanation for this paradoxical observation [25]. We identified select genes whose transcriptional activities are differential among cells in each cluster across all keratinocytes comprising each cluster. Figure 3B shows multiple differentially-expressed collagen and MMP target genes identified this way. Of interest, expression of integrin α-6 (*Itga6*) and integrin α-8 (*Itga8*) were regulated in an opposite manner in the absence of epithelial FAK. Both integrin subtypes can modulate keratinocyte migration and cell-ECM interactions and are implicated in wound healing. Expression profiles of integrin α-3 (*Itga3*), an integrin subtype involved in epithelial-mesenchymal transition, and *Cd44* cell adhesion molecule were both significantly up-regulated in FAK KO cells [36]. In addition, there was a trend that an array of differentially-expressed MMP proteins was up-regulated with FAK deletion, suggesting that loss of FAK can modulate ECM-remodeling activities and can potentially lead to altered wound healing profiles seen in FAK KO mice.

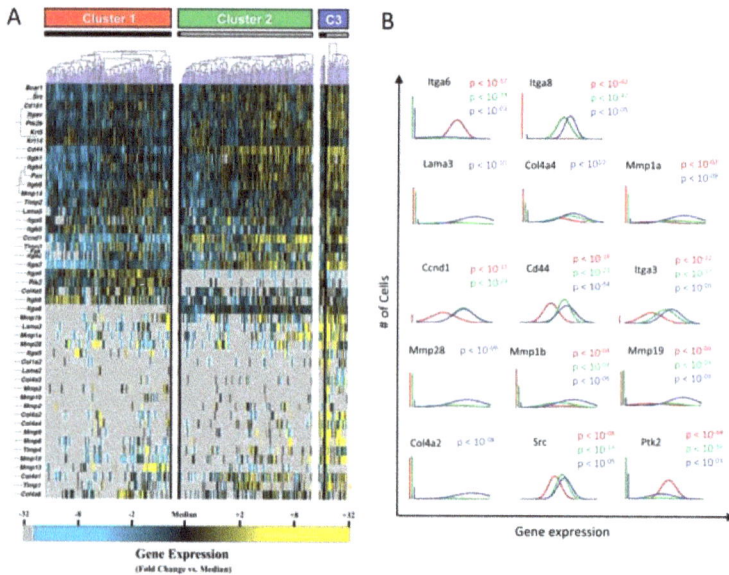

Figure 3. Partitional cluster analysis of single-cell transcriptional data. (**A**) K-means clustering of WT (black bar) and FAK-deleted keratinocytes (grey bar). Gene expression is presented as fold change from median on a color scale from yellow (high expression, 32-fold above median) to blue (low expression, 32-fold below median); (**B**) Differentially-expressed genes among cells in each cluster using non-parametric two sample Kolmogorov-Smimov testing across all (both WT and FAK KO) keratinocytes comprising each cluster, illustrated here using median-centered Gaussian curve fits. The left bar for each panel represents the fraction of qPCR reactions that failed to amplify in each group.

We next applied enrichment analysis to the gene sets up-regulated in cells from each putative subpopulation. Cluster 1 cells were characterized by comparatively increased integrin signaling, FAK signaling, and ERK/MAPK signaling (Figure S3A), consistent with intact FAK expression in these predominantly WT cells. By contrast, cluster 2 cells were associated with ILK signaling and inhibition of MMPs, as well as suppression of paxillin signaling (Figure S3B). Corresponding transcriptional networks were subsequently generated for each subgroup using IPA as described above. Analysis of signaling pathways implicated in cluster 1 resulted in a network defined by increased expression of numerous *Timp* genes (Figure 4A). Of interest, the network of cluster 2 was characterized by elevated expression of various collagen genes (Figure 4B), which was similar to cluster 3 (Figure 4C). Given the complex molecular interactions orchestrated by mechanical signaling [10], we explored intracellular pathways affected by FAK modulation by combining regulatory networks from each keratinocyte subgroup using known relationships among commonly expressed genes (Figure 5). The pathway analysis indicates that a complex signaling cascade centered on the FAK-AKT-P13K axis, may regulate expression of collagen, integrin, and *Mmp* genes. These data suggest that FAK activation/deactivation in keratinocytes has widespread downstream implications and represents a key mechanism underlying the regulation of cellular mechanotransduction, extracellular matrix deposition, and tissue remodeling.

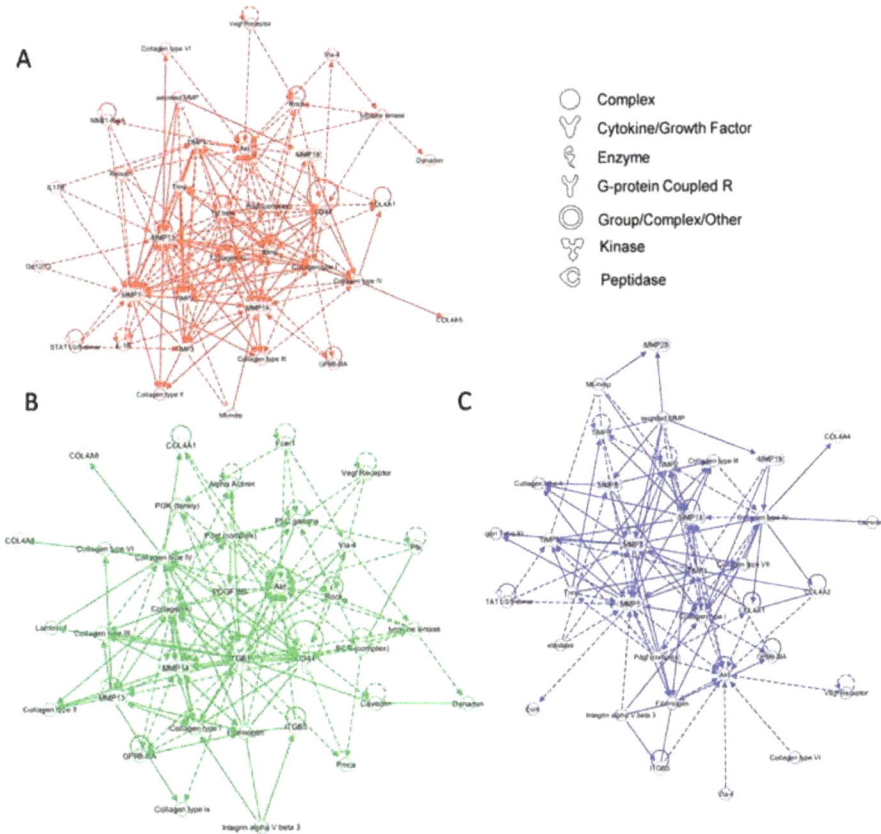

Figure 4. Network analysis of keratinocyte subpopulations. Top scoring Ingenuity Pathway Analysis (IPA)-constructed transcriptome networks based genes that were significantly up-regulated in cluster 1 (**A**; red), cluster 2 (**B**; green), and cluster 3 (**C**; blue). Direct relationships are indicated by solid lines, and dashed lines represent indirect relationships.

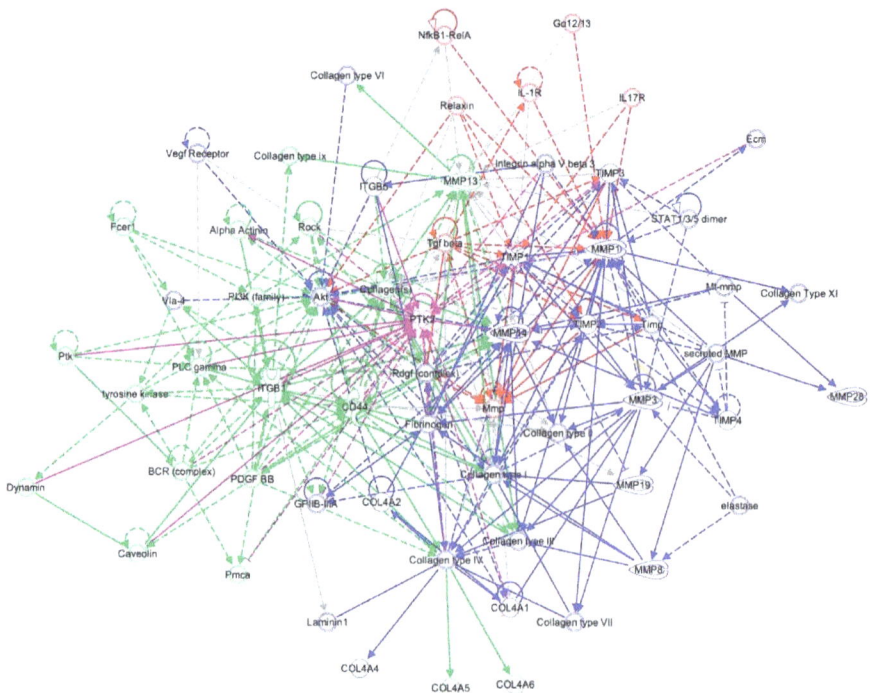

Figure 5. Merged network analysis of keratinocyte subpopulations. Merged network analysis of the three clusters in Figure 4. The top scoring networks generated from genes significantly up-regulated in each of the three keratinocyte subpopulations (cluster 1 (red), cluster 2 (green), and cluster 3 (blue)) were merged using Ingenuity Pathway Analysis (IPA), creating a super-network centered on the FAK-AKT-PI3K axis. Known relationships among molecules across the original three networks are represented in magenta. * denotes FAK (PTK2).

3. Discussion

Mechanotransduction pathways have long been shown to regulate human skin biology, scar formation, and wound repair [37]. Previously, our laboratory has identified FAK as a key mechanosensor in both the epidermal and dermal skin compartments with an important role in the cutaneous response to injury [24,25]. Biomechanical aberrations in the wound environment may predispose patients to either hypertrophic scarring or non-healing wounds, both of which are growing public health burdens awaiting effective therapeutics.

In a recent study, we demonstrated that deletion of FAK from epithelial keratinocytes resulted in accelerated dermal proteolysis and delayed wound healing in a mouse model [25]. Here we utilized single-cell microfluidic techniques to identify distinct subgroups of FAK-deleted keratinocytes that may drive this pathologic response. Specifically, one subgroup appears primed to activate matrix degradation and remodeling. Multiple genes linked to wound mechanotransduction were shown to be regulated by FAK, including integrins, *Mmps*, and matrix components, supporting a primary role for epithelial mechanosensing in wound repair. Furthermore, these results suggest that distinct subpopulations of dysfunctional keratinocytes may have the potential to impair normal cutaneous tissue repair.

These findings also confirm the importance of FAK-MAPK intracellular signaling networks in skin mechanobiology [24,25]. The FAK-MAPK pathway has been shown to regulate mechanosensing in numerous tissues throughout the body, including endothelium, bone, heart, and lung [38–41]. These intracellular pathways may serve as molecular targets to modulate aberrant wound repair. Current device-based approaches to modulate mechanical wound signaling, such as negative pressure wound therapy, are believed to work in part via these intracellular networks [42,43]. An alternative strategy would be to directly target these mechanotransduction pathways pharmacologically, as has been successfully demonstrated in a murine hypertrophic scar model in which small molecule FAK inhibitors were employed [24].

Focal adhesion complexes are dynamic multi-component mechanosensors that influence multiple aspects of cell biology [44,45]. The focus of our study is to investigate the role of epithelial FAK in regulating the transcriptional profiles of mechanically unstimulated keratinocytes at baseline. With loss of epithelial FAK, we observe a corresponding decrease in expression of integrin α-6 (*Itga6*) and increased expression of integrin α-8 (*Itga8*). Integrins are transmembrane receptor complexes that dimerize and transmit extracellular cues (such as mechanical force) into cells. The differential expression and activation of integrins can modify diverse cell behaviors and is highly implicated in wound healing [46]. Itga6 has been closely linked to keratinocyte migration and cell-matrix interactions with the basement membrane component laminin-111 [47]. Itga8 has been less frequently examined in keratinocytes, but studies suggest a role for this protein in epithelial-mesenchymal interactions, matrix remodeling, and cell migration [48,49]. These findings demonstrate that loss of FAK alters the surface integrin profile of keratinocytes and potentially disrupts how the epithelial layer senses the wound environment. It is also interesting to note that expression of *Cd44*, a cell adhesion molecule that has been shown to stimulate activation of FAK via Rho-dependent mechanisms, was up-regulated with loss of epithelial FAK. Since Cd44 has a direct functional link to FAK-mediated signaling and plays a role in suppressing apoptosis, its importance in the context of wound healing may be of future interest. In a similar context, expression of *Ccnd1* was also increased with epithelial FAK deletion, suggesting that cellular proliferation profiles can also be altered with loss of FAK in keratinocytes.

Further, we found that WT keratinocytes demonstrated considerable expression of integrin-FAK-MAPK pathways, whereas FAK-deleted keratinocytes preferentially expressed ILK and suppressed paxillin signaling - known components of focal adhesion complexes [50]. The significance of this is unclear, as both ILK and paxillin activate many of the same targets as FAK. Of interest, increased activation of ILK and dysregulation of paxillin signaling have been linked to pathologic expression of the matrix-degrading enzyme MMP9 [51,52]. This concept of altered biomechanical sensing has also been applied to cancer metastasis, as tumor cells invade adjacent tissues by secreting MMPs and bypassing physical contact inhibition signals. Our current study will form the basis for future studies on characterization of keratinocyte subpopulations under mechanical stimuli.

4. Materials and Methods

4.1. Animals

All procedures were approved by the Administrative Panel on Laboratory Animal and Care committee (Protocol number APLAC-12080 approved on 21 December 2012) at Stanford University. Wildtype (WT) mice (C57BL/6) were purchased from Jackson Laboratories (Bar Harbor, ME, USA). Keratinocyte-restricted FAK knockout mice were produced as previously described [25]. Transgenic mice were generated by crossing keratinocyte-specific K5-Cre recombinase mice (FVB.Cg-Tg(KRT1-5-cre)5132JIj, purchased from University of North Carolina Mutant Mouse Regional Resource Center) to homozygous floxed FAK mice (B6.129-Ptk2^{tm1Lfr}/Mmcd, purchased from University of California Davis Mutant Mouse Regional Resource Center (Davis, CA, USA).

4.2. Keratinocyte Harvest and Culture

Primary epithelial keratinocytes were isolated from WT and keratinocyte-specific FAK knockout mice as previously published [53]. Briefly, full thickness skin was harvested from the mouse dorsum using sterile techniques. Specimens were floated in trypsin (Gibco/Invitrogen, Carlsbad, CA, USA) before adhering the epidermal surface to a culture dish lid. After removing the dermis with forceps, the epidermal specimens were minced and placed into suspension. Cells were grown on plates coated with Coating Matrix Kit (Gibco/Invitrogen, Waltham, MA, USA) and keratinocyte basal media (Lonza, Walkersville, MD, USA) supplemented with KGM Gold Bullet Kit (Lonza) was used to maintain cells. After a short-term culture for approximately 7 days without subculturing, passage 0 keratinocytes were used for all experiments.

4.3. Single-Cell Transcriptional Analysis

Cultured keratinocytes were sorted as single cells into lysis buffer using a Becton Dickinson FACSAria flow cytometer. Live/dead gating was performed based on propidium iodide exclusion. Reverse transcription and low cycle pre-amplification were performed following addition of Superscript Ill reverse transcriptase enzyme (Invitrogen, Waltham, MA, USA), Cells Direct reaction mix (Invitrogen), TaqMan assay primer sets (Applied Biosystems, Foster City, CA, USA) (Figure S1). The resulting cDNA was loaded onto a 48.48 Dynamic Array (Fluidigm, South San Francisco, CA, USA) for qPCR amplification using Universal PCR Master Mix (Applied Biosystems) with TaqMan assay primer sets, and products were analyzed on the BioMark reader system (Fluidigm).

Analysis of single-cell data was performed as previously described [54]. Briefly, expression data from all chips (both WT and FAK KO) were normalized relative to the median expression for each gene in the pooled sample and converted to base 2 logarithms. Absolute bounds of ± 5 cycle thresholds from the median were applied, and non-expressers were assigned to this floor. Heatmaps were generated and organized using hierarchical clustering in order to facilitate data visualization using MATLAB (R2011 b, MathWorks, Natick, MA, USA).

Partitional clustering of gene expression data for each group of keratinocytes was performed using a modified k-means algorithm to identify subpopulations based on transcriptional profiles [55]. Gene-wise comparisons between groups and across clusters was achieved using a standard two sample Kolmogorov-Smirnov test with a strict cutoff of $p < 0.05$ following Bonferroni correction for multiple hypothesis testing. For comparisons among subgroups, the empirical distribution of cells from each cluster was evaluated against that of the remaining cells in the experiment.

Canonical pathway calculations and network analyses were performed using Ingenuity Pathway Analysis (IPA, Ingenuity Systems, Redwood City, CA, USA). For these analyses, the gene targets interrogated in the corresponding single-cell analysis (rather than the entire transcriptome) were used as the reference set in order to avoid biasing the associated enrichment and network calculations.

5. Conclusions

Our study contributes to improved understanding of FAK-mediated keratinocyte gene transcription by dissecting and highlighting the effects of this mediator and related pathway components in the epidermis. Global suppression of FAK not only restricted to keratinocytes (e.g., using pharmacological FAK inhibitors), however, may result in different pathophysiological outcomes and remains to be studied. In addition, we have shown the existence of a novel subpopulation of keratinocytes that may respond differentially to fibrogenic mechanical cues during the course of wound healing and scar formation. It also remains to be studied how FAK alters the transcriptional profiles of mechanically-strained keratinocytes. The challenge of targeting these cells in vivo to modulate this process remains an ongoing effort, and future studies will clarify the clinical relevance of these single-cell transcriptional findings.

In conclusion, the findings of our study will enhance our understanding of the epidermal mechanotransduction important for scar formation and will have future implications for developing fibrosis and scar therapies targeting FAK-mediated mechanotransduction pathways.

Supplementary Materials: Supplementary materials can be found at www.mdpi.com/1422-0067/18/9/1915/s1.

Acknowledgments: Funding for this research was provided by the National Institutes of Health (LM007033) and the Department of Defense (Armed Forces Institute for Regenerative Medicine (AFIRM; W81XWH-08-2-0032)).

Author Contributions: Michael Januszyk designed the study, acquired, and analyzed data. Michael Januszyk, Sun Hyung Kwon, Victor W. Wong, and Jagannath Padmanabhan wrote the article. Sun Hyung Kwon, Victor W. Wong, Jagannath Padmanabhan, and Zeshaan N. Maan provided substantial contributions to data interpretation and manuscript preparation. Alexander J. Whittam and Melanie R. Major analyzed data and assisted with manuscript preparation. Geoffrey C. Gurtner designed the study, analyzed data, and wrote the article.

Conflicts of Interest: The authors declare no conflict of interest.

References

1. Gurtner, G.C.; Werner, S.; Barrandon, Y.; Longaker, M.T. Wound repair and regeneration. *Nature* **2008**, *453*, 314–321. [CrossRef] [PubMed]
2. Horch, R.E.; Popescu, L.M.; Polykandriotis, E. History of Regenerative Medicine. In *Regenerative Medicine*, 1st ed.; Steinhoff, G., Ed.; Springer: Dordrecht, The Netherlands, 2011; Volume 1, p. 1.
3. Walmsley, G.G.; Maan, Z.N.; Wong, V.W.; Duscher, D.; Hu, M.S.; Zielins, E.R.; Wearda, T.; Muhonen, E.; McArdle, A.; Tevlin, R.; et al. Scarless wound healing: Chasing the holy grail. *Plast. Reconstr. Surg.* **2015**, *135*, 907–917. [CrossRef] [PubMed]
4. Zielins, E.R.; Atashroo, D.A.; Maan, Z.N.; Duscher, D.; Walmsley, G.G.; Hu, M.; Senarath-Yapa, K.; McArdle, A.; Tevlin, R.; Wearda, T.; et al. Wound healing: An update. *Regen. Med.* **2014**, *9*, 817–830. [CrossRef] [PubMed]
5. Demidova-Rice, T.N.; Hamblin, M.R.; Herman, I.M. Acute and impaired wound healing: Pathophysiology and current methods for drug delivery, part 1: Normal and chronic wounds: Biology, causes, and approaches to care. *Adv. Skin Wound Care* **2012**, *25*, 304–314. [CrossRef] [PubMed]
6. Sen, C.K.; Gordillo, G.M.; Roy, S.; Kirsner, R.; Lambert, L.; Hunt, T.K.; Gottrup, F.; Gurtner, G.C.; Longaker, M.T. Human skin wounds: A major and snowballing threat to public health and the economy. *Wound Repair Regen.* **2009**, *17*, 763–771. [CrossRef] [PubMed]
7. Van De Water, L.; Varney, S.; Tomasek, J.J. Mechanoregulation of the myofibroblast in wound contraction, scarring, and fibrosis: Opportunities for new therapeutic intervention. *Adv. Wound Care* **2013**, *2*, 122–141. [CrossRef] [PubMed]
8. Ingber, D.E. Cellular mechanotransduction: Putting all the pieces together again. *FASEB J.* **2006**, *20*, 811–827. [CrossRef] [PubMed]
9. Wang, N.; Tytell, J.D.; Ingber, D.E. Mechanotransduction at a distance: Mechanically coupling the extracellular matrix with the nucleus. *Nat. Rev. Mol. Cell Biol.* **2009**, *10*, 75–82. [CrossRef] [PubMed]
10. Duscher, D.; Maan, Z.N.; Wong, V.W.; Rennert, R.C.; Januszyk, M.; Rodrigues, M.; Hu, M.; Whitmore, A.J.; Whittam, A.J.; Longaker, M.T.; et al. Mechanotransduction and fibrosis. *J. Biomech.* **2014**, *47*, 1997–2005. [CrossRef] [PubMed]
11. Edlich, R.F.; Carl, B.A. Predicting scar formation: From ritual practice (langer's lines) to scientific discipline (static and dynamic skin tensions). *J. Emerg. Med.* **1998**, *16*, 759–760. [PubMed]
12. Evans, N.D.; Oreffo, R.O.; Healy, E.; Thurner, P.J.; Man, Y.H. Epithelial mechanobiology, skin wound healing, and the stem cell niche. *J. Mech. Behav. Biomed. Mater.* **2013**, *28*, 397–409. [CrossRef] [PubMed]
13. Jiang, H.Q.; Zhang, X.L.; Liu, L.; Yang, C.C. Relationship between focal adhesion kinase and hepatic stellate cell proliferation during rat hepatic fibrogenesis. *World J. Gastroenterol.* **2004**, *10*, 3001–3005. [CrossRef] [PubMed]
14. Clemente, C.F.; Tornatore, T.F.; Theizen, T.H.; Deckmann, A.C.; Pereira, T.C.; Lopes-Cendes, I.; Souza, J.R.; Franchini, K.G. Targeting focal adhesion kinase with small interfering RNA prevents and reverses load-induced cardiac hypertrophy in mice. *Circ. Res.* **2007**, *101*, 1339–1348. [CrossRef] [PubMed]
15. Morla, A.O.; Mogford, J.E. Control of smooth muscle cell proliferation and phenotype by integrin signaling through focal adhesion kinase. *Biochem. Biophys. Res. Commun.* **2000**, *272*, 298–302. [CrossRef] [PubMed]
16. Garneau-Tsodikova, S.; Thannickal, V.J. Protein kinase inhibitors in the treatment of pulmonary fibrosis. *Curr. Med. Chem.* **2008**, *15*, 2632–2640. [CrossRef] [PubMed]

17. Song, G.; Ouyang, G.; Bao, S. The activation of akt/pkb signaling pathway and cell survival. *J. Cell. Mol. Med.* **2005**, *9*, 59–71. [CrossRef] [PubMed]

18. Chin, Y.R.; Toker, A. Function of akt/pkb signaling to cell motility, invasion and the tumor stroma in cancer. *Cell. Signal.* **2009**, *21*, 470–476. [CrossRef] [PubMed]

19. Nishimura, K.; Li, W.; Hoshino, Y.; Kadohama, T.; Asada, H.; Ohgi, S.; Sumpio, B.E. Role of akt in cyclic strain-induced endothelial cell proliferation and survival. *Am. J. Physiol. Cell Physiol.* **2006**, *290*, C812–C821. [CrossRef] [PubMed]

20. Potter, C.J.; Pedraza, L.G.; Xu, T. Akt regulates growth by directly phosphorylating tsc2. *Nat. Cell Biol.* **2002**, *4*, 658–665. [CrossRef] [PubMed]

21. Tian, B.; Lessan, K.; Kahm, J.; Kleidon, J.; Henke, C. Beta 1 integrin regulates fibroblast viability during collagen matrix contraction through a phosphatidylinositol 3-kinase/AKT/protein kinase b signaling pathway. *J. Biol. Chem.* **2002**, *277*, 24667–24675. [CrossRef] [PubMed]

22. Ding, Q.; Gladson, C.L.; Wu, H.; Hayasaka, H.; Olman, M.A. Focal adhesion kinase (FAK)-related non-kinase inhibits myofibroblast differentiation through differential mapk activation in a FAK-dependent manner. *J. Biol. Chem.* **2008**, *283*, 26839–26849. [CrossRef] [PubMed]

23. Hayashida, T.; Wu, M.H.; Pierce, A.; Poncelet, A.C.; Varga, J.; Schnaper, H.W. Map-kinase activity necessary for tgfbeta1-stimulated mesangial cell type I collagen expression requires adhesion-dependent phosphorylation of FAK tyrosine 397. *J. Cell Sci.* **2007**, *120*, 4230–4240. [CrossRef] [PubMed]

24. Wong, V.W.; Rustad, K.C.; Akaishi, S.; Sorkin, M.; Glotzbach, J.P.; Januszyk, M.; Nelson, E.R.; Levi, K.; Paterno, J.; Vial, I.N.; et al. Focal adhesion kinase links mechanical force to skin fibrosis via inflammatory signaling. *Nat. Med.* **2012**, *18*, 148–152. [CrossRef] [PubMed]

25. Wong, V.W.; Garg, R.K.; Sorkin, M.; Rustad, K.C.; Akaishi, S.; Levi, K.; Nelson, E.R.; Tran, M.; Rennert, R.; Liu, W.; et al. Loss of keratinocyte focal adhesion kinase stimulates dermal proteolysis through upregulation of mmp9 in wound healing. *Ann. Surg.* **2014**, *260*, 1138–1146. [CrossRef] [PubMed]

26. Bernard, F.X.; Pedretti, N.; Rosdy, M.; Deguercy, A. Comparison of gene expression profiles in human keratinocyte mono-layer cultures, reconstituted epidermis and normal human skin; transcriptional effects of retinoid treatments in reconstituted human epidermis. *Exp. Dermatol.* **2002**, *11*, 59–74. [CrossRef] [PubMed]

27. Hu, M.S.; Januszyk, M.; Hong, W.X.; Walmsley, G.G.; Zielins, E.R.; Atashroo, D.A.; Maan, Z.N.; McArdle, A.; Takanishi, D.M., Jr.; Gurtner, G.C.; et al. Gene expression in fetal murine keratinocytes and fibroblasts. *J. Surg. Res.* **2014**, *190*, 344–357. [CrossRef] [PubMed]

28. Januszyk, M.; Gurtner, G.C. High-throughput single-cell analysis for wound healing applications. *Adv. Wound Care* **2013**, *2*, 457–469. [CrossRef] [PubMed]

29. Glotzbach, J.P.; Januszyk, M.; Vial, I.N.; Wong, V.W.; Gelbard, A.; Kalisky, T.; Thangarajah, H.; Longaker, M.T.; Quake, S.R.; Chu, G.; et al. An information theoretic, microfluidic-based single cell analysis permits identification of subpopulations among putatively homogeneous stem cells. *PLoS ONE* **2011**, *6*, e21211. [CrossRef] [PubMed]

30. Varkey, M.; Ding, J.; Tredget, E.E. Fibrotic remodeling of tissue-engineered skin with deep dermal fibroblasts is reduced by keratinocytes. *Tissue Eng. Part A* **2014**, *20*, 716–727. [CrossRef] [PubMed]

31. Kippenberger, S.; Loitsch, S.; Guschel, M.; Muller, J.; Knies, Y.; Kaufmann, R.; Bernd, A. Mechanical stretch stimulates protein kinase b/akt phosphorylation in epidermal cells via angiotensin ii type 1 receptor and epidermal growth factor receptor. *J. Biol. Chem.* **2005**, *280*, 3060–3067. [CrossRef] [PubMed]

32. Kippenberger, S.; Loitsch, S.; Guschel, M.; Muller, J.; Kaufmann, R.; Bernd, A. Hypotonic stress induces e-cadherin expression in cultured human keratinocytes. *FEBS Lett.* **2005**, *579*, 207–214. [CrossRef] [PubMed]

33. Knies, Y.; Bernd, A.; Kaufmann, R.; Bereiter-Hahn, J.; Kippenberger, S. Mechanical stretch induces clustering of beta1-integrins and facilitates adhesion. *Exp. Dermatol.* **2006**, *15*, 347–355. [CrossRef] [PubMed]

34. Lavker, R.M.; Sun, T.T. Heterogeneity in epidermal basal keratinocytes: Morphological and functional correlations. *Science* **1982**, *215*, 1239–1241. [CrossRef] [PubMed]

35. Haustead, D.J.; Stevenson, A.; Saxena, V.; Marriage, F.; Firth, M.; Silla, R.; Martin, L.; Adcroft, K.F.; Rea, S.; Day, P.J.; et al. Transcriptome analysis of human ageing in male skin shows mid-life period of variability and central role of nf-kappab. *Sci. Rep.* **2016**, *6*, 26846. [CrossRef] [PubMed]

36. Shirakihara, T.; Kawasaki, T.; Fukagawa, A.; Semba, K.; Sakai, R.; Miyazono, K.; Miyazawa, K.; Saitoh, M. Identification of integrin $\alpha 3$ as a molecular marker of cells undergoing epithelial-mesenchymal transition and of cancer cells with aggressive phenotypes. *Cancer Sci.* **2013**, *104*, 1189–1197. [CrossRef] [PubMed]

37. Wong, V.W.; Longaker, M.T.; Gurtner, G.C. Soft tissue mechanotransduction in wound healing and fibrosis. *Semin. Cell Dev. Biol.* **2012**, *23*, 981–986. [CrossRef] [PubMed]

38. Young, S.R.; Gerard-O'Riley, R.; Kim, J.B.; Pavalko, F.M. Focal adhesion kinase is important for fluid shear stress-induced mechanotransduction in osteoblasts. *J. Bone Miner. Res.* **2009**, *24*, 411–424. [CrossRef] [PubMed]

39. Chaturvedi, L.S.; Marsh, H.M.; Basson, M.D. Src and focal adhesion kinase mediate mechanical strain-induced proliferation and erk1/2 phosphorylation in human h441 pulmonary epithelial cells. *Am. J. Physiol. Cell Physiol.* **2007**, *292*, C1701–1713. [CrossRef] [PubMed]

40. Dalla Costa, A.P.; Clemente, C.F.; Carvalho, H.F.; Carvalheira, J.B.; Nadruz, W., Jr.; Franchini, K.G. Fak mediates the activation of cardiac fibroblasts induced by mechanical stress through regulation of the mtor complex. *Cardiovasc. Res.* **2010**, *86*, 421–431. [CrossRef] [PubMed]

41. Zebda, N.; Dubrovskyi, O.; Birukov, K.G. Focal adhesion kinase regulation of mechanotransduction and its impact on endothelial cell functions. *Microvasc. Res.* **2012**, *83*, 71–81. [CrossRef] [PubMed]

42. Wong, V.W.; Akaishi, S.; Longaker, M.T.; Gurtner, G.C. Pushing back: Wound mechanotransduction in repair and regeneration. *J. Investig. Dermatol.* **2011**, *131*, 2186–2196. [CrossRef] [PubMed]

43. Huang, C.; Leavitt, T.; Bayer, L.R.; Orgill, D.P. Effect of negative pressure wound therapy on wound healing. *Curr. Probl. Surg.* **2014**, *51*, 301–331. [CrossRef] [PubMed]

44. Hall, J.E.; Fu, W.; Schaller, M.D. Focal adhesion kinase: Exploring fak structure to gain insight into function. *Int. Rev. Cell Mol. Biol.* **2011**, *288*, 185–225. [PubMed]

45. Jahed, Z.; Shams, H.; Mehrbod, M.; Mofrad, M.R. Mechanotransduction pathways linking the extracellular matrix to the nucleus. *Int. Rev. Cell Mol. Biol.* **2014**, *310*, 171–220. [PubMed]

46. Margadant, C.; Sonnenberg, A. Integrin-tgf-beta crosstalk in fibrosis, cancer and wound healing. *EMBO Rep.* **2010**, *11*, 97–105. [CrossRef] [PubMed]

47. Golbert, D.C.; Correa-de-Santana, E.; Ribeiro-Alves, M.; de Vasconcelos, A.T.; Savino, W. Itga6 gene silencing by rna interference modulates the expression of a large number of cell migration-related genes in human thymic epithelial cells. *BMC Genom.* **2013**, *14* (Suppl. S6), S3.

48. Muller, U.; Wang, D.; Denda, S.; Meneses, J.J.; Pedersen, R.A.; Reichardt, L.F. Integrin α8β1 is critically important for epithelial-mesenchymal interactions during kidney morphogenesis. *Cell* **1997**, *88*, 603–613. [CrossRef]

49. Bieritz, B.; Spessotto, P.; Colombatti, A.; Jahn, A.; Prols, F.; Hartner, A. Role of α8 integrin in mesangial cell adhesion, migration, and proliferation. *Kidney Int.* **2003**, *64*, 119–127. [CrossRef] [PubMed]

50. Geiger, B.; Spatz, J.P.; Bershadsky, A.D. Environmental sensing through focal adhesions. *Nat. Rev. Mol. Cell Biol.* **2009**, *10*, 21–33. [CrossRef] [PubMed]

51. Bedal, K.B.; Grassel, S.; Oefner, P.J.; Reinders, J.; Reichert, T.E.; Bauer, R. Collagen xvi induces expression of mmp9 via modulation of ap-1 transcription factors and facilitates invasion of oral squamous cell carcinoma. *PLoS ONE* **2014**, *9*, e86777. [CrossRef] [PubMed]

52. Chen, J.Y.; Tang, Y.A.; Huang, S.M.; Juan, H.F.; Wu, L.W.; Sun, Y.C.; Wang, S.C.; Wu, K.W.; Balraj, G.; Chang, T.T.; et al. A novel sialyltransferase inhibitor suppresses fak/paxillin signaling and cancer angiogenesis and metastasis pathways. *Cancer Res.* **2011**, *71*, 473–483. [CrossRef] [PubMed]

53. Lichti, U.; Anders, J.; Yuspa, S.H. Isolation and short-term culture of primary keratinocytes, hair follicle populations and dermal cells from newborn mice and keratinocytes from adult mice for in vitro analysis and for grafting to immunodeficient mice. *Nat. Protoc.* **2008**, *3*, 799–810. [CrossRef] [PubMed]

54. Rennert, R.C.; Januszyk, M.; Sorkin, M.; Rodrigues, M.; Maan, Z.N.; Duscher, D.; Whittam, A.J.; Kosaraju, R.; Chung, M.T.; Paik, K.; et al. Microfluidic single-cell transcriptional analysis rationally identifies novel surface marker profiles to enhance cell-based therapies. *Nat. Commun.* **2016**, *7*, 11945. [CrossRef] [PubMed]

55. Januszyk, M.; Sorkin, M.; Glotzbach, J.P.; Vial, I.N.; Maan, Z.N.; Rennert, R.C.; Duscher, D.; Thangarajah, H.; Longaker, M.T.; Butte, A.J.; et al. Diabetes irreversibly depletes bone marrow-derived mesenchymal progenitor cell subpopulations. *Diabetes* **2014**, *63*, 3047–3056. [CrossRef] [PubMed]

International Journal of
Molecular Sciences

MDPI

Review

Estrogen Effects on Wound Healing

Huann-Cheng Horng [1,2,3,†], Wen-Hsun Chang [1,4,5,†], Chang-Ching Yeh [1,2,6],
Ben-Shian Huang [1,2,6], Chia-Pei Chang [1,2], Yi-Jen Chen [1,2,6], Kuan-Hao Tsui [2,7,8]
and Peng-Hui Wang [1,2,6,9,*]

1   Department of Obstetrics and Gynecology, Taipei Veterans General Hospital, Taipei 112, Taiwan;
     hchorng@vghtpe.gov.tw (H.-C.H.); whchang@vghtpe.gov.tw (W.-H.C.); ccyeh39@gmail.com (C.-C.Y.);
     benshianhuang@gmail.com (B.-S.H.); jpchang2@vghtpe.gov.tw (C.-P.C.); chenyj@vghtpe.gov.tw (Y.-J.C.)
2   Department of Obstetrics and Gynecology, National Yang-Ming University, Taipei 112, Taiwan;
     khtsui60@gmail.com
3   Institute of BioMedical Informatics, National Yang-Ming University, Taipei 112, Taiwan
4   Department of Nursing, Taipei Veterans General Hospital, Taipei 112, Taiwan
5   Department of Nursing, National Yang-Ming University, Taipei 112, Taiwan
6   Institute of Clinical Medicine, National Yang-Ming University, Taipei 112, Taiwan
7   Department of Obstetrics and Gynecology, Kaohsiung Veterans General Hospital, Kaohsiung 813, Taiwan
8   Department of Pharmacy and Graduate Institute of Pharmaceutical Technology, Tajen University,
     Pingtung County 900, Taiwan
9   Department of Medical Research, China Medical University Hospital, Taichung 404, Taiwan
*   Correspondence: phwang@vghtpe.gov.tw or pongpongwang@gmail.com; Tel.: +886-2-28757566;
     Fax: +886-2-5570-2788
†   These authors contributed equally to this work.

Received: 29 September 2017; Accepted: 2 November 2017; Published: 3 November 2017

Abstract: Wound healing is a physiological process, involving three successive and overlapping
phases—hemostasis/inflammation, proliferation, and remodeling—to maintain the integrity of skin
after trauma, either by accident or by procedure. Any disruption or unbalanced distribution of these
processes might result in abnormal wound healing. Many molecular and clinical data support the
effects of estrogen on normal skin homeostasis and wound healing. Estrogen deficiency, for example
in postmenopausal women, is detrimental to wound healing processes, notably inflammation and
re-granulation, while exogenous estrogen treatment may reverse these effects. Understanding the
role of estrogen on skin might provide further opportunities to develop estrogen-related therapy for
assistance in wound healing.

Keywords: estrogen; estrogen receptor; wound healing

1. Introduction

The epidermis and dermis of normal skin maintain a steady-state equilibrium to ensure
a protective barrier against the external environment [1–3]. When the protective barrier is
broken by trauma, injury, radiation, chemical injury, and/or burns, the wound healing process is
immediately set in motion and complex biochemical events take place to repair the damage [4–8].
The normal wound healing process involves three successive and overlapping stages, including the
hemostasis/inflammation phase, the proliferation phase, and the remodeling phase [9–15].

After an injury to skin, hemostasis activates immediately. The exposed sub-endothelium,
collagen, and tissue factor will activate platelet aggregation, which leads to degranulation and
releasing chemotactic and growth factors (GF), such as proteases, platelet-derived GF, and vasoactive
agents (histamine and serotonin) [16]. Clot formation is obtained, thus achieving successful
hemostasis. Chemokines released by platelet activation arouse chemotaxis of neutrophil granulocytes,

macrophages, and T lymphocytes into the wound to cleanse debris and bacteria to provide a desired environment for wound healing [16]. Neutrophils are the first cells to appear at the injury site, and create an environment to prevent infection [10,11]. Then, macrophages accumulate within the injury sites and facilitate the phagocytosis of bacteria and damaged tissue. The hemostasis/inflammatory phase proceeds for 48 to 72 h and then the proliferative phase follows.

The proliferative phase, represented by profuse fibroblasts, keratinocytes, endothelial cells, and an accumulation of extracellular matrix (ECM), attenuates gradually and lasts for three to six weeks [10,11]. During the proliferation phase, granulation tissue is formed to replace the clot formation originally created after injury. The major components of granulation tissue are macrophages, fibroblasts, proteoglycans, hyaluronic acid, collagen, and elastin. Additionally, the recruited fibroblasts differentiate into myofibroblasts to achieve wound contraction by cytokines and/or molecules and GFs, including platelet-derived GF, fibroblast GF, nerve GF, transforming GF-β (TGF-β, including TGF-β1, TGF-β2, and TGF-β3), connective tissue GF (CTGF), cysteine-rich 61 (Cyr61), interleukins (IL, including IL-6, IL-8, and IL-10), homeobox 13, the Wnt signaling pathway, osteopontin, and early growth response protein 1 [6–8,17–24]. The upregulation of vascular endothelial GFs all provide a structural framework for endothelial cells to proliferate and lay down new vessels by the process of angiogenesis during this phase [10,22,23]. Re-epithelialization is essential for the re-establishment of tissue integrity. After the completion of re-epithelialization, represented by keratinocytes from wound edge and neighboring adnexal structures migrating across the wound surface, apoptosis of myofibroblasts then occurs, and the skin defect is closed.

The subsequent remodeling phase often takes six to nine months or up to a year to complete [7–10,13]. During the remodeling phase, a precise balance between the apoptosis of existing cells and the production of new cells should be fulfilled. Profuse ECM is degraded gradually and the immature type III collagen is amended into mature type I collagen, resulting in stronger scar tissue. The following are critical for normal wound healing: the orientation of type I collagen, the regression of the immature vessels, and the activation of myofibroblasts [10,25–27]. The final actin-rich cells and reorganization of mature blood contribute to normal wound healing.

Many factors play a role in wound healing [3,4,7–24,28–30], including endocrine factors. The aims of the current review are limited mainly to the discussion of the effects of estrogen on wound healing.

2. Estrogen and Estrogen Signaling

The three major forms of estrogens in humans are estradiol, estrone, and estriol, with estradiol being the predominant form [31–35]. The ovary is the main source of estrogen in the premenopausal period, and granulosa cells are the key cells to produce estrogen [36–40]. Estrogen is also synthesized in skin tissue [41], presenting a concept termed "intracrinology" [42]. Crucially, estrogen regulation depends on estrogen-metabolism enzymes, which include aromatase, catalyzing the conversion of androgens (including dehydroepiandrosterone (DHEA) and testosterone) to estrogens, as well as 17β-hydroxysteroid dehydrogenase 1 and 17β-hydroxysteroid dehydrogenase 2 converting estrone to estradiol [42].

Many natural or chemical synthetic compounds imitating estrogen have been found, and we called these compounds estrogen-like compounds, such as selective estrogen receptor modulators (SERMs), phytoestrogens, and others [31,43–45]. The action of these estrogen-like compounds on cells or tissues is far more complicated, although the main pathway might follow the classical ligand-mediated nuclear receptor signaling, such as estrogen receptors α and β (ERα and ERβ) located within the nuclei of target cells [46,47]. The absence of a ligand (estrogen) shows that ERs are associated with an inhibitory heat shock protein [41].

Estrogen receptors, class I members of the superfamily of nuclear receptors characterized as ligand-inducible transcription factors, have considerable sequence identity in the DNA-binding domains, which permits both receptor types (homo- or hetero-dimers) to interact with the ER elements of various genes. Sequence differences between the two receptors occur primarily in the N- and C-terminal regions. Estrogen receptors can also act as coregulators through binding to other transcription factors already attached to the gene regulatory region and by ligand-independent mechanisms [31,43]. Furthermore, each ER has several known isoforms; at least two isoforms of the gene product relating to ERα and six isoforms of the gene product relating to ERβ are known [31]. ERs are identified in skin, but the expression of ERs on skin depends on the location in the body and cell type [42].

In addition of classical estrogen and ER action, membrane-bound ERs show a distinct role [48]. G-protein-coupled ER1 (GPER1) or G-protein-coupled receptor 30 (GRP30) is a seven-transmembrane receptor protein that binds specifically to estrogen and plays an important role in rapid non-genomic cell signaling [48,49]. Taken together, estrogen signaling can be mediated either through a rapid non-genomic effect or a classical genomic effect. The intricate biochemistry of the mode of action of ERs can be separated into three major aspects, including: (1) the interaction of ERs with other membrane-bound receptors as a form of non-genomic action to regulate the activation or suppression of different transcription factors; (2) the mechanism of ER action alone or in combination of other transcription factors as a form of genomic action to modulate DNA-binding ability to influence gene expression; (3) specific amino acid sequences and structural configurations of ERs [48].

3. Estrogen Signaling and Wound Healing

To explore the relationship between estrogen and wound healing, the term "wound healing and oestrogen" (from 1947 to 30 July 2017) was used to search PubMed for relevant English-language articles (Available online: https://www.ncbi.nlm.nih.gov/pubmed/?term=wound+healing%2C+estrogen) [50], which identified 532 articles. The following is a brief summary.

The influence of estrogen on wound healing was investigated in the animal model in 1947 [50] and in humans in 1953 [51]. Afterwards, much evidence supports that the estrogen is important on wound healing [31,41,42,52–61]. It has been shown that primary estrogen deficiency contributes to cutaneous aging and delayed wound repair or impaired wound healing [41,42]. A decrease in the collagen content (both type I and especially type III) of thigh skin in postmenopausal women without hormone replacement therapy has been found at a rate of 2% (skin wrinkling) per postmenopausal year [56]. Compared with premenopausal women, postmenopausal women demonstrate a decrease in collagen types I and III and a reduction in the type III/type I ratio within the dermis, and this change is much more related with the period of estrogen deficiency rather than chronological age [57].

Skin elasticity correlates negatively with years since menopause [56]. Estrogen therapy can rescue the above with an increasing elasticity of 5% over a year [56]. The daily administration of topical estradiol to the skin of postmenopausal women has shown that the amount of hydroxyproline is significantly increased and a morphologic examination by electron microscopy showed a significant improvement in elastic and collagen fibers [58]. Systemic oral estrogen replacement therapy also showed the similar positive effect on skin contour due to the significantly increasing amount of collagen fibers [59].

Microarray-based profiling of genes differentially expressed in rapid and slow wound healing identified 83% of downregulated gene sets and 80% of upregulated probe sets that were estrogen-regulated [60]. Among these, many factors have been investigated, including estrogen-regulated protease inhibitors, arginase 1 (Arg1), and macrophage migration inhibitory factor (MIF) [60–62]. These protease inhibitors, including SERPINB4 and SERPINB7, which act to protect against the inappropriate activation of cathepsins, are important for wound healing [60]. Deprivation of these protease inhibitors will result in tissue breakdown [60]. Arg1-expressing cells were significantly decreased in the wound granulation tissue of mice after oophorectomy. Arg1 metabolizes L-arginine to generate proline, a substrate for collagen synthesis. Hence, Arg1 is central to modulating the balance between inflammation and matrix deposition during wound healing [60]. MIF is further discussed in the next section (estrogen signaling on hemostasis/inflammation process) [62]. All of this hints that estrogen has a more profound influence on aging than previously thought.

Int. J. Mol. Sci. **2017**, *18*, 2325

Additionally, estrogen can protect against the development of a chronic wound [52,60]. Furthermore, estrogen replacement and topical application of estrogen or its precursor accelerate wound healing [53–55,61], since systemic DHEA levels are strongly associated with protection against chronic venous ulceration in humans. Women who received hormone replacement therapy were less likely to develop a venous leg ulcer (age-adjusted relative risk (RR) 0.65, 95% confidence interval (CI) 0.61–0.69) or a pressure ulcer (RR 0.68, 95% CI 0.62–0.75) than those who did not use hormone replacement therapy, according to a case-cohort study in the UK General Practice Research Database [63]. Moreover, anti-estrogen therapy, such as tamoxifen or aromatase inhibitors, which plays a major role in hormone receptor-positive breast cancer management [64–66], is also correlated with poor wound healing [66]. A study found that women who received anti-estrogen therapy experienced more wound healing complications (61% vs. 28%, $p < 0.001$), fat necrosis (26% vs. 8.3%, $p < 0.001$), infections (15% vs. 2.8%, $p < 0.001$), delayed wound healing (49% vs. 13%, $p < 0.001$), and grade III/IV capsular contracture (55% vs. 9.1%, $p = 0.001$) than those who did not [67], suggesting the important role of estrogen in the wound healing process.

The new insights on the molecular mechanisms mediating the estrogen protective function on wound healing are discussed and topics addressing the involvement of estrogen signaling on hemostasis/inflammation process, proliferation process, and final remodeling process are shown in the following sections.

4. Estrogen Signaling on Hemostasis/Inflammation Process

The recognition of the importance of estrogen in skin physiology would suggest it that may have an important role in wound healing, and a number of studies are available to provide evidence that estrogen might have an important role in all phases of wound healing by modifying the inflammatory reaction, accelerating re-epithelialization, stimulating granulation formation, regulating proteolysis, and balancing collagen synthesis and degradation [68–70].

The first step of wound healing is immediate hemostasis and a prompt initiation of the inflammatory process. These conditions are mediated by local activation of the coagulation system, hematopoietic system, inflammatory cells, and immune system. Coagulation systems, including coagulation factors, are significantly influenced by estrogen signaling transduction [71–73]. However, current knowledge on the effects of estrogen on hemostasis shows the different effect on the coagulation system when the administration route of estrogen is different [73]. A hemostasis imbalance was found among oral estrogen users with a decrease in coagulation inhibitors and an increase in markers of activation coagulation, leading to a global enhanced thrombin formation; by contrast, transdermal estrogen use (avoiding the first-pass effect of liver metabolism) was associated with less change in hemostasis variables and did not activate coagulation and fibrinolysis [73]. Therefore, estrogen deficiency might slow down the activation of coagulation and subsequently impair the immediate hemostasis, which initiates the wound healing process, suggesting the important role of estrogen in hemostasis during wound healing.

During the following inflammation process, the aggregation of megakaryocytes, leukocytes, monocytes, macrophages, lymphocytes, and mast cells is needed. These immune cell populations, including monocytes, neutrophils, macrophages, lymphocytes, and mast cells, as well as hematopoietic progenitors in bone marrow express ERs, suggesting that estrogen directly affects the functions of these cells, including cytokines, in growth factor production [74–79]. Even CD34+ hematopoietic stem cells in human adult bone marrow, but not in hematopoietic stem cells from cord blood, also express ERs, suggesting that ER expression is highly regulated in hematopoietic precursor development [77]. Estrogen receptor activity augments and dampens innate immune signaling pathways in dendritic cells and macrophages [74].

Monocyte-derived dendritic cells express high levels of ERα and low levels of ERβ [77]. B lymphocytes and plasmacytoid dendritic cells display the highest levels of ERβ in comparison with any other cell type, although B lymphocytes also express the highest levels of ERα compared to other leukocytes [77]. Furthermore, macrophages predominantly express the N-terminal truncated ERα46 protein, and this response is mainly dependent on estrogen induction [78,79]. Since ERα acts directly on hematopoietic stem cells, lymphoid progenitors and myeloid progenitors promote development pathways [78].

Excessive neutrophil recruitment and protease production is often associated with impaired wound healing [68]. Estrogen therapy can reduce the number of wound neutrophils, and reduce the neutrophil adhesion molecule L-selectin, leading to diminished neutrophil localization at sites of inflammation [41]. Administration of estrogen can increase wound fibronectin levels, as well as reduce elastase activity and lessen the degradation of fibronectin in wound tissue [79–81].

Estrogen signaling pro-inflammatory cytokine production is varied by the cell type and local estrogen concentration (environmental factors). One report showed that treatment with estrogen can induce ERα expression in macrophages, but the ERβ expression is not changed, and the increasing ERα expression is time-dependent during monocyte-to-macrophage differentiation [82]. The C-X-C motif chemokine ligand 8 (CXCL-8)—one of the chemokines known to cause allergic reactions and profound inflammation [83]—production of macrophages is suppressed in estrogen-mediated ER-dependent pathways.

Additionally, MIF—one of the pro-inflammatory cytokines, which was established by micro-array techniques to identify more than 600 differently regulating repair/inflammation-associated gene targets—is a key player in skin biology and wound healing [60–62]. MIF has a direct effect on the expression of genes involved in all aspects of the wound healing process, in addition to genes associated with cellular proliferation, differentiation, and apoptosis in a wide range of cell types [62]. Impaired healing in the absence of estrogen via elevated MIF is associated with dysregulated differentiation, cell contractile machinery, and altered signaling and transcription, coupled with a proteolytic and a pro-inflammatory state [62]. Estrogen treatment can downregulate MIF expression, resulting in an improvement in wound healing [62]. Furthermore, in vitro studies suggested a direct effect on specific pro-inflammatory cytokine production by macrophages via mitogen-activated kinase (MAP), phosphatidylinositol 3 (PI3) kinase pathways, and a nuclear factor κB-dependent mechanism [60–62]. This information suggests the importance of ERα expression and regulation in the ability of estrogen to modulate innate immune response.

There are several possible mechanisms to address the relationship between the function of these innate immune cells and estrogen, including estrogen signaling epigenetic changes of immune cells, especially for precursor cells, which might influence downstream developmental pathways or functional responses in mature immune cells. For example, estrogen signaling gene activation promotes a developmental pathway, or altered activity of estrogen signaling transduction in immune cells within a pathway when they are activated [74].

To summarize during the inflammatory phase of wound healing, neutrophils are the first responders of the inflammatory response, and function mainly by clearing debris and pathogens; subsequently, activated macrophages join and continue to clear debris and pathogens, as well as clear apoptotic neutrophils that remain at the wound site. All of them are influenced by estrogen, supporting the important roles of estrogen-mediated signaling transduction in the first step of the wound healing process. Estrogen results in dampening purulent inflammation, decreasing neutrophil numbers as shown above, promoting alternative macrophage polarization (promoting a shift from M1 to M2 subtypes), reducing the expression of pro-inflammatory cytokines, such as tissue necrotic factor (TNF) α, and decreasing elastase production, contributing to the beneficial roles of aiding wound closure and collagen deposition. The role of macrophage phenotypes on wound healing has been extensively reviewed recently [83,84].

In brief, monocytes can be classically or alternatively activated to form M1 and M2 macrophages, respectively [83]. Conventionally, M1 macrophages, which are activated by pro-inflammatory mediators interferon-γ (IFN-γ), TNF, and damage-associated pattern molecules, have a known scavenger function [84]. These classically activated M1 macrophages prolifically produce pro-inflammatory cytokines, such as TNF and IL-6 and other mediators, enabling them to phagocytose neutrophils that have undergone apoptosis and remove any pathogens or debris in the wound. All of this facilitates the initial stages of wound healing. M2 macrophages are typically anti-inflammatory and regulate re-vascularization and wound closure [83]. M2 macrophages can been further divided into four discrete types: M2a, M2b, M2c, and M2d, based on their function and key markers [83]. Collectively, alterations in macrophage number and phenotype will disrupt the wound healing.

Although estrogen is important for wound healing, the effects of estrogen on wounds should be a balancing act, since both low and high levels of estrogen will slow inflammation, and over-inhibition of inflammation is not good for wound healing.

5. Estrogen Signaling on the Proliferation Process

The proliferative process of wound healing involves profuse fibroblasts, keratinocytes, and endothelial cells, as well as an accumulation of ECM [10,11]. Therefore, the relationship between estrogen and targeted cells, including fibroblasts, keratinocytes, and endothelial cells, is discussed. Al these cells contain ERs.

Cells with low-level differentiation potential, for example mesenchymal cells, have the ability to stimulate tissue renewal. The fibroblast is the key mesenchymal cell type in connective tissue and deposits the collagen and elastic fibers of the ECM, critical for wound healing. Aging and estrogen deficiency result in defects in fibroblast differentiation and functionality associated with impaired hyaluronan synthase 2 and epidermal growth factor receptor function, as a result of upregulated microRNA-7 expression, which mediates the over-activation of JAK/STAT1 [53].

Estrogen signaling interacts with the microRNA-7 promoter, suppresses microRNA-7 expression, and further attenuates STAT1 expression and activity. Estrogen is important for the proliferation, migration, and differentiation of fibroblasts [53], which is primarily mediated by ERα [67]. Estrogen induces a rapid re-organization of the cytoskeleton in dermal fibroblast via the non-classical receptor GPR30, MAPK, PI3 K/Akt, and ERK1/2 activation [48,49,85–87]. Finally, the function of fibroblasts may be markedly varied even within a single tissue [87], which results from different autocrine (intracrine) signals. Aromatase activity and Wnt-regulated signals form the overlying epidermis can act both locally, via ECM deposition, and via secreted factors. All impact the behavior of fibroblasts in different dermal locations.

Additionally, human dermal fibroblasts can be "transiently" activated to go forward to myofibroblast differentiation (αSMA expression) [88,89]. Permanent or sequential presence of TGF-β1 and IL-10 might modify the proliferation and migration of fibroblasts and their activated form-myofibroblasts [90]. Study has shown that the removal of TGF-β1 after initial stimulation resulted in an increase of apoptosis of myofibroblasts [90]. TGF-β1 stimulation followed by IL-10 treatment did not result in increased cell apoptosis, but instead led to a significant increase of cell motility and a reduction of myofibroblasts [90]. All of this hints at the precise and dynamic function of cytokines—such as TGF-β1 and IL-10—as an important cue for the completion of wound healing [90].

Keratinocytes proliferate and migrate over the wound to create a barrier between the outer and inner environments, mediated through re-epithelialization. Similar to the impact of estrogen on fibroblasts, estrogen also imparts potent mitogenic effect on keratocytes, promoting in vitro and in vivo migration [41], which might be affected by the estrogen-mediated ERβ interaction with keratocytes. One study showed that the pharmacological activation of ERβ, but not that of ERα, led to a significant alteration in the pattern of differentiation and the proliferation activity of keratinocytes, suggesting that the stimulation of epidermal regeneration may ensue after treating a wound with a targeting to ERβ [91]. However, although the activation of the TGF-β pathway is critical for fibroblast transformation to myofibroblasts [90], the TGF-β pathway is downstream of ERα and ERβ; thus, the ER-mediated function might be opposite [92].

The relation of estrogen to endothelial cells has been investigated before [93,94]. Mature endothelial cells are derived from endothelial progenitor cells from bone marrow, and these endothelial progenitor cells are defined as cells that have a positive receptor to hematopoietic stem cell marker-CD34, endothelial cell marker-vascular endothelial growth factor receptor 2, and an immature hematopoietic stem cell marker CD133 [95]. Estrogen signaling induces the mobilization of circulating endothelial progenitor cells from bone marrow, and these cells help to build and restore injured and/or damaged endothelium. Estrogen also induces proliferation and migration, and inhibits the apoptosis of endothelial progeny cells. The main mechanism of estrogen-mediated re-reendothelialization is a nitric oxide-dependent pathway [93,95]. Endothelial cells constitute an essential cornerstone in the building and maintenance of endothelial blood vessels (lining the lumen of every blood vessel). Endothelial cells are involved in filtration, hemostasis, barrier function, inflammation, and angiogenesis, serving as gatekeepers, preventing degeneration, and assisting in wound healing [95,96].

6. Estrogen Signaling on the Remodeling Process

The last step of the wound healing is the remodeling phase, which relies on a controlled balance between the synthesis and degradation of the ECM, and estrogen is thought to involve both [67]. Collagens, composing three α chains of repeating Gly-Xaa-Yaa triplets, which not only induce each α chain to adopt a left-handed PolyPro II helix, but also represent a base by which to classify collagen types, are the most abundant proteins in the human body and the main components of the ECM [96]. Thus far, at least 28 types of collagens have been reported [97]. Collagens can be classified as homotrimeric and heterotrimetric types. Homotrimeric collagens (e.g., type II and type III) have three α chains of identical sequence, and heterotrimetric collagens have two α chains of identical sequence (α1) and one α chain of differing sequence (α2) (i.e., type I), or three α chains with different sequences (α1, α2, and α3) (i.e., type VI) [96].

Among these collagens, type I and type III collagens are important for wound healing [10,27], and both have fibrillar structures [97]. Type I collagens are the most profuse and ubiquitous of the collagens in the connective tissues [97]. Evidence has shown that the deposition of different types of collagens may cause diverse wound healing [7–10]. Type I collagen is dominant in normal wounds; by contrast, in hypertrophic scar or keloid scar, type III is dominant [4].

A myriad of factors stringently control the formation and degradation of collagens [98–105]. Among these, matrix metalloproteinases (MMPs), a family of zinc dependent proteases, are crucial [97]. These MMPs involve the tight control of ECM not only for an initial step of wound healing—inflammation—but also for the end step of wound healing, remodeling over time [99]. The interaction of the keratinocytes with collagen I, in animal models, triggers MMP-1 expression immediately when tissue insult occurs, and the basal keratinocytes at the epithelial front secretes MMP-1, which cleaves the provisional matrix, paving the path for the proper migration of these rapidly proliferating cells at the distal end [98]. This highest level of MMP-1 at day one gradually will decrease to basal level towards the completion of re-epithelialization. Persistent high levels of MMP-1 (collagenase 1) result in impairing wound healing [98]. This abnormal over-activity of MMP-1 is also supported by other studies [99].

One study showed that a magnitude of a 116-fold increase in the average protease activity is found in abnormal wound exudates when compared to normal acute wound exudates [97]. The expressions of the other MMPs, such as MMP-8 and MMP-13, are also disrupted in poor wound healing. MMP-8 (collagenase 2) has a stronger affinity toward type I collagen; therefore, the overexpression and activation of MMP-8 is directly involved in the pathogenesis of chronic non-healing wounds [98–104]. Absent expression of MMP-13 (collagenase 3) in the epidermis and overexpression of MMP-13 in non-healed wounds also suggests that the expression of several MMPs are derailed both at mRNA and protein levels in abnormal wound healing [98].

MMPs also regulate cell-cell and cell-matrix interactions through modulating and releasing cytokines, growth factors, and other biological active fragments that are sequestered in the ECM [100]. Growing evidence has convincingly identified select MMPs in membrane-type and intracellular compartments with unexpected physiological and pathological roles [98,99]. Membrane-type MMPs involve sheddase activities, collagenolysis, bacterial killing, and intracellular trafficking reaching as far as the nucleus. These membrane-type MMPs may also support pericellular proteolysis and endocytosis [100]. The above described research supports the important role of MMPs in wound healing, because MMPs are not only responsible for the direct degradation of ECM molecules, but are also key modulators of cardinal bioactive factors [102,103].

Coinciding with the dysregulation of MMPs, the tight regulation of endogenous inhibitors, tissue inhibitors of metalloproteinases (TIMPs), is also reported to be involved in wound healing [101,102]. These TIMPs directly regulate MMP activity and the MMPs/TIMPs balance can determine the net MMP activity, ECM turnover, and tissue remodeling, including wound healing.

The successful remodeling phase of wound healing depends on type I collagen in place of the original type III collagen. Type III collagens have more "flexible" potential cleavage sites than type I, and thus are more susceptible to hydrolysis by a variety of MMPs [97]. The classical collagenases include MMP-1, MMP-8, and MMP-13, based on efficiently catalyzing collagen hydrolysis ability [94]. Membrane-type I MMP prefers type I collagen, and the activity against type I collagen is 6.5 times that of type III collagens [97]. Since clear evidence shows that a single MMP cannot be unequivocally labeled as "good" or "bad", because the net result of proteolytic activity should be dependent on situations when considering wound healing in general [104], it is not surprising that dysregulation in any protease function that affects ECM homeostasis might contribute to poor wound healing.

7. Conclusions

A full understanding of cutaneous estrogen synthesis and signaling is essential for future estrogen-based pharmacological manipulations of wound healing. Existing clinical estrogen-like compounds (e.g., tamoxifen, raloxifene, phytoestrogens, and genistein) are also known to influence the wound healing process [105], suggesting that the effects of estrogen on wound healing are complicated and worthy of further investigation. Translation of these estrogen and estrogen-like compound-mediated pathways from the bench to the clinic remains a promising proposition.

Acknowledgments: This work was supported by grants from the Ministry of Science and Technology (MOST 103-2314-B-010-043-MY3, and MOST 106-2314-B-075-061-MY3) and the Taipei Veterans General Hospital (Grant V104C-095, V105C-096, V106C-129; V106D23-001-MY2-1; and V106A-012), Taipei, Taiwan. The funders had no role in the study design, data collection and analysis, decision to publish, or preparation of the manuscript. We thank the Clinical Research Core Laboratory and the Medical Science and Technology Building of the Taipei Veterans General Hospital for providing experimental space and facilities.

Author Contributions: Huann-Cheng Horng, Wen-Hsun Chang, Chang-Ching Yeh and Ben-Shian Huang designed the study, performed the experiments, analyzed the data, and drafted the manuscript. Chang-Ching Yeh, Chia-Pei Chang, Yi-Jen Chen and Kuan-Hao Tsui revised the paper. Peng-Hui Wang designed the study, analyzed the data, drafted the manuscript, edited the paper, supervised research, and made a final revision. All authors read and approved the final manuscript.

Conflicts of Interest: The authors declare no conflict of interest.

Abbreviations

The following abbreviations are used in this manuscript:

GF	growth factor
ECM	extracellular matrix
TGF-β	transforming growth factor-β
DHEA	dehydroepiandrosterone
SERMs	selective estrogen receptor modulators
ER	estrogen receptor
Cyr61	cysteine-rich 61
IL	interleukin
GPER1	G-protein-coupled ER1
GRP30	G-protein-coupled receptor 30
Arg1	arginase 1

TNF	tumor necrosis factor
MIF	macrophage migration inhibitory factor
RR	relative risk
CI	confidence interval
MMPs	metalloproteinases
TIMPs	tissue inhibitors of metalloproteinases
siRNA	small interfering RNA
MAP	mitogen-activated kinase (MAP)
PI3K	phosphatidylinositol-3-kinase
miRNAs	microRNAs
CXCL-8	C-X-C motif chemokine ligand 8
IFN-γ	interferon-γ

References

1. Su, W.H.; Cheng, M.H.; Lee, W.L.; Tsou, T.S.; Chang, W.H.; Chen, C.S.; Wang, P.H. Nonsteroidal anti-inflammatory drugs for wounds: Pain relief or excessive scar formation? *Mediat. Inflamm.* **2010**, *2010*, 413238. [CrossRef] [PubMed]

2. Wang, Y.W.; Liou, N.H.; Cherng, J.H.; Chang, S.J.; Ma, K.H.; Fu, E.; Liu, J.C.; Dai, N.T. siRNA-targeting transforming growth factor-β type I receptor reduces wound scarring and extracellular matrix deposition of scar tissue. *J. Investig. Dermatol.* **2014**, *134*, 2016–2025. [CrossRef] [PubMed]

3. Ogawa, R. Keloid and hypertrophic scars are the result of chronic inflammation in the reticular dermis. *Int. J. Mol. Sci.* **2017**, *18*, 606. [CrossRef] [PubMed]

4. Butzelaar, L.; Ulrich, M.M.; Mink van der Molen, A.B.; Niessen, F.B.; Beelen, R.H. Currently known risk factors for hypertrophic skin scarring: A review. *J. Plast. Reconstr. Aesthet. Surg.* **2016**, *69*, 163–169. [CrossRef] [PubMed]

5. Hofer, M.; Hoferová, Z.; Falk, M. Pharmacological modulation of radiation damage. Does it exist a chance for other substances than hematopoietic growth factors and cytokines? *Int. J. Mol. Sci.* **2017**, *18*, 1385. [CrossRef]

6. Zhang, K.; Si, X.-P.; Huang, J.; Han, J.; Liang, X.; Xu, X.-B.; Wang, Y.-T.; Li, G.-Y.; Wang, H.-Y.; Wang, J.-H. Preventive effects of *Rhodiola rosea* L. on bleomycin-induced pulmonary fibrosis in rats. *Int. J. Mol. Sci.* **2016**, *17*, 879. [CrossRef] [PubMed]

7. Janis, J.E.; Harrison, B. Wound healing: Part I. Basic science. *Plast. Reconstr. Surg.* **2016**, *138*, 9S–17S. [CrossRef] [PubMed]

8. Profyris, C.; Tziotzios, C.; do Vale, I. Cutaneous scarring: Pathophysiology, molecular mechanisms, and scar reduction therapeutics Part I. The molecular basis of scar formation. *J. Am. Acad. Dermatol.* **2012**, *66*, 1–10. [CrossRef] [PubMed]

9. Gauglitz, G.G.; Korting, H.C.; Pavicic, T.; Ruzicka, T.; Jeschke, M.G. Hypertrophic scarring and keloids: Pathomechanisms and current and emerging treatment strategies. *Mol. Med.* **2011**, *17*, 113–125. [CrossRef] [PubMed]

10. Berman, B.; Maderal, A.; Raphael, B. Keloids and hypertrophic scars: Pathophysiology, classification, and treatment. *Dermatol. Surg.* **2017**, *43*, S3–S18. [CrossRef] [PubMed]

11. Plikus, M.V.; Guerrero-Juarez, C.F.; Ito, M.; Li, Y.R.; Dedhia, P.H.; Zheng, Y.; Shao, M.; Gay, D.L.; Ramos, R.; Hsi, T.C.; et al. Regeneration of fat cells from myofibroblasts during wound healing. *Science* **2017**, *355*, 748–752. [CrossRef] [PubMed]

12. Yannas, I.V.; Tzeranis, D.S.; So, P.T.C. Regeneration of injured skin and peripheral nerves requires control of wound contraction, not scar formation. *Wound Repair. Regen.* **2017**, *25*, 177–191. [CrossRef] [PubMed]

13. Walmsley, G.G.; Maan, Z.N.; Wong, V.W.; Duscher, D.; Hu, M.S.; Zielins, E.R.; Wearda, T.; Muhonen, E.; McArdle, A.; Tevlin, R.; et al. Scarless wound healing: Chasing the holy grail. *Plast. Reconstr. Surg.* **2015**, *135*, 907–917. [CrossRef] [PubMed]

14. Tsai, H.W.; Wang, P.H.; Tsui, K.H. Mesenchymal stem cell in wound healing and regeneration. *J. Chin. Med. Assoc.* **2017**. [CrossRef] [PubMed]

15. Hrosley, V.; Watt, F. Repeal and replace: Adipocyte regeneration in wound repair. *Cell Stem Cell* **2017**, *20*, 424–426. [CrossRef]

16. Brockmann, L.; Giannou, A.D.; Gagliani, N.; Huber, S. Regulation of TH17 cells and associated cytokines in wound healing, tissue regeneration, and carcinogenesis. *Int. J. Mol. Sci.* **2017**, *18*, 33. [CrossRef] [PubMed]

17. Driskell, R.R.; Jahoda, C.A.; Chuong, C.M.; Watt, F.M.; Horsley, V. Defining dermal adipose tissue. *Exp. Dermatol.* **2014**, *23*, 629–631. [CrossRef] [PubMed]

18. Mateu, R.; Živicová, V.; Krejčí, E.D.; Grim, M.; Strnad, H.; Vlček, Č.; Kolář, M.; Lacina, L.; Gál, P.; Borský, J.; et al. Functional differences between neonatal and adult fibroblasts and keratinocytes: Donor age affects epithelial-mesenchymal crosstalk in vitro. *Int. J. Mol. Med.* **2016**, *38*, 1063–1074. [CrossRef] [PubMed]

19. Gerarduzzi, C.; di Battista, J.A. Myofibroblast repair mechanisms post-inflammatory response: A fibrotic perspective. *Inflamm. Res.* **2017**, *66*, 451–465. [CrossRef] [PubMed]

20. Ojeh, N.; Pastar, I.; Tomic-Canic, M.; Stojadinovic, O. Stem cells in skin regeneration, wound healing, and their clinical applications. *Int. J. Mol. Sci.* **2015**, *16*, 25476–25501. [CrossRef] [PubMed]

21. Lindley, L.E.; Stojadinovic, O.; Pastar, I.; Tomic-Canic, M. Biology and biomarkers for wound healing. *Plast. Reconstr. Surg.* **2016**, *138*, 18S–28S. [CrossRef] [PubMed]

22. Syeda, M.M.; Jing, X.; Mirza, R.H.; Yu, H.; Sellers, R.S.; Chi, Y. Prostaglandin transporter modulates wound healing in diabetes by regulating prostaglandin-induced angiogenesis. *Am. J. Pathol.* **2012**, *181*, 334–346. [CrossRef] [PubMed]

23. Cohen, B.E.; Geronemus, R.G.; McDaniel, D.H.; Brauer, J.A. The role of elastic fibers in scar formation and treatment. *Dermatol. Surg.* **2017**, *43*, S19–S24. [CrossRef] [PubMed]

24. Darby, I.A.; Weller, C.D. Aspirin treatment for chronic wounds: Potential beneficial and inhibitory effects. *Wound. Repair. Regen.* **2017**, *25*, 7–12. [CrossRef] [PubMed]

25. Ashcroft, K.J.; Syed, F.; Bayat, A. Site-specific keloid fibroblasts alter the behaviour of normal skin and normal scar fibroblasts through paracrine signalling. *PLoS ONE* **2013**, *8*, e75600. [CrossRef] [PubMed]

26. Huang, C.; Akaishi, S.; Hyakusoku, H.; Ogawa, R. Are keloid and hypertrophic scar different forms of the same disorder? A fibroproliferative skin disorder hypothesis based on keloid findings. *Int. Wound. J.* **2014**, *11*, 517–522. [CrossRef] [PubMed]

27. Cheng, J.; Wang, Y.; Wang, D.; Wu, Y. Identification of collagen 1 as a post-transcriptional target of miR-29b in skin fibroblasts: Therapeutic implication for scar reduction. *Am. J. Med. Sci.* **2013**, *346*, 98–103. [CrossRef] [PubMed]

28. Ding, S.L.S.; Kumar, S.; Mok, P.L. Cellular reparative mechanisms of mesenchymal stem cells for retinal diseases. *Int. J. Mol. Sci.* **2017**, *18*, 1406. [CrossRef] [PubMed]

29. Jeong, W.; Yang, C.E.; Roh, T.S.; Kim, J.H.; Lee, J.H.; Lee, W.J. Scar prevention and enhanced wound healing induced by polydeoxyribonucleotide in a rat incisional wound-healing model. *Int. J. Mol. Sci.* **2017**, *18*, 1698. [CrossRef] [PubMed]

30. Yang, S.Y.; Yang, J.Y.; Hsiao, Y.H.; Chuang, S.S. A comparison of gene expression of decorin and MMP13 in hypertrophic scars treated with calcium channel blocker, steroid, and interferon: A human-scar-carrying animal model study. *Dermatol. Surg.* **2017**, *43*, S37–S46. [CrossRef] [PubMed]

31. Leblanc, D.R.; Schneider, M.; Angele, P.; Vollmer, G.; Docheva, D. The effect of estrogen on tendon and ligament metabolism and function. *J. Steroid. Biochem. Mol. Biol.* **2017**, *172*, 106–116. [CrossRef] [PubMed]

32. Casarini, L.; Riccetti, L.; de Pascali, F.; Gilioli, L.; Marino, M.; Vecchi, E.; Morini, D.; Nicoli, A.; la Sala, G.B.; Simoni, M. Estrogen modulates specific life and death signals induced by LH and hCG in human primary granulosa cells in vitro. *Int. J. Mol. Sci.* **2017**, *18*, 926. [CrossRef] [PubMed]

33. Parodi, D.A.; Greenfield, M.; Evans, C.; Chichura, A.; Alpaugh, A.; Williams, J.; Cyrus, K.C.; Martin, M.B. Alteration of mammary gland development and gene expression by in utero exposure to cadmium. *Int. J. Mol. Sci.* **2017**, *18*, 1939. [CrossRef] [PubMed]

34. Tsai, H.W.; Wang, P.H.; Huang, B.S.; Twu, N.F.; Yen, M.S.; Chen, Y.J. Low-dose add-back therapy during postoperative GnRH agonist treatment. *Taiwan. J. Obstet. Gynecol.* **2016**, *55*, 55–59. [CrossRef] [PubMed]

35. Tsui, K.H.; Huang, B.S.; Wang, P.H. Kisspeptin system in female reproduction: A next-generation target in the manipulation of sex hormones. *J. Chin. Med. Assoc.* **2016**, *79*, 519–520. [CrossRef] [PubMed]

36. Liu, C.H.; Horng, H.C.; Chang, W.H.; Wang, P.H. Granulosa cell tumor of ovary: Perspective of Taiwan. *Taiwan. J. Obstet. Gynecol.* **2017**, *56*. [CrossRef] [PubMed]

37. Tsui, K.H.; Wang, P.H.; Lin, L.T.; Li, C.J. DHEA protects mitochondria against dual modes of apoptosis and necroptosis in human granulosa HO23 cells. *Reproduction* **2017**, *154*, 101–110. [CrossRef] [PubMed]

38. Lai, W.A.; Yeh, Y.T.; Fang, W.L.; Wu, L.S.; Harada, N.; Wang, P.H.; Ke, F.C.; Lee, W.L.; Hwang, J.J. Calcineurin and CRTC2 mediate FSH and TGFβ1 upregulation of Cyp19a1 and Nr5a in ovary granulosa cells. *J. Mol. Endocrinol.* **2014**, *53*, 259–270. [CrossRef] [PubMed]

39. Wang, N.; Zhao, F.; Lin, P.; Zhang, G.; Tang, K.; Wang, A.; Jin, Y. Knockdown of XBP1 by RNAi in mouse granulosa cells promotes apoptosis, inhibits cell cycle, and decreases estradiol synthesis. *Int. J. Mol. Sci.* **2017**, *18*, 1152. [CrossRef] [PubMed]

40. Worku, T.; Rehman, Z.U.; Talpur, H.S.; Bhattarai, D.; Ullah, F.; Malobi, N.; Kebede, T.; Yang, L. MicroRNAs: New insight in modulating follicular atresia: A review. *Int. J. Mol. Sci.* **2017**, *18*, 333. [CrossRef] [PubMed]

41. Mukai, K.; Urai, T.; Asano, K.; Nakajima, Y.; Nakatani, T. Evaluation of effects of topical estradiol benzoate application on cutaneous wound healing in ovariectomized female mice. *PLoS ONE* **2016**, *11*, e0163560. [CrossRef] [PubMed]

42. Wilkinson, H.N.; Hardman, M.J. The role of estrogen in cutaneous ageing and repair. *Maturitas* **2017**, *103*, 60–64. [CrossRef] [PubMed]

43. Lee, W.L.; Cheng, M.H.; Tarng, D.C.; Yang, W.C.; Lee, F.K.; Wang, P.H. The benefits of estrogen or selective estrogen receptor modulator on kidney and its related disease-chronic kidney disease-mineral and bone disorder: Osteoporosis. *J. Chin. Med. Assoc.* **2013**, *76*, 365–371. [CrossRef] [PubMed]

44. Ahangarpour, A.; Najimi, S.A.; Farbood, Y. Effects of Vitex agnus-castus fruit on sex hormones and antioxidant indices in a D-galactose-induced aging female mouse model. *J. Chin. Med. Assoc.* **2016**, *79*, 589–596. [CrossRef] [PubMed]

45. Chen, F.P.; Chang, C.J.; Chao, A.S.; Huang, H.Y.; Huang, J.P.; Wu, M.H.; Tsai, C.C.; Kung, F.T.; Chang, C.W.; Tsai, Y.C. Efficacy of Femarelle for the treatment of climacteric syndrome in postmenopausal women: An open label trial. *Taiwan. J. Obstet. Gynecol.* **2016**, *55*, 336–340. [CrossRef] [PubMed]

46. Huang, B.S.; Lee, W.L.; Wang, P.H. The slow down of renal deterioration but acceleration of cardiac hypertrophy: Is the estrogen receptor-α a hero or villain? *Am. J. Physiol. Renal. Physiol.* **2014**, *307*, F1352. [CrossRef] [PubMed]

47. Huang, B.S.; Chang, W.H.; Wang, K.C.; Huang, N.; Guo, C.Y.; Chou, Y.J.; Huang, H.Y.; Chen, T.J.; Lee, W.L.; Wang, P.H. Endometriosis might be inversely associated with developing chronic kidney disease: A population-based cohort study in Taiwan. *Int. J. Mol. Sci.* **2016**, *17*, 1079. [CrossRef] [PubMed]

48. Seto, K.; Hoang, M.; Santos, T.; Bandyopadhyay, M.; Kindy, M.S.; Dasgupta, S. Non-genomic oestrogen receptor signal in B lymphocytes: An approach towards therapeutic interventions for infection, autoimmunity and cancer. *Int. J. Biochem. Cell. Biol.* **2016**, *76*, 115–118. [CrossRef] [PubMed]

49. Zhou, T.; Yang, Z.; Chen, Y.; Chen, Y.; Huang, Z.; You, B.; Peng, Y.; Chen, J. Estrogen accelerates cutaneous wound healing by promoting proliferation of epidermal keratinocytes via Erk/Akt signaling pathway. *Cell. Physiol. Biochem.* **2016**, *38*, 959–968. [CrossRef] [PubMed]

50. Sjovall, A. The influence of oestrogen upon the healing of vaginal wounds in rats. *Acta Obstet. Gynecol. Scand.* **1947**, *27*, 1–10. [CrossRef] [PubMed]

51. Sjostedt, S. The effect of diethylstilbenediol on the healing of wounds in the human vagina. *Acta Endocrinol.* **1953**, *12*, 260–263. [CrossRef] [PubMed]

52. Chenu, C.; Adlanmerini, M.; Boudou, F.; Chantalat, E.; Guihot, A.L.; Toutain, C.; Raymond-Letron, I.; Vicendo, P.; Gadeau, A.P.; Henrion, D.; et al. Testosterone prevents cutaneous ischemia and necrosis in males through complementary estrogenic and androgenic actions. *Arterioscler. Thromb. Vasc. Biol.* **2017**, *37*, 909–919. [CrossRef] [PubMed]

53. Brufani, M.; Rizzi, N.; Meda, C.; Filocamo, L.; Ceccacci, F.; D'Aiuto, V.; Bartoli, G.; Bella, A.; Migneco, L.M.; Bettolo, R.M.; et al. Novel locally active estrogens accelerate cutaneous wound healing—Part 2. *Sci. Rep.* **2017**, *7*, 2510. [CrossRef] [PubMed]

54. Midgley, A.C.; Morris, G.; Phillips, A.O.; Steadman, R. 17β-estradiol ameliorates age-associated loss of fibroblast function by attenuating IFN-γ/STAT1-dependent miR-7 upregulation. *Aging Cell* **2016**, *15*, 531–541. [CrossRef] [PubMed]

55. Pepe, G.; Braga, D.; Renzi, T.A.; Villa, A.; Bolego, C.; D'Avila, F.; Barlassina, C.; Maggi, A.; Locati, M.; Vegeto, E. Self-renewal and phenotypic conversion are the main physiological responses of macrophages to the endogenous estrogen surge. *Sci. Rep.* **2017**, *7*, 44270. [CrossRef] [PubMed]

56. Brincat, M.; Versi, E.; Moniz, C.F.; Magos, A.; de Trafford, J.; Studd, J.W. Skin collagen changes in postmenopausal women receiving different regimens of estrogen therapy. *Obstet. Gynecol.* **1987**, *70*, 123–127. [CrossRef]

57. Affinito, P.; Palomba, S.; Sorrentino, C.; di Carlo, C. Bifulco, G. Arienzo, M.P.; Nappi, C. Effects of postmenopausal hypoestrogenism on skin collagen. *Maturitas.* **1999**, *33*, 239–247. [CrossRef]

58. Varila, E.; Rantala, I.; Oikarinen, A.; Risteli, J.; Reunala, T.; Oksanen, H.; Punnonen, R. The effect of topical oestradiol on skin collagen of postmenopausal women. *Br. J. Obstet. Gynaecol.* **1995**, *102*, 985–989. [CrossRef] [PubMed]

59. Sauerbronn, A.D.V.; Fonseca, A.M.; Bagnoli, V.R.; Saldiva, P.H.; Pinotti, J.A. The effects of systemic hormonal replacement therapy of the skin of postmenopausal women. *Int. J. Gynaecol. Obstet.* **2000**, *68*, 35–41. [CrossRef]

60. Hardman, M.J.; Ashcroft, G.S. Estrogen, not intrinsic aging, is the major regulator of delayed human wound healing in the elderly. *Genome Biol.* **2008**, *9*, R80. [CrossRef] [PubMed]

61. Mills, S.J.; Ashworth, J.J.; Gilliver, S.C.; Hardman, M.J.; Ashcroft, G.S. The sex steroid precursor DHEA accelerates cutaneous wound healing via the estrogen receptors. *J. Investig. Dermatol.* **2005**, *125*, 1053–1062. [CrossRef] [PubMed]

62. Hardman, M.J.; Waite, A.; Zeef, L.; Burow, M.; Nakayama, T.; Ashcroft, G.S. Macrophage migration inhibitory factor: A central regulator of wound healing. *Am. J. Pathol.* **2005**, *167*, 1561–1574. [CrossRef]

63. Margolis, D.J.; Knauss, J.; Bilker, W. Hormone replacement therapy and prevention of pressure ulcers and venous leg ulcers. *Lancet* **2002**, *359*, 675–677. [CrossRef]

64. Rondón-Lagos, M.; Villegas, V.E.; Rangel, N.; Sánchez, M.C.; Zaphiropoulos, P.G. Tamoxifen resistance: Emerging molecular targets. *Int. J. Mol. Sci.* **2016**, *17*, 1357. [CrossRef] [PubMed]

65. Villegas, V.E.; Rondón-Lagos, M.; Annaratone, L.; Castellano, I.; Grismaldo, A.; Sapino, A.; Zaphiropoulos, P.G. Tamoxifen treatment of breast cancer cells: Impact on Hedgehog/GLI1 signaling. *Int. J. Mol. Sci.* **2016**, *17*, 308. [CrossRef] [PubMed]

66. Wang, P.H.; Chao, H.T. To switch or not to switch: Should the study of tamoxifen and raloxifene (STAR) trial alter our decision? *Taiwan. J. Obstet. Gynecol.* **2008**, *47*, 372–374. [CrossRef]

67. Billon, R.; Bosc, R.; Belkacemi, Y.; Assaf, E.; SidAhmed-Mezi, M.; Hersant, B.; Meningaud, J.P. Impact of adjuvant anti-estrogen therapies (tamoxifen and aromatase inhibitors) on perioperative outcomes of breast reconstruction. *J. Plast. Reconstr. Aesthet. Surg.* **2017**. [CrossRef] [PubMed]

68. Thornton, M.J. Estrogens and aging skin. *Dermato-endocrinology* **2013**, *5*, 264–270. [CrossRef] [PubMed]

69. Archer, D.F. Postmenopausal skin and estrogen. *Gynecol. Endocrinol.* **2012**, *28*, 2–6. [CrossRef] [PubMed]

70. Stevenson, S.; Thornton, J. Effect of estrogens on skin aging and the potential role of SERMs. *Clin. Interv. Aging* **2007**, *2*, 283–297. [PubMed]

71. Chen, C.H.; Chin, H.Y.; Chen, H.H.; Chang, H.Y.; Liu, W.M. Pills-related severe adverse events: A case report in Taiwan. *Taiwan. J. Obstet. Gynecol.* **2016**, *55*, 588–590. [CrossRef] [PubMed]

72. Koh, K.K.; Mincemoyer, R.; Bui, M.N.; Csako, G.; Pucino, F.; Guetta, V.; Waclawiw, M.; Cannon, R.O., 3rd. Effects of hormone-replacement therapy on fibrinolysis in postmenopausal women. *N. Engl. J. Med.* **1997**, *336*, 683–690. [CrossRef] [PubMed]

73. Canonico, M. Hormone therapy and hemostasis among postmenopausal women: A review. *Menopause* **2014**, *21*, 753–762. [CrossRef] [PubMed]

74. Kovats, S. Estrogen receptors regulate innate immune cells and signaling pathways. *Cell. Immunol.* **2015**, *294*, 63–69. [CrossRef] [PubMed]

75. Gubbels Bupp, M.R. Sex, the aging immune system, and chronic disease. *Cell. Immunol.* **2015**, *294*, 102–110. [CrossRef] [PubMed]

76. Khan, D.; Ansar Ahmed, S. The immune system is a natural target for estrogen action: Opposing effects of estrogen in two prototypical autoimmune diseases. *Front. Immunol.* **2016**, *6*, 635. [CrossRef] [PubMed]

77. Laffont, S.; Seillet, C.; Guéry, J.C. Estrogen receptor-dependent regulation of dendritic cell development and function. *Front. Immunol.* **2017**, *8*, 108. [CrossRef] [PubMed]

78. Murphy, A.J.; Guyre, P.M.; Wira, C.R.; Pioli, P.A. Estradiol regulates expression of estrogen receptor ERα46 in human macrophages. *PLoS ONE* **2009**, *4*, e5539. [CrossRef] [PubMed]

79. Cooper, R.L.; Segal, R.A.; Diegelmann, R.F.; Reynolds, A.M. Modeling the effects of systemic mediators on the inflammatory phase of wound healing. *J. Theor. Biol.* **2015**, *367*, 86–99. [CrossRef] [PubMed]

80. Ashcroft, G.S.; Greenwell-Wild, T.; Horan, M.A.; Wahl, S.M.; Ferguson, M.W. Topical estrogen accelerates cutaneous wound healing in aged humans associated with an altered inflammatory response. *Am. J. Pathol.* **1999**, *155*, 1137–1146. [CrossRef]

81. Ashcroft, G.S.; Lei, K.; Jin, W.; Longenecker, G.; Kulkarni, A.B.; Greenwell-Wild, T.; Hale-Donze, H.; McGrady, G.; Song, X.Y.; Wahl, S.M. Secretory leukocyte protease inhibitor mediates non-redundant functions necessary for normal wound healing. *Nat. Med.* **2000**, *6*, 1147–1153. [CrossRef] [PubMed]

82. Murphy, A.J.; Guyre, P.M.; Pioli, P.A. Estradiol suppresses NF-κB activation through coordinated regulation of let-7a and miR-125b in primary human macrophages. *J. Immunol.* **2010**, *184*, 5029–5037. [CrossRef] [PubMed]

83. Hesketh, M.; Sahin, K.B.; West, Z.E.; Murray, R.Z. Macrophage phenotypes regulate scar formation and chronic wound healing. *Int. J. Mol. Sci.* **2017**, *18*, 1545. [CrossRef] [PubMed]

84. Kotwal, G.J.; Chien, S. Macrophage differentiation in normal and accelerated wound healing. *Results Probl. Cell Differ.* **2017**, *62*, 353–364. [CrossRef] [PubMed]

85. Yeh, C.C.; Horng, H.C.; Chou, H.; Tai, H.Y.; Shen, H.D.; Hsieh, S.L.; Wang, P.H. Dectin-1-mediated pathway contributes to *Fusarium proliferatum*-induced CXCL-8 release from human respiratory epithelial cells. *Int. J. Mol. Sci.* **2017**, *18*, 624. [CrossRef] [PubMed]

86. Carnesecchi, J.; Malbouyres, M.; de Mets, R.; Balland, M.; Beauchef, G.; Vié, K.; Chamot, C.; Lionnet, C.; Ruggiero, F.; Vanacker, J.M. Estrogens induce rapid cytoskeleton re-organization in human dermal fibroblasts via the non-classical receptor GPR30. *PLoS ONE* **2015**, *10*, e0120672. [CrossRef] [PubMed]

87. Tsui, K.H.; Wang, P.H.; Chen, C.K.; Chen, Y.J.; Chiou, S.H.; Sung, Y.J.; Li, H.Y. Non-classical estrogen receptors action on human dermal fibroblasts. *Taiwan. J. Obstet. Gynecol.* **2011**, *50*, 474–478. [CrossRef] [PubMed]

88. Driskell, R.R.; Lichtenberger, B.M.; Hoste, E.; Kretzschmar, K.; Simons, B.D.; Charalambous, M.; Ferron, S.R.; Herault, Y.; Pavlovic, G.; Ferguson-Smith, A.C.; et al. Distinct fibroblast lineages determine dermal architecture in skin development and repair. *Nature* **2013**, *504*, 277–281. [CrossRef] [PubMed]

89. Driskell, R.R.; Watt, F.M. Understanding fibroblast heterogeneity in the skin. *Trends Cell Biol.* **2015**, *25*, 92–99. [CrossRef] [PubMed]

90. Sapudom, J.; Wu, X.; Chkolnikov, M.; Ansorge, M.; Anderegg, U.; Pompe, T. Fibroblast fate regulation by time dependent TGF-β1 and IL-10 stimulation in biomimetic 3D matrices. *Biomater. Sci.* **2017**, *5*, 1858–1867. [CrossRef] [PubMed]

91. Peržeľová, V.; Sabol, F.; Vasilenko, T.; Novotný, M.; Kováč, I.; Slezák, M.; Ďurkáč, J.; Hollý, M.; Pilátová, M.; Szabo, P.; et al. Pharmacological activation of estrogen receptors-α and -β differentially modulates keratinocyte differentiation with functional impact on wound healing. *Int. J. Mol. Med.* **2016**, *37*, 21–28. [CrossRef] [PubMed]

92. Wang, Y.X.; Li, M.; Zhang, H.Q.; Tang, M.X.; Guo, C.F.; Deng, A.; Chen, Y.; Xiao, L.G. Opposite function of ERα and ERβ in controlling 17β-estradiol-mediated osteogenesis in osteoblasts. *Arch. Med. Res.* **2016**, *47*, 255–261. [CrossRef] [PubMed]

93. Yang, K.; Zhang, H.; Luo, Y.; Zhang, J.; Wang, M.; Liao, P.; Cao, L.; Guo, P.; Sun, G.; Sun, X. Gypenoside XVII Prevents atherosclerosis by attenuating endothelial apoptosis and oxidative stress: Insight into the ERα-mediated PI3K/Akt pathway. *Int. J. Mol. Sci.* **2017**, *18*, 77. [CrossRef] [PubMed]

94. Tsui, K.H.; Li, H.Y.; Cheng, J.T.; Sung, Y.J.; Yen, M.S.; Hsieh, S.L.; Wang, P.H. The role of nitric oxide in the outgrowth of trophoblast cells on human umbilical vein endothelial cells. *Taiwan. J. Obstet. Gynecol.* **2015**, *54*, 227–231. [CrossRef] [PubMed]

95. Huang, B.S.; Yang, M.H.; Wang, P.H.; Li, H.Y.; Chou, T.Y.; Chen, Y.J. Oestrogen-induced angiogenesis and implantation contribute to the development of parasitic myomas after laparoscopic morcellation. *Reprod. Biol. Endocrinol.* **2016**, *14*, 64. [CrossRef] [PubMed]

96. Blum, A. Endoethelial progenitor cells are affected by medications and estrogen. *Isr. Med. Assoc. J.* **2015**, *17*, 578–580. [PubMed]

97. Okonkwo, U.A.; DiPietro, L.A. Diabetes and wound angiogenesis. *Int. J. Mol. Sci.* **2017**, *18*, 1419. [CrossRef] [PubMed]

98. Amar, S.; Smith, L.; Fields, G.B. Matrix metalloproteinase collagenolysis in health and disease. *Biochim. Biophys. Acta* **2017**, *1864*, 1940–1951. [CrossRef] [PubMed]

99. Krishnaswamy, V.R.; Mintz, D.; Sagi, I. Matrix metalloproteinases: The sculptors of chronic cutaneous wounds. *Biochim. Biophys. Acta* **2017**. [CrossRef] [PubMed]

100. Freitas-Rodríguez, S.; Folgueras, A.R.; López-Otín, C. The role of matrix metalloproteinases in aging: Tissue remodeling and beyond. *Biochim. Biophys. Acta* **2017**, *1864*, 2015–2025. [CrossRef] [PubMed]

101. Jobin, P.G.; Butler, G.S.; Overall, C.M. New intracellular activities of matrix metalloproteinases shine in the moonlight. *Biochim. Biophys. Acta* **2017**, *1864*, 2043–2055. [CrossRef] [PubMed]

102. Van Doren, S.R.; Marcink, T.C.; Koppisetti, R.K.; Jurkevich, A.; Fulcher, Y.G. Peripheral membrane associations of matrix metalloproteinases. *Biochim. Biophys. Acta* **2017**, *1864*, 1964–1973. [CrossRef] [PubMed]

103. Levin, M.; Udi, Y.; Solomonov, I.; Sagi, I. Next generation matrix metalloproteinase inhibitors—Novel strategies bring new prospects. *Biochim. Biophys. Acta* **2017**, *1864*, 1927–1939. [CrossRef] [PubMed]

104. Fingleton, B. Matrix metalloproteinases as regulators of inflammatory processes. *Biochim. Biophys. Acta* **2017**, *1864*, 2036–2042. [CrossRef] [PubMed]

105. Irrera, N.; Pizzino, G.; D'Anna, R.; Vaccaro, M.; Arcoraci, V.; Squadrito, F.; Altavilla, D.; Bitto, A. Dietary management of skin health: The role of genistein. *Nutrients* **2017**, *9*, 622. [CrossRef] [PubMed]

International Journal of
Molecular Sciences

MDPI

Article

TRPV3 Channel in Keratinocytes in Scars with Post-Burn Pruritus

Chun Wook Park [1,†], Hyun Ji Kim [1,†], Yong Won Choi [1,†], Bo Young Chung [1], So-Youn Woo [2], Dong-Keun Song [3] and Hye One Kim [1,*]

[1]   Department of Dermatology, Kangnam Sacred Heart Hospital, Hallym University, 1 Singil-ro, Seoul 07441, Korea; dermap@daum.net (C.W.P.); rlatofhal90@hanmail.net (H.J.K.); henric@naver.com (Y.W.C.); victoryby@naver.com (B.Y.C.)
[2]   Department of Microbiology, School of Medicine, Ewha Womans University, 911-1 Mok-Dong, Yang Cheon-Gu, Seoul 158-710, Korea; soyounwoo@ewha.ac.kr
[3]   Department of Pharmacology, College of Medicine, Institute of Natural Medicine, Hallym University, Chunchon 200-702, Korea; dksong@hallym.ac.kr
*   Correspondence: hyeonekim@hallym.or.kr; Tel.: +82-2-829-5221; Fax: +82-2-832-3237
†   These authors contributed equally to this work.

Received: 4 October 2017; Accepted: 11 November 2017; Published: 15 November 2017

Abstract: Post-burn pruritus is a common and distressing sequela of burn scars. Empirical antipruritic treatments usually fail to have a satisfactory outcome because of their limited selectivity and possible side effects. Therefore, novel drug targets need to be identified. Here, we aimed to investigate the possible role of protease-activated receptor 2 (PAR2) and transient receptor potential vanilloid 3 (TRPV3), along with the relation of TRPV3 to thymic stromal lymphopoietin (TSLP). Specimens from normal (unscarred) or burn-scarred (with or without pruritus) tissue were obtained from burn patients for this study. In each sample, the keratinocytes were isolated and cultured, and the intracellular Ca^{2+} level at the time of stimulation of each factor was quantified and the interaction was screened. PAR2 function was reduced by antagonism of TRPV3. Inhibiting protein kinase A (PKA) and protein kinase C (PKC) reduced TRPV3 function. TSLP mRNA and protein, and TSLPR protein expressions, increased in scars with post-burn pruritus, compared to scars without it or to normal tissues. In addition, TRPV1 or TRPV3 activation induced increased TSLP expression. Conclusively, TRPV3 may contribute to pruritus in burn scars through TSLP, and can be considered a potential therapeutic target for post-burn pruritus.

Keywords: protease-activated receptor 2; transient receptor potential vanilloid 3; thymic stromal lymphopoietin; post-burn pruritus

1. Introduction

Transient receptor potential (TRP) channels are an ion channel group located in the plasma membrane of numerous types of cells. It was shown that certain thermosensitive TRP channels, especially TRP vanilloid 1 (TRPV1), TRPV3, TRPV4 and TRP ankyrin 1 (TRPA1), have important roles in the pathogenesis of pruritus as well as pain [1]. In a few of the early studies, the absence of TRPV3 in the TRP channels of skin cells was reported, but this finding was later reversed [2,3]. Besides skin and neuronal cells, TRPV3 is also expressed in the nasal and oral cavities [4]. In our previous study, we evaluated the clinical and histopathological characteristics of patients with post-burn pruritus and discovered increased expression of TRPV3, TRPV4 and TRPA1 [5]. Among them, TRPV3, in particular, is predominantly expressed in the epidermis of the tissue of pruritic burn scars. Moreover, Ca^{2+} influx via TRPV3 is markedly greater in the keratinocytes isolated from pruritic burn scars than in nonpruritic burn scars or normal skin [6]. In our previous study, protease-activated receptor 2 (PAR2) and neurokinin receptor 1 (NK1R) also were more abundant in pruritic burn scars than in nonpruritic burn scars [6].

Despite these previous studies, there is still much more to be studied about the pathway. For example, how TRPV3, which is highly expressed in keratinocytes, is involved in the itching of burn patients. We hypothesized that the PAR2 pathway might have the effect on TRPV3 of inducing itching in burn scars. This is suggested because of previous discoveries showing that PAR2 activation sensitizes thermoTRPs like TRPV1, TRPV4 and TRPA1 [7–9]. Moreover, the effect of PAR2 on TRPV3 channel has not been established.

Thymic stromal lympopoietin (TSLP) is a cytokine involved in allergic diseases and fibrotic conditions such as asthma, allergic rhinitis, atopic dermatitis, systemic sclerosis and scleroderma [10–14]. Intracellular increase of Ca^{2+} promotes TSLP expression through the calmodulin–calcineurin pathway [15]. PAR2 and TRPV1 activation reportedly increased TSLP in airway epithelium [15,16], but the evidence for TRPV3 inducing TSLP expression in epidermal keratinocyte is lacking. Previous study revealed that TSLP derived from keratinocytes act directly via TSLP receptors on sensory nerve terminals to induce pruritus [17]. We hypothesized that epithelial TSLP might have a role in pruritus of burn scar patients and that the TSLP expression level might be increased in burn-scar tissues.

The aims of this study were (1) to determine whether PAR2 downstream pathways have relations with TRPV3 function; (2) to elucidate the mechanism of it and (3) to compare the expression of TSLP in pruritic burn scars with that in nonpruritic burn scars and normal tissue.

2. Results

2.1. Characteristics of Patients

The 27 burn patients included 19 males and 8 females, ages between 8 and 60 years (36.9 ± 14.2). The average time after burn injury for all patients was 103.2 ± 125.0 months (range, 6–487 months). The patients had total body surface area (TBSA) from 2% to 70% (22.7 ± 22.6). There was no statistical difference on age, proportion of males, TBSA, and time after burn injury (Table 1).

Table 1. The clinical differences between burn patients with and without pruritus.

Characteristic	Nonpruritic Burn Scar Patients ($n = 15$)	Pruritic Burn Scar Patients ($n = 12$)	p-Value
Age, years	34.3 ± 23.6	40.1 ± 13.8	0.419
Gender (male), %	80	75	0.6374
Burned area (% of TBSA)	19.3 ± 23.6	26.9 ± 21.4	0.123
Time after burn injury, months	135.6 ± 147.5	62.9 ± 77.9	0.188
VAS for pruritus	0	6.4 ± 0.8	<0.001 *

All data shown are means ± SD or percentage. Significantly different at * $p < 0.05$. VAS, visual analogue scale; TBSA, total body surface area. p-values were calculated by Wilcoxon signed rank test for continuous variables and chi-square test for categorical variable.

2.2. PAR2 Activation Induces Intracellular Ca^{2+} Influx in Cultured Human Keratinocytes and Amplifies TRPV3 Activation Induced Intra + Cellular Ca^{2+} Influx in Burn Scars

To compare intracellular Ca^{2+} influx after TRPV3 activation between keratinocytes cultured from normal tissue and burn-scarred tissue with or without pruritus, TRPV3 agonist (500 μM Carvacrol and 200 μM 2-APB mixture) was used to treat each group.

The Ca^{2+} influx in the cultured keratinocytes decreased slowly with time (Figure 1A–C). The keratinocytes from pruritic burn scars showed higher peak level of intracellular Ca^{2+} influx than normal and nonpruritic burn scars, at the time of TRPV3 agonist treatment (Figure 1D).

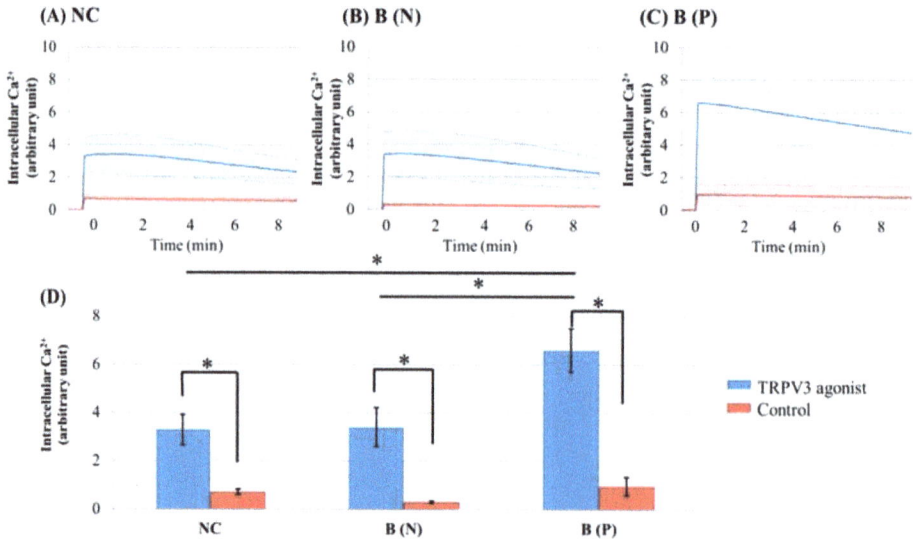

Figure 1. Ca^{2+} influx in cultured keratinocytes of: (**A**) Normal control; (**B**) Nonpruritic burn scar and (**C**) pruritic burn scar; (**D**) Intracellular Ca^{2+} levels at the time of TRPV3 agonist treatment. Each group was treated with TRPV3 agonist (500 μM Carvacrol and 200 μM 2-APB mixture) and Ca^{2+} buffer with 0.5 M ethylenediaminetetraacetic acid (EDTA) was used for control. The Ca^{2+} influx in cultured keratinocytes decreased slowly with time. NC: keratinocytes from normal control; B (N): keratinocytes from a nonpruritic burn scar; B (P): keratinocytes from a pruritic burn scar; Error bars in (**A–C**): standard deviation of the mean value obtained from three experiments; Error bars in (**D**): standard error, each performed in triplicate.* $p < 0.001$.

To investigate the effects of PAR2 on TRPV3 activation, keratinocytes cultured from normal tissue and burn-scarred tissue with or without pruritus, were treated with PAR2 agonist (100 uM SLIGRL-NH2). After that, they were treated with TRPV3 or TRPV1 agonist. The intracellular Ca^{2+} level was recorded continuously with a multimode detector.

PAR2 agonist induced intracellular Ca^{2+} influx in cultured human keratinocytes of normal tissue, and in nonpruritic and pruritic burn-scar tissues (Figures 2 and 3). The keratinocytes from pruritic burn scars showed higher peak level of intracellular Ca^{2+} influx than normal and nonpruritic burn scars, at the time of TRPV3 agonist treatment (Figure 2G). Unlike normal tissue, keratinocytes from scar tissue showed an increasing pattern of intracellular Ca^{2+} level (Figure 2A–C). In keratinocytes transfected by TRPV3 siRNA, PAR2 agonist increased intracellular Ca^{2+} level; whereas TRPV3 agonist decreased it (Figure 2D–F).

After treatment with TRPV1 agonist, intracellular Ca^{2+} slowly increased in all keratinocyte groups except for TRPV3 siRNA transfected normal control keratinocyte (Figure 3A–C,E,F). The keratinocytes from pruritic burn scars showed higher peak level of intracellular Ca^{2+} influx than normal and nonpruritic burn scars, at the time of TRPV1 agonist treatment (Figure 3G).

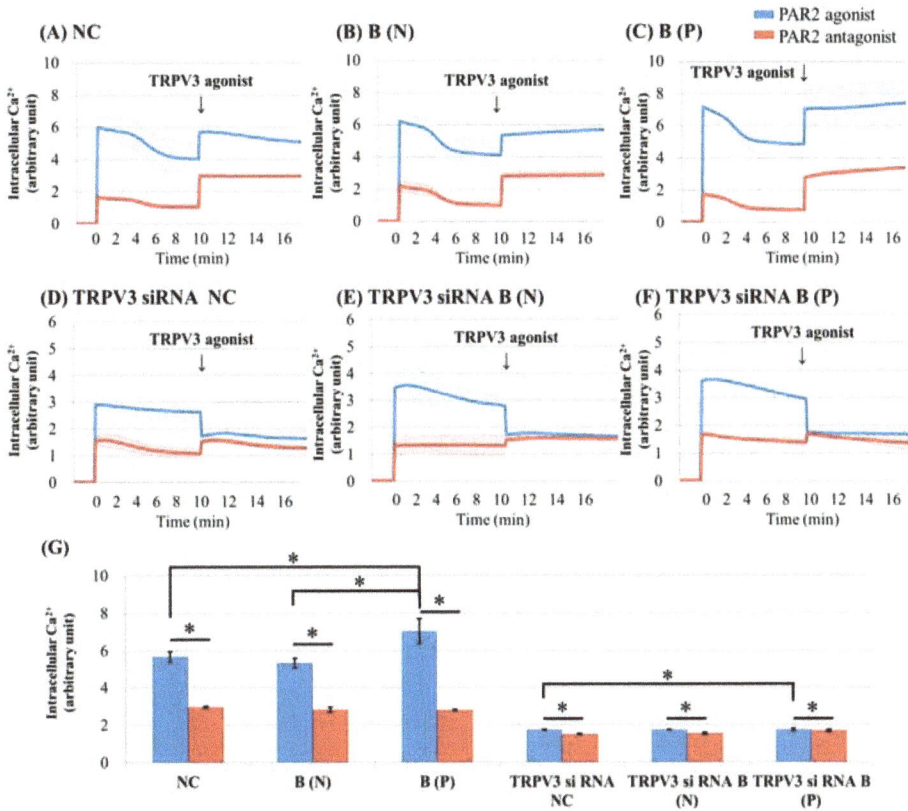

Figure 2. Ca^{2+} influx in cultured keratinocytes of: (**A**) Normal control; (**B**) nonpruritic burn scar; (**C**) pruritic burn scar; (**D**) TRPV3 siRNA transfected normal control; (**E**) TRPV3 siRNA transfected nonpruritic burn scar and (**F**) TRPV3 siRNA transfected pruritic burn scar; (**G**) Intracellular Ca^{2+} levels at the time of TRPV3 agonist treatment. Each group was pretreated with protease-activated receptor 2 (PAR2) agonist (100 µM SLIGRL-NH2) or PAR2 antagonist (100 uM LRGILS-NH2); then TRPV3 agonist (500 µM Carvacrol and 200 µM 2-APB mixture) was added. PAR2 agonist induced intracellular Ca^{2+} influx in cultured human keratinocytes of normal (unscarred), and in nonpruritic and pruritic burn-scar tissues. The pruritic burn scars showed the highest level of intracellular Ca^{2+} influx. Unlike normal tissue, keratinocytes from scar tissue showed a pattern of increasing intracellular Ca^{2+}. In keratinocytes transfected with TRPV3 siRNA, PAR2 agonist increased the level of intracellular Ca^{2+}, but TRPV3 agonist decreased it. NC: keratinocytes from normal control; B (N): keratinocytes from a nonpruritic burn scar; B (P): keratinocytes from a pruritic burn scar; Error bars in (**A–F**): standard deviation of the mean value obtained from three experiments; Error bars in (**G**): standard error, each performed in triplicate.* $p < 0.001$.

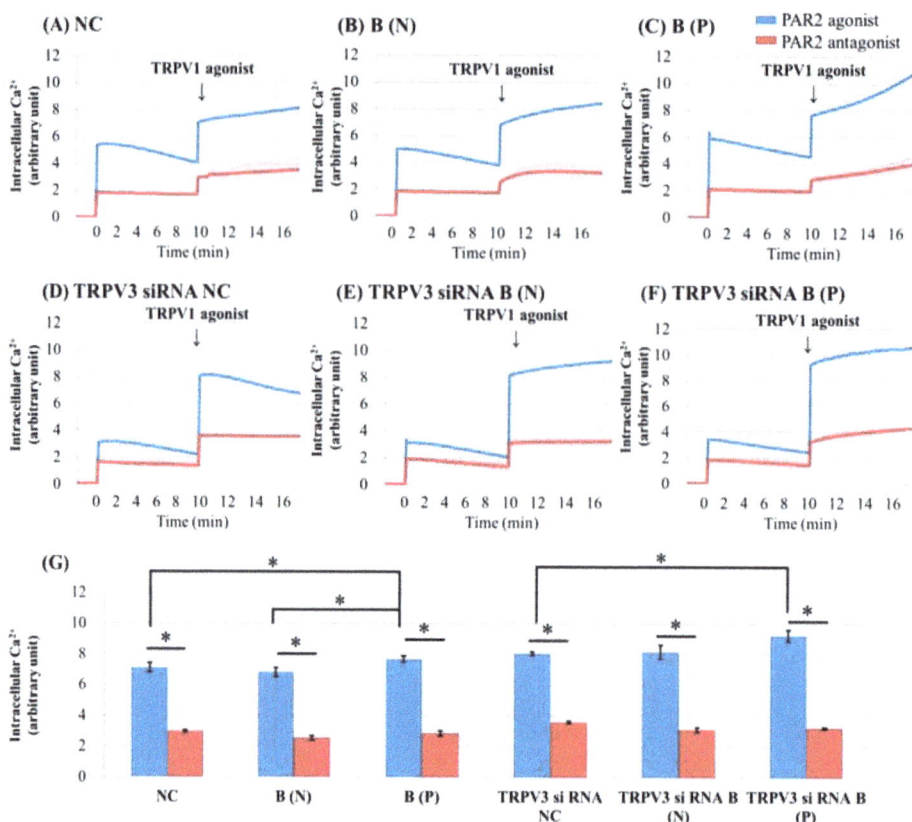

Figure 3. Ca^{2+} influx in cultured keratinocytes of: (**A**) Normal control; (**B**) nonpruritic burn scar; (**C**) pruritic burn scar; (**D**) TRPV3 siRNA transfected normal control; (**E**) TRPV3 siRNA transfected nonpruritic burn scar and (**F**) TRPV3 siRNA transfected pruritic burn scar; (**G**) intracellular Ca^{2+} levels at the time of TRPV3 agonist treatment. Each group was pretreated with PAR2 agonist (100 uM SLIGRL-NH2) or PAR2 antagonist (100 uM LRGILS-NH2); then TRPV1 agonist (1 uM Capsaicin) was added. PAR2 agonist induced intracellular Ca^{2+} influx in cultured human keratinocytes of normal (unscarred), and in nonpruritic and pruritic burn-scar tissues. After treatment with TRPV3 agonist, intracellular Ca^{2+} slowly increased in all keratinocyte groups except for TRPV3 siRNA transfected normal control keratinocyte. NC: keratinocytes from normal control; B (N): keratinocytes from a nonpruritic burn scar; B (P): keratinocytes from a pruritic burn scar; Error bars in (**A–F**): standard deviation of the mean value obtained from three experiments; Error bars in (**G**): standard error, each performed in triplicate.* $p < 0.001$.

2.3. PAR2 Amplification of TRPV3 Agonist Effects Via PKA, PKC and PLC-β Related Mechanism

To examine the mechanism of amplified TRPV3 activation in PAR2 agonist pretreated keratinocytes of burn-scar tissue, we evaluated the level of intracellular Ca^{2+} of cultured keratinocytes from normal tissue or burn-scarred tissue (with or without pruritus) after inhibiting PLC-β, protein kinase A (PKA) and protein kinase C (PKC), which are essential components of intracellular Ca^{2+} influx signalling by PAR2.

The blocking of PKA and PKC is associated with reduced intracellular Ca^{2+} influx compared to the control when treated with PAR2 agonist or with additional TRPV3 agonist treatments (Figure 4).

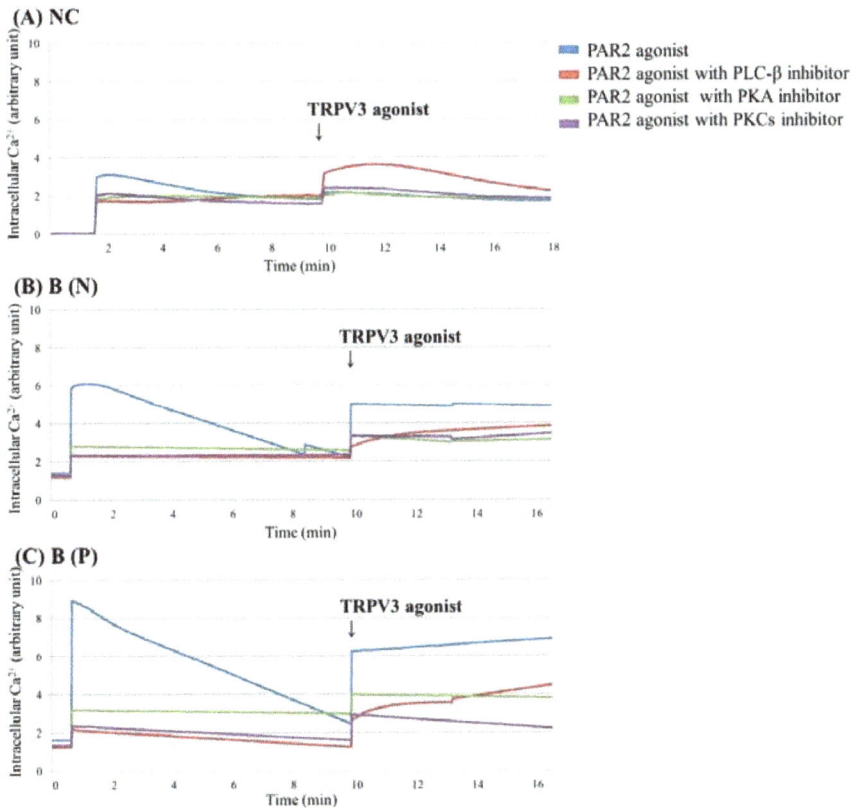

Figure 4. Ca^{2+} influx in cultured keratinocytes of: (**A**) Normal control, (**B**) Nonpruritic burn scar and (**C**) Pruritic burn scar. Each group was pretreated with PAR2 agonist (100 μM SLIGRL-NH2), PAR2 agonist with phospholipase C (PLC)-β inhibitor (10 μM U73122), PAR2 agonist with protein kinase A (PKA) inhibitor (10 μM H-89) or PAR2 agonist with and protein kinase C (PKCs) (α, β, γ, δ, ε and ζ) inhibitor (10 μM GF109203X); then TRPV3 agonist (500 μM Carvacrol and 200 μM 2-APB mixture) was added. Inhibition of PKA and PKC, which are essential components of intracellular Ca^{2+} influx signalling by PAR2, is associated with reduced intracellular Ca^{2+} influx compared to the control, even when treated with PAR2 agonist or with additional TRPV3 agonist treatments. NC: keratinocytes from normal control; B (N): keratinocytes from a nonpruritic burn scar; B (P): keratinocytes from a pruritic burn scar.

2.4. Inhibition of TRPV3 Channels Attenuated the Action of PAR2

To evaluate the role of the TRPV3 channel on PAR2 function, we compared the intracellular Ca^{2+} level with Fluo-3, with PAR2 agonist in TRPV3 pretreated keratinocytes cultured from normal tissue or burn-scarred tissue (with or without pruritus).

Intracellular Ca^{2+} influx by PAR2 agonist in keratinocytes was markedly reduced after pre-treatment with TRPV3 antagonist (Figure 5). The keratinocytes from pruritic burn scars showed higher peak level of intracellular Ca^{2+} influx than normal and nonpruritic burn scars, at the time of PAR2 agonist treatment (Figure 5D).

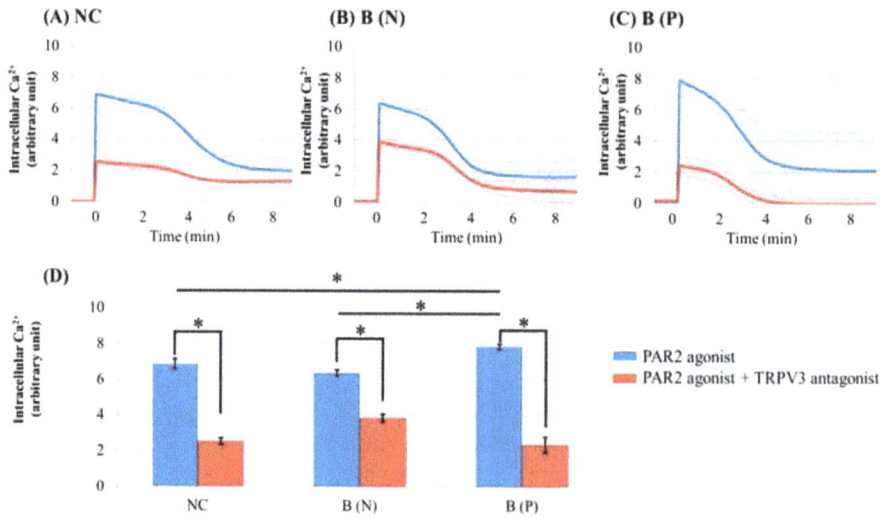

Figure 5. Ca^{2+} influx in cultured keratinocytes of: (**A**) Normal control, (**B**) Nonpruritic burn scar and (**C**) pruritic burn scar. (**D**) Intracellular Ca^{2+} levels at the time of PAR2 agonist treatment. Control group was treated only with PAR2 agonist (100 uM SLIGRL-NH2), while test group was treated with PAR2 agonist and TRPV3 antagonist (125 µM DPTHF). TRPV3 antagonist pretreated group showed lower intracellular Ca^{2+} influx induced by PAR2 agonist than did control group. NC: keratinocytes from normal control; B (N): keratinocytes from nonpruritic burn scar; B (P): keratinocytes from pruritic burn scar; Error bars in (**A–C**): standard deviation of the mean value obtained from three experiments, Error bars in (**D**): standard error, each performed in triplicate.* $p < 0.001$.

2.5. Thymic Stromal Lymphopoietin (TSLP) and TSLPR Expression Increased in Burn-Scar Tissues, Especially in Pruritic Burn Scars

In order to compare the expression of TSLP and TSLPR in normal and burn-scar tissues with or without pruritus, we measured the mRNA level of TSLP in the tissues using qPCR. Then, western blotting was performed to measure the protein level of TSLP and TSLPR in the tissue.

The highest TSLP expression of mRNA and protein was observed in pruritic burn-scar tissue by qPCR and western blotting. Expression of TSLPR proteins was also greatest in pruritic burn-scar tissue subjected to western blotting (Figure 6).

To visualize the expression pattern of TRPV3, TSLP and TSLPR, immunohistochemistry was performed on normal and burn-scar tissues with or without pruritus, each in twelve sets, using DAPI (4′,6-diamidino-2-phenylindole), TRPV3 antibody, TSLP antibody, TSLPR antibody and PGP 9.5 antibody.

Immunohistochemistry showed markedly increased TRPV3, TSLP and TSLPR expression in pruritic burn-scar tissue compared to normal (unscarred) tissue. TRPV and TSLP were highly expressed in the spinous layer and TSLPR was mainly observed in the lower spinous layer and basal layer (Figures 7–10).

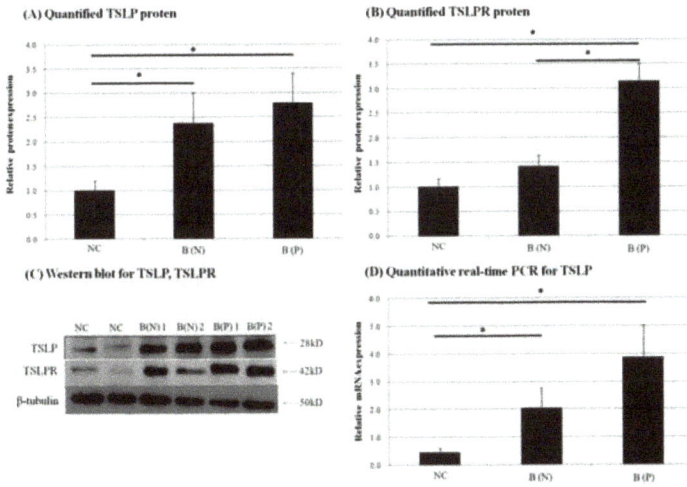

Figure 6. Expression of (**A**) thymic stromal lymphopoietin (TSLP) protein and (**B**) TSLP receptor (TSLPR) protein in normal and burn-scar tissues with or without pruritus was measured by (**C**) western blotting (WB). Expression of (**D**) TSLP mRNA was measured using quantitative real-time PCR in normal and burn-scar tissues with or without pruritus. The greatest TSLP expression of mRNA and protein was observed in pruritic burn-scar tissue using qPCR and western blotting (**A–C**). TSLPR proteins were also highest in pruritic burn-scar tissue via western blotting results (**D**). NC: keratinocytes from normal control; B (N): keratinocytes from nonpruritic burn scar; B (P): keratinocytes from pruritic burn scar; TSLP: thymic stromal lymphopoietin; TSLPR: thymic stromal lymphopoietin receptor; Error bars: standard deviation of the mean value obtained from three experiments, each performed in triplicate. * $p < 0.001$.

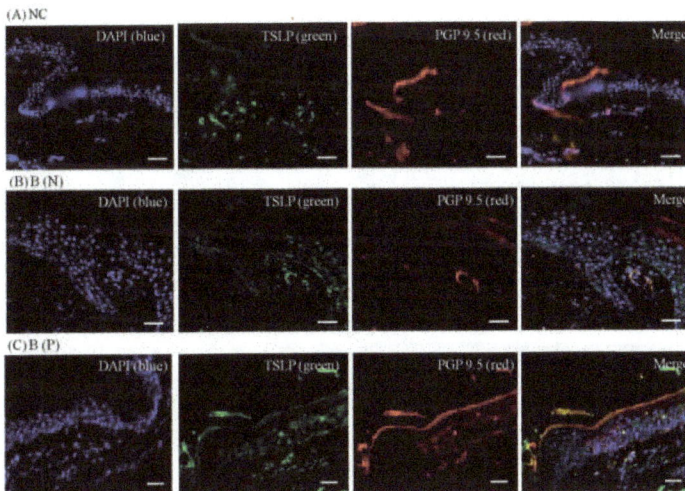

Figure 7. Three-color immunofluorescence confocal images were obtained for 4′,6-diamidino-2-phenylindole (DAPI), **blue**), TSLP (**green**) and protein gene product 9.5 (PGP 9.5, **red**) from tissues of (**A**) normal; (**B**) nonpruritic and (**C**) pruritic burn scars. The staining of TSLP, well observed in the spinous layer of the epidermis, was least in (**A**) normal control and greatest in (**C**) pruritic burn-scar tissue. NC: keratinocytes from normal control; B (N): keratinocytes from a nonpruritic burn scar; B (P): keratinocytes from a pruritic burn scar; TSLP: thymic stromal lymphopoietin. Scale bars = 50 μm.

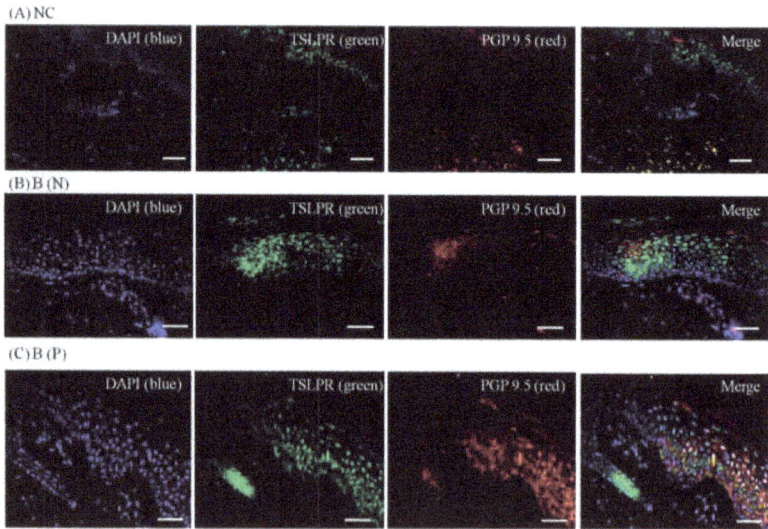

Figure 8. Three-color immunofluorescence confocal images were obtained for DAPI (**blue**), TSLPR (**green**) and PGP 9.5 (**red**) from tissues of (**A**) Normal, (**B**) Nonpruritic and (**C**) Pruritic burn scars. The staining of TSLPR, mainly observed in the lower spinous layer and basal layer of the epidermis, was least in (**A**) Normal control and greater in (**B,C**) Burn-scar tissues with or without pruritus. NC: keratinocytes from normal control; B (N): keratinocytes from nonpruritic burn scar; B (P): keratinocytes from pruritic burn scar; TSLPR: thymic stromal lymphopoietin receptor. Scale bars = 50 μm.

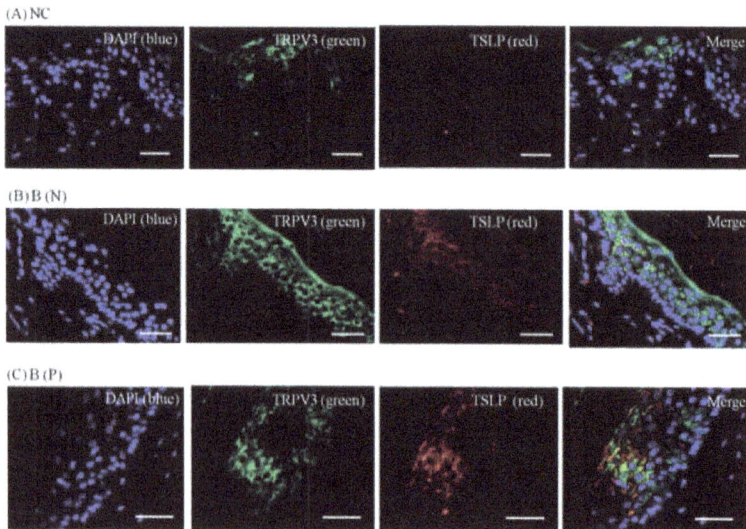

Figure 9. Three-color immunofluorescence confocal images were obtained for DAPI (**blue**), TRPV3 (**green**) and TSLP (**red**) from tissues of (**A**) normal, (**B**) nonpruritic and (**C**) pruritic burn scars. The staining of TRPV3 and TSLP, well observed in spinous layer of the epidermis, was least in (**A**) normal control and greater in (**B,C**) burn-scar tissues with or without pruritus. NC: keratinocytes from normal control; B (N): keratinocytes from nonpruritic burn scar; B (P): keratinocytes from pruritic burn scar; TSLP: thymic stromal lymphopoietin. Scale bars = 50 μm.

(A) NC

(B) B (N)

(C) B (P)

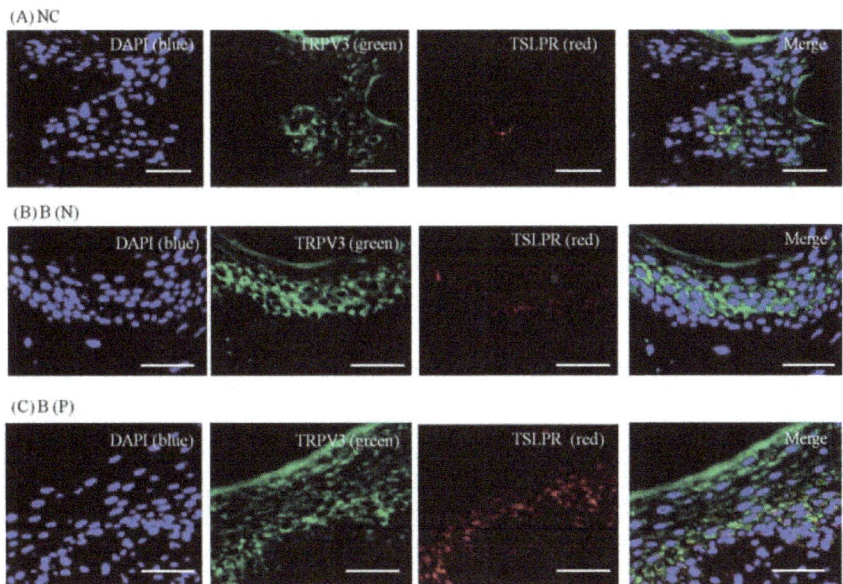

Figure 10. Three-color immunofluorescence confocal images were obtained for DAPI (**blue**), TRPV3 (**green**) and TSLPR (**red**) from skin tissues of (**A**) normal, (**B**) nonpruritic and (**C**) pruritic burn scars. The staining of TRPV3, well observed in spinous layer of the epidermis, and TSLPR, in lower spinous and basal layer, was least in (**A**) normal control and higher in (**B,C**) burn-scar tissues with or without pruritus. Staining of TSLPR was greatest in burn-scar tissue with pruritus. NC: keratinocytes from normal control; B (N): keratinocytes from nonpruritic burn scar; B (P): keratinocytes from pruritic burn scar; TSLPR: thymic stromal lymphopoietin receptor. Scale bars = 50 μm.

2.6. TRPV1 and TRPV3 Activation Induce TSLP Expression in Normal Human Epidermal Keratinocytes

Finally, to confirm that TRPV3 activation induces TSLP production in keratinocytes, normal human epidermal keratinocytes (NHEKs) were transfected with TRPV3 cDNA containing vectors and TRPV3 siRNA. TRPV3 agonist was used to treat MOCK, TRPV3 overexpressing, and TRPV3 blocked NHEKs. TRPV1 agonist was also treated as control. The mRNA level of TSLP was quantified with qPCR, and the protein level was quantified using western blot.

In normal human epidermal keratinocytes treated with TRPV3 agonist, TSLP mRNA and protein expression increased. The expression of TSLP mRNA was significantly increased by TRPV3 agonist in TRPV3 overexpressed NHEK, but not in the *TRPV3* knockout NHEK with siRNA. The expression of TSLP mRNA and protein was also increased by TRPV1 agonist treatment. There was no significant increase in TSLP protein expression in TRPV1 agonist treatment with TRPV3 overexpressed NHEK, and no increase in the *TRPV3* knockout NHEK (Figure 11).

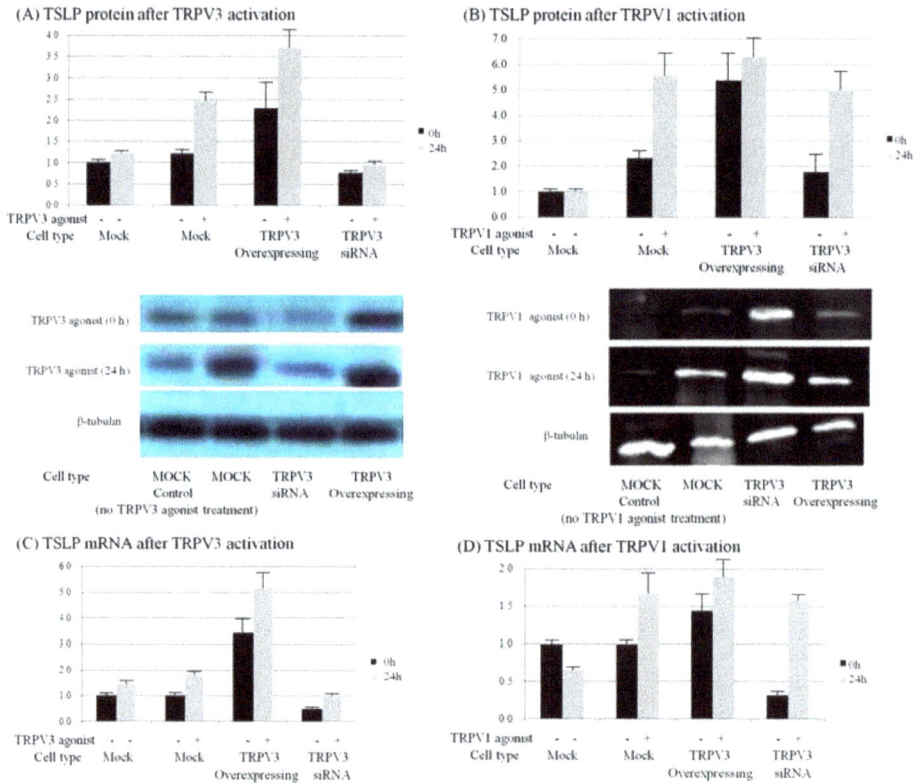

(A) TSLP protein after TRPV3 activation

(B) TSLP protein after TRPV1 activation

(C) TSLP mRNA after TRPV3 activation

(D) TSLP mRNA after TRPV1 activation

Figure 11. Normal human epidermal keratinocytes (NHEKs) were transfected with TRPV3 overexpressing vector or with siRNA, to analyse the effect of TRPV3 and TRPV1 channels on TSLP expression. Western blotting was conducted to measure the TSLP protein level in response to (**A**) TRPV3 agonist (500 μM Carvacrol + 200 μM mixture 2-APB) and (**B**) TRPV1 agonist (1 μM Capsaicin). Quantitative real-time PCR was performed to measure the TSLP mRNA level in response to (**C**) TRPV3 agonist and (**D**) TRPV1 agonist. In normal human epidermal keratinocytes treated with TRPV3 agonist, TSLP mRNA and protein expression increased. The expression of TSLP mRNA was significantly increased by TRPV3 agonist treatment with TRPV3 overexpressed NHEK, but not with *TRPV3* knockout NHEK with siRNA. The expression of TSLP mRNA and protein was also increased by TRPV1 agonist treatment. There was no significant increase in TSLP protein expression in TRPV1 agonist treatment with TRPV3 overexpressed NHEK and no increase in *TRPV3* knockout NHEK. NC: keratinocytes from normal control; B (N): keratinocytes from nonpruritic burn scar; B (P): keratinocytes from pruritic burn scar; NHEK: normal human epidermal keratinocyte; TSLP: thymic stromal lymphopoietin; TSLPR: thymic stromal lymphopoietin receptor; Error bars: standard deviation of the mean value obtained from three experiments, each performed in triplicate.

3. Discussion

In this study we demonstrated the relationships of PAR2, TRPV3 and TSLP in keratinocytes from scars with or without pruritus.

PARs are widely expressed G-protein coupled receptors, differentially activated by various physiological factors such as thrombin (PAR1, PAR3 and PAR4), mast cell-derived tryptase (PAR2) and trypsin (PAR2 and PAR4), which mediate various signals [18,19]. PAR2 can be activated by proteases from circulation, inflammatory cells, neurons and keratinocytes; and controls inflammation, pain and neuronal excitability [20–22]. Its agonists also cause thermal and mechanical somatic hyperalgesia [23].

In contrast with TRPV1, TRPV4 and TRPA1, which are reportedly sensitized by PAR2 activation in keratinocytes and neurons, there has been no report on the relationship of TRPV3 with PAR2 [7–9]. The increase in intracellular Ca^{2+} by PAR2 was greatest in burn scars with pruritus (Figures 2C and 3C). In addition, the effect of PAR2 agonist on intracellular Ca^{2+} (increase) in keratinocytes was not decreased by TRPV3 siRNA treatment (Figure 2D–F); while in the presence of TRPV3 antagonist, the PAR2 agonist was less effective in increasing intracellular Ca^{2+} levels (Figure 5). This suggests that the TRPV3 channel is involved but not necessarily required for PAR2 to increase intracellular Ca^{2+}. In the presence of the pretreated PAR2 antagonist in keratinocytes cultured in normal and burn scars, the TRPV3 agonist did not induce marked intracellular Ca^{2+} influx (Figure 2). PLC-β, PKA, and PKC are the subsequent signalling pathways after PAR2 activation. In this study, the action of PAR2 on intracellular Ca^{2+} was decreased by inhibition of PKA and PKC, as expected. Moreover, the action of TRPV3 agonist was also decreased by inhibiting PKA and PKC (Figure 4). In the context of the fact that PKA and PKC modulate the activation of TRPV1, this result implies that PKA and PKC are necessary for TRPV3 to function [24–26]. It can be suggested that PKA, PKC and/or other downstream signalling pathways of PAR2 affect TRPV3 inducing Ca^{2+} influx.

On the other hand, the decrease of intracellular Ca^{2+} after treating TRPV3 agonist (carvacrol + 2-APB mixture) on PAR2 agonist pretreated si-RNA transfected keratinocytes (Figure 2D–F) may be caused by the inhibitory effect of 2-APB on inositol phosphate 3 (IP_3) receptor [27]. The IP_3 receptor itself is a calcium channel that is linked to intracellular space like endoplasmic reticulum, and when activated, transfers intraluminal calcium to intracellular space to increase intracellular Ca^{2+} level. Inhibition of the IP_3 receptor, a calcium channel, may have reduced intracellular Ca^{2+}. Despite the fact that 2-APB can act on other TRP channels such as TRPV1/2/3/4/6, TRPM7/8, TRPC6, and store-operated calcium channels and show a different pattern in Ca^{2+} regulation in a dose-dependent manner, 2-APB is widely used as TRPV3 agonist since there are few alternatives [28–31].

Thymic stromal lymphopoietin (TSLP) is known to have an important role in the pathogenesis of atopic dermatitis by mediating an immune response to Th2 [32]. It is released via the Ca^{2+}-calmodulin/nuclear factor of activated T cell NFAT pathway in multiple cells including keratinocytes and dendritic cells [33]. Keratinocyte-derived TSLP expression is increased in acute and chronic lesions of atopic dermatitis patients. Genetic variants of TSLP can either induce or protect against the cutaneous inflammation of atopic dermatitis, which may be the result of corresponding increase or decrease in TSLP protein activity [32,34,35]. TSLPR is a heterodimer of the IL-7 receptor alpha chain and TSLP-specific receptor chain, and is expressed in nerve tissue and dorsal root ganglia. We found that TRPV3 and TSLP were increased in burn-scar tissues, especially in burn scars with pruritus (Figures 6 and 7). TRPV3 is responsible for the epidermal layer differentiation and proliferation, which is important to the skin barrier function as well as the pathogenesis of xerotic eczema and the process of wound healing [36,37]. Increased expression of TRPV3 may have a role in the healing of burn injury, and it is possible that increased TRPV3 induced the expression of TSLP. TSLP secreted by keratinocytes may acts directly on TSLP receptors on sensory nerve terminals to induce pruritus [17]. Recently, it has been reported that TSLP expression increased in idiopathic pulmonary fibrosis, atopic dermatitis fibrosis and keloid scar tissue, and it has been reported that it induces transforming growth factor-β, and affects collagen synthesis and fibrocyte infiltration [13]. The findings of our study could be of great significance in explaining the role of TSLP because it might play an important role in the pathogenesis of post-burn pruritus and, although not focused on in this study, burn-scar formation.

The TRPV family of ion channels has six members [38]. Among them, TRPV3 is abundantly expressed in the skin, especially in epidermal and hair follicular keratinocytes, as well as in the cornea, the distal colon, human larynx and inner ear [39–46]. Activation of the epidermal growth factor receptor (EGFR) induces TRPV3 channel activity, which eventually stimulates the release of the growth factor (TGF) [36]. In addition, temperature stimulation activates TRPV3 (and TRPV4) and promotes barrier recovery after mechanical damage [47]. With respect to pain sensation, it is noteworthy that the naturally existing TRPV3 agonists (camphor, eugenol, carvacrol, etc.) function as skin sensitizers, because they induce irritation and pain on topical application [48].

In the present study, expression of TSLP at the protein level and mRNA level increased after TRPV1 and TRPV3 channel activation in normal human epidermal keratinocytes (Figure 11). It has been reported that the activation of the TRPV1 channel in bronchial epithelial cells increases the secretion of TSLP [13]. However, there has been no study of TSLP and TSLPR increase after TRPV1 activation in human epidermal keratinocytes. Furthermore, it is significant that we found that TSLP expression increases after TRPV3 stimulation. Burn-scar tissues had greater TSLP protein and mRNA expression, and greater TSLPR protein expression, than did normal skin tissue (Figure 6). Also, immunofluorescence confocal microscopy images of pruritic burn-scar tissue showed increased staining for TRPV3, TSLP and TSLPR as well (Figures 7–10). From the results of our previous study, it appeared that pruritic burn-scar tissue had greater TRPV3 mRNA expression than did normal and nonpruritic burn-scar tissue [5]. Therefore, upregulation of the TRPV3 channel in the scar tissue with pruritus may lead to an increase in TSLP expression, which may be one of the mechanisms of post-burn pruritus.

There have been reports that TRP channels other than TRPV3, such as TRPV1/4, TRPA1, and TRPM6/7 contribute to pruritus. Authors noted at the location of expression of TRP channels, since they are all different. Although TRPV1 and TRPV3 are highly expressed in skin, TRPV1 is also highly expressed in and nerve tissues while TRPV3 is more restricted to the skin, being more selective for itching or pain. Other TRP channels also may have some role in pruritus. However, practically, selective agonists and antagonists have not been found on many TRP channels and further investigation is necessary. Given that TRPV3 is mainly expressed in epithelial tissues, it now appears that agents selectively inhibiting TRPV3 may be beneficial in controlling pruritus and may have fewer side effects.

4. Materials and Methods

4.1. Patient Selection

Patients with burn scars treated in Hangang Sacred Heart Hospital, Hallym University (Seoul, Korea) were included. Inclusion criteria was burn patients who were 18 years or older who agreed to volunteer. Exclusion criteria were patients with (i) pre-existing chronic systemic or dermatologic diseases which cause pruritus, (ii) concurrent systemic medications affecting pruritus symptoms (such as antihistamines or immunomodulators like systemic steroids and cyclosporine), (iii) concurrent psychotic disease, (iv) pregnancy, and (v) patients who could not provide clinical information about their pruritus. Each patient received a thorough dermatologic examination and underwent a complete examination to exclude other cutaneous or systemic causes of pruritus. Laboratory tests were done to assess blood sugar, liver function and kidney function. The Institutional Review Board of Hangang Sacred Heart Hospital approved the study protocol. Informed consent was obtained from each patient or their parents/legal representative.

4.2. Pruritus in Burn Scars

All patients were asked whether they had pruritus on their burn scars. The patients were divided into two groups: those with pruritus and those without.

4.3. Tissue Collection

Tissue samples were taken from two different sites from 27 patients undergoing plastic surgery to correct excessive scar tissue: one from a hypertrophic burn scar and the other from normal skin of the inguinal area, which skin is rarely exposed to the sun. In the patients with post-burn pruritus, the specimens were obtained from the most pruritic areas.

4.4. Isolation and Cultivation of Keratinocytes

Post-burn hypertrophic scars and normal skin biopsy specimens were obtained. Dispase (Boehringer Mannheim, Mannheim, Germany) (2.4 U/mL for 14 h at 4 °C) was treated to them to mechanically separate the dermis from epidermis. The epidermis part was rinsed with phosphate-buffered saline (PBS) and a single-cell suspension was obtained by treatment with 1 mM EDTA (Gibco, Grand Island, NY, USA) for 30 min in a water bath at 37 °C. The dermis was also washed with PBS and a single-cell suspension was obtained by treatment with 500 U/mL collagenase (Gibco) for 1 h in a water bath at 37 °C. For the cultivation of keratinocytes, isolated cells were rinsed with PBS and were then overlaid (10×10^5/mL) on a culture of adherent, nonproliferating NIH 3T3 mouse fibroblast cells (2×10^4/cm^2) which is treated with mitomycin C. Cells were cultivated in vitro under standard conditions (5% CO_2 conditioned atmosphere, 99% humidity and 37 °C) for various time periods before RNA extraction and qRT-PCR were performed. Culture medium of 0.15 mM calcium chloride (3:4 Dulbecco's modified Eagle's medium, 1:4 Ham's F12; Gibco) was supplemented with 10% fetal calf serum (FCS), an antibiotic mixture, 1% L-glutamine, 0.5 µg/mL hydrocortisone, 5 µg/mL insulin, 2.4 µg/mL adenine, 5 µg/mL transferrin, 2 nM triiodothyronine, 1 nM cholera toxin and 10 ng/mL of epidermal growth factor (all from Sigma, St. Louis, MO, USA). Half of the exhausted medium was replaced with fresh medium every 4 days.

4.5. Transfecting Normal Human Epidermal Keratinocytes

For TRPV3 siRNA transfection, keratinocytes were cultured for a day in serum-free media. Then, a TRPV3 siRNA reagent system (Santa Cruz Biotechnology, Dallas, CA, USA) was applied. TRPV3 siRNA and the transfection reagent were mixed and incubated for 45 min at room temperature. The TRPV3 siRNA mixture was overlaid onto the washed cell and incubated for 5 h in a CO_2 incubator. Two-fold serum medium was added and the mixture was cultured for one day.

For TRPV3 overexpression, TRPV3 DNA (TrueORF Gold, Rockville, MD, USA) and Lipofectamine 2000 (Thermo Fisher, Carlsbad, CA, USA) were gently mixed and incubated for 10 min at room temperature. DNA-lipid complex was dripped onto keratinocytes and then cultured for a day in a CO_2 incubator.

4.6. RNA Extraction and Quantitative Real-Time PCR (qRT-PCR)

The total RNA of tissues and cultured cells were extracted and purified using an RNeasy mini kit (#74106, Qiagen, Hilden, Germany) according to the manufacturer's instructions. A total of 500 ng RNA, and RT premix solution from High Capacity cDNA RT kits (#4374966; Applied Biosystems, Foster City, CA, USA), were added in steps at 25 °C for 10 min, 37 °C for 2 h and 85 °C for 5 min. For the real-time PCR, a reaction mix was prepared containing TaqMan Mastermix (Applied Biosystems), TaqMan probe (Hs00263639_m1 QuantiFast Probe Assay, TSLP; Hs00845692_m1 QuantiFast Probe Assay, TSLPR).

In order to measure TSLP expression in keratinocytes cultured from pruritic, nonpruritic burn scars and normal skin, data was collected using a Light-Cycler480II (Roche, Rotkreuz, Switzerland) under the following conditions: 45 cycles of 95° for 30 s, 60° for 30 s, and 72° for 1 s.

4.7. Western Blot

Tissue samples were homogenized in 400 µL RIPA buffer (50 mM Tris-HCl (pH 7.4), 150 mM NaCl, 1 mM PMSF, 1 mM EDTA, 1% Triton X-100, 0.5% sodium deoxycholate and 0.1% sodium dodecyl sulfate (SDS)). The protein was separated on SDS-polyacrylamide gels (20 µg per sample) and transferred to a nitrocellulose membrane using standard procedure (Protran, Schleicher & Schuell, city, Germany). The membranes were put in 5% non-fat milk for 1 h at room temperature and then incubated with primary anti-TSLPR antibody (1:1000; sc-83871, Santa Cruz Biotechnology, Dallas, CA, USA) and anti-TSLP antibody (1:10,000; ab47943, Abcam, Cambridge, MA, USA), overnight at 4 °C. After three washes of 5 minutes each in PBS–Tween 20 (0.1%, *v/v*), membranes were incubated with the secondary antibody (1:5000; AP307P, Millipore, Burlington, CA, USA) conjugated with horseradish peroxidase, anti-rabbit IgG for 2 h at 4 °C and then three 5-min washes were performed. The signal was detected with an enhanced chemiluminescence system according to the manufacturer's manual (Amersham Pharmacia Biotech, Milan, Italy).

4.8. Intracellular Ca^{2+} Measurement

Cells (3×10^4/200 mL/well) were cultured in a 96-well plate (NalgeNunc, Naperville, IL, USA) for 3–4 days in supplemented media with 20 μM calcium. Cells were washed using 4-(2-hydroxyethyl) -1-piperazineethanesulfonic acid (HEPES)-buffered saline solution (mM: 121 NaCl, 5.4 KCl, 0.8 $MgCl_2$, 25 HEPES, 1.8 $CaCl_2$ and 6.0 $NaHCO_3$ at pH 7.3) and then loaded with 4 μM Fluo-3/AM (Invitrogen, Carlsbad, CA, USA) with an equivalent volume of 20% Pluronic F127 (Invitrogen) for 45 min in an incubator at 37 °C. Thereafter, the cells were washed twice with HEPES-buffered saline solution for 10 min at room temperature in the dark. Following transfer to EDTA solution with 2 mM calcium at room temperature, the cells were measured using a multimode detector (DTX880, Beckman-Coulter, Brea, CA, USA). Fluorescence was excited by an argon laser at 488 nm and emissions were collected using a 515 filter. The concentrations of materials used for activating and blocking TRP channels in this study were: Carvacrol 500 μM + 2-Aminoethoxydiphenyl borate (2-APB) 200 μM mixture, as TRPV3 agonist; DPTHF 125 μM, as TRPV3 antagonist; Capsacin 1 uM as TRPV1 agonist; SLIGRL-NH2 100 uM, as PAR2 agonist; LRGILS-NH2 100 uM, as PAR2 antagonist; U73122 10 μM, as phospholipase C inhibitor; H-89 10 μM, as PKA inhibitor; GF109203X 10 μM, as PKC-α, β, γ, δ, ε and ζ inhibitor. Chemicals were obtained from Santa Cruz Biotechnology, Sigma, or Abcam.

4.9. Confocal Microscopy

Zeiss Axiovert and Bio-Rad MRC1000 confocal microscopes with Zeiss Plan Apo ×40 (NA 1.4) or ×100 (NA 1.3) objectives were used to observe specimens. Image collection was done at a zoom of 1–2, iris of <3 μm and typically 5–10 optical sections were taken at intervals of 0.5 μm. Images were coloured to represent the appropriate fluorophores, and processed using Adobe Photoshop 7.0 (Adobe Systems, Mountain View, CA, USA) to modulate contrast and brightness. Images of stained and control slides were collected and processed identically. The fluorescence intensity of the cells was checked by selecting a straight line across the neuronal soma and using the plot profile function of ImageJ software (version 10.2, NIH image).

4.10. Statistical Analysis

Data were expressed as mean ± SD. All statistical analyses were conducted using PASW Statistics 18 (SPSS, Inc., Chicago, IL, USA). Statistical comparisons were made using paired *t*-test or McNemar's test. Statistical comparisons between two groups were made using the Fisher's exact test and the Mann–Whitney *U* test.

Acknowledgments: This study was supported by grants of the National Research Foundation of Korea (NRF), funded by the Ministry of Science, ICT & Future Planning (NRF-2017R1A2B4006252), Korea Healthcare technology R&D project, funded by Ministry of Health & Welfare, Republic of Korea (HI17C0597), and the Hallym University Research Fund (HURF-2017-35).

Author Contributions: Chun Wook Park, Hyun Ji Kim, Yong Won Choi contributed equally to this work as co-first authors; Chun Wook Park, Yong Won Choi, and Hye One Kim conceived the idea of the article, designed the experimental plan and wrote the paper; Bo Young Chung, So-Youn Woo and Dong-keun Song performed the experiments, wrote the paper and reviewed scientific literature. All the authors gave final approval for the version to be submitted.

Conflicts of Interest: The authors declare no conflicts of interest.

References

1. Tóth, B.I.; Bíró, T. TRP channels and pruritus. *Open Pain J.* **2013**, *6*, 62–80. [CrossRef]
2. Smith, G.D.; Gunthorpe, M.J.; Kelsell, R.E.; Hayes, P.D.; Reilly, P.; Facer, P.; Wright, J.E.; Jerman, J.C.; Walhin, J.P.; Ooi, L.; et al. TRPV3 is a temperature-sensitive vanilloid receptor-like protein. *Nature* **2002**, *418*, 186–190. [CrossRef] [PubMed]
3. Peier, A.M.; Reeve, A.J.; Andersson, D.A.; Moqrich, A.; Earley, T.J.; Hergarden, A.C.; Story, G.M.; Colley, S.; Hogenesch, J.B.; McIntyre, P.; et al. A heat-sensitive trp channel expressed in keratinocytes. *Science* **2002**, *296*, 2046–2049. [CrossRef] [PubMed]
4. Xu, H.; Delling, M.; Jun, J.C.; Clapham, D.E. Oregano, thyme and clove-derived flavors and skin sensitizers activate specific trp channels. *Nat. Neurosci.* **2006**, *9*, 628–635. [CrossRef] [PubMed]
5. Yang, Y.S.; Cho, S.I.; Choi, M.G.; Choi, Y.H.; Kwak, I.S.; Park, C.W.; Kim, H.O. Increased expression of three types of transient receptor potential channels (TRPA1, TRPV4 and TRPV3) in burn scars with post-burn pruritus. *Acta Dermato-Venereol.* **2015**, *95*, 20–24. [CrossRef] [PubMed]
6. Kim, H.O.; Cho, Y.S.; Park, S.Y.; Kwak, I.S.; Choi, M.G.; Chung, B.Y.; Park, C.W.; Lee, J.Y. Increased activity of TRPV3 in keratinocytes in hypertrophic burn scars with postburn pruritus. *Wound Repair Regen.* **2016**, *24*, 841–850. [CrossRef] [PubMed]
7. Amadesi, S.; Cottrell, G.S.; Divino, L.; Chapman, K.; Grady, E.F.; Bautista, F.; Karanjia, R.; Barajas-Lopez, C.; Vanner, S.; Vergnolle, N.; et al. Protease-activated receptor 2 sensitizes TRPV1 by protein kinase cepsilon- and a-dependent mechanisms in rats and mice. *J. Physiol.* **2006**, *575*, 555–571. [CrossRef] [PubMed]
8. Grant, A.D.; Cottrell, G.S.; Amadesi, S.; Trevisani, M.; Nicoletti, P.; Materazzi, S.; Altier, C.; Cenac, N.; Zamponi, G.W.; Bautista-Cruz, F.; et al. Protease-activated receptor 2 sensitizes the transient receptor potential vanilloid 4 ion channel to cause mechanical hyperalgesia in mice. *J. Physiol.* **2007**, *578*, 715–733. [CrossRef] [PubMed]
9. Dai, Y.; Wang, S.; Tominaga, M.; Yamamoto, S.; Fukuoka, T.; Higashi, T.; Kobayashi, K.; Obata, K.; Yamanaka, H.; Noguchi, K. Sensitization of trpa1 by par2 contributes to the sensation of inflammatory pain. *J. Clin. Investig.* **2007**, *117*, 1979–1987. [CrossRef] [PubMed]
10. Al-Shami, A.; Spolski, R.; Kelly, J.; Keane-Myers, A.; Leonard, W.J. A role for tslp in the development of inflammation in an asthma model. *J. Exp. Med.* **2005**, *202*, 829–839. [CrossRef] [PubMed]
11. Ziegler, S.F.; Roan, F.; Bell, B.D.; Stoklasek, T.A.; Kitajima, M.; Han, H. The biology of thymic stromal lymphopoietin (TSLP). *Adv. Pharmacol.* **2013**, *66*, 129–155. [PubMed]
12. Yoo, J.; Omori, M.; Gyarmati, D.; Zhou, B.; Aye, T.; Brewer, A.; Comeau, M.R.; Campbell, D.J.; Ziegler, S.F. Spontaneous atopic dermatitis in mice expressing an inducible thymic stromal lymphopoietin transgene specifically in the skin. *J. Exp. Med.* **2005**, *202*, 541–549. [CrossRef] [PubMed]
13. Shin, J.U.; Kim, S.H.; Kim, H.; Noh, J.Y.; Jin, S.; Park, C.O.; Lee, W.J.; Lee, D.W.; Lee, J.H.; Lee, K.H. Tslp is a potential initiator of collagen synthesis and an activator of CXCR4/SDF-1 axis in keloid pathogenesis. *J. Investig. Dermatol.* **2016**, *136*, 507–515. [CrossRef] [PubMed]
14. Usategui, A.; Criado, G.; Izquierdo, E.; Del Rey, M.J.; Carreira, P.E.; Ortiz, P.; Leonard, W.J.; Pablos, J.L. A profibrotic role for thymic stromal lymphopoietin in systemic sclerosis. *Ann. Rheum. Dis.* **2013**, *72*, 2018–2023. [CrossRef] [PubMed]
15. Jia, X.; Zhang, H.; Cao, X.; Yin, Y.; Zhang, B. Activation of TRPV1 mediates thymic stromal lymphopoietin release via the Ca^{2+}/NFAT pathway in airway epithelial cells. *FEBS Lett.* **2014**, *588*, 3047–3054. [CrossRef] [PubMed]
16. Kouzaki, H.; O'Grady, S.M.; Lawrence, C.B.; Kita, H. Proteases induce production of thymic stromal lymphopoietin by airway epithelial cells through protease-activated receptor-2. *J. Immunol.* **2009**, *183*, 1427–1434. [CrossRef] [PubMed]
17. Turner, M.J.; Zhou, B. A new itch to scratch for TSLP. *Trends Immunol.* **2014**, *35*, 49–50. [CrossRef] [PubMed]
18. Ossovskaya, V.S.; Bunnett, N.W. Protease-activated receptors: Contribution to physiology and disease. *Physiol. Rev.* **2004**, *84*, 579–621. [CrossRef] [PubMed]
19. Cottrell, G.S.; Amadesi, S.; Schmidlin, F.; Bunnett, N. Protease-activated receptor 2: Activation, signalling and function. *Biochem. Soc. Trans.* **2003**, *31*, 1191–1197. [CrossRef] [PubMed]
20. Camerer, E.; Huang, W.; Coughlin, S.R. Tissue factor- and factor x-dependent activation of protease-activated receptor 2 by factor viia. *Proc. Natl. Acad. Sci. USA* **2000**, *97*, 5255–5260. [CrossRef] [PubMed]

21. Corvera, C.U.; Dery, O.; McConalogue, K.; Bohm, S.K.; Khitin, L.M.; Caughey, G.H.; Payan, D.G.; Bunnett, N.W. Mast cell tryptase regulates rat colonic myocytes through proteinase-activated receptor 2. *J. Clin. Investig.* **1997**, *100*, 1383–1393. [CrossRef] [PubMed]

22. Molino, M.; Barnathan, E.S.; Numerof, R.; Clark, J.; Dreyer, M.; Cumashi, A.; Hoxie, J.A.; Schechter, N.; Woolkalis, M.; Brass, L.F. Interactions of mast cell tryptase with thrombin receptors and PAR-2. *J. Biol. Chem.* **1997**, *272*, 4043–4049. [CrossRef] [PubMed]

23. Kawabata, A.; Kawao, N.; Kuroda, R.; Tanaka, A.; Itoh, H.; Nishikawa, H. Peripheral PAR-2 triggers thermal hyperalgesia and nociceptive responses in rats. *Neuroreport* **2001**, *12*, 715–719. [CrossRef] [PubMed]

24. Liu, L.; Chen, L.; Liedtke, W.; Simon, S.A. Changes in osmolality sensitize the response to capsaicin in trigeminal sensory neurons. *J. Neurophysiol.* **2007**, *97*, 2001–2015. [CrossRef] [PubMed]

25. Premkumar, L.S.; Ahern, G.P. Induction of vanilloid receptor channel activity by protein kinase C. *Nature* **2000**, *408*, 985–990. [CrossRef] [PubMed]

26. Mohapatra, D.P.; Nau, C. Desensitization of capsaicin-activated currents in the vanilloid receptor TRPV1 is decreased by the cyclic AMP-dependent protein kinase pathway. *J. Biol. Chem.* **2003**, *278*, 50080–50090. [CrossRef] [PubMed]

27. Diver, J.M.; Sage, S.O.; Rosado, J.A. The inositol trisphosphate receptor antagonist 2-aminoethoxydiphenylborate (2-apb) blocks Ca2+ entry channels in human platelets: Cautions for its use in studying Ca2+ influx. *Cell Calcium* **2001**, *30*, 323–329. [CrossRef] [PubMed]

28. Bootman, M.D.; Collins, T.J.; Mackenzie, L.; Roderick, H.L.; Berridge, M.J.; Peppiatt, C.M. 2-aminoethoxydiphenyl borate (2-APB) is a reliable blocker of store-operated Ca2+ entry but an inconsistent inhibitor of insp3-induced ca2+ release. *FASEB J.* **2002**, *16*, 1145–1150. [CrossRef] [PubMed]

29. Chokshi, R.; Fruasaha, P.; Kozak, J.A. 2-aminoethyl diphenyl borinate (2-APB) inhibits TRPM7 channels through an intracellular acidification mechanism. *Channels* **2012**, *6*, 362–369. [CrossRef] [PubMed]

30. Kovacs, G.; Montalbetti, N.; Simonin, A.; Danko, T.; Balazs, B.; Zsembery, A.; Hediger, M.A. Inhibition of the human epithelial calcium channel TRPV6 by 2-aminoethoxydiphenyl borate (2-APB). *Cell Calcium* **2012**, *52*, 468–480. [CrossRef] [PubMed]

31. Khairatkar Joshi, N.K.; Maharaj, N.; Thomas, A. The TRPV3 receptor as a pain target: A therapeutic promise or just some more new biology. *Open Drug Discov. J.* **2010**, *2*, 89–97. [CrossRef]

32. Zhang, Y.; Zhou, B. Functions of thymic stromal lymphopoietin in immunity and disease. *Immunol. Res.* **2012**, *52*, 211–223. [CrossRef] [PubMed]

33. Wilson, S.R.; The, L.; Batia, L.M.; Beattie, K.; Katibah, G.E.; McClain, S.P.; Pellegrino, M.; Estandian, D.M.; Bautista, D.M. The epithelial cell-derived atopic dermatitis cytokine TSLP activates neurons to induce itch. *Cell* **2013**, *155*, 285–295. [CrossRef] [PubMed]

34. Gao, P.S.; Rafaels, N.M.; Mu, D.; Hand, T.; Murray, T.; Boguniewicz, M.; Hata, T.; Schneider, L.; Hanifin, J.M.; Gallo, R.L.; et al. Genetic variants in thymic stromal lymphopoietin are associated with atopic dermatitis and eczema herpeticum. *J. Allergy Clin. Immunol.* **2010**, *125*, 1403–1407. [CrossRef] [PubMed]

35. Margolis, D.J.; Kim, B.; Apter, A.J.; Gupta, J.; Hoffstad, O.; Papadopoulos, M.; Mitra, N. Thymic stromal lymphopoietin variation, filaggrin loss of function, and the persistence of atopic dermatitis. *JAMA Dermatol.* **2014**, *150*, 254–259. [CrossRef] [PubMed]

36. Cheng, X.; Jin, J.; Hu, L.; Shen, D.; Dong, X.P.; Samie, M.A.; Knoff, J.; Eisinger, B.; Liu, M.L.; Huang, S.M.; et al. TRP channel regulates EGFR signaling in hair morphogenesis and skin barrier formation. *Cell* **2010**, *141*, 331–343. [CrossRef] [PubMed]

37. Yamamoto-Kasai, E.; Imura, K.; Yasui, K.; Shichijou, M.; Oshima, I.; Hirasawa, T.; Sakata, T.; Yoshioka, T. TRPV3 as a therapeutic target for itch. *J. Investig. Dermatol.* **2012**, *132*, 2109–2112. [CrossRef] [PubMed]

38. Vriens, J.; Appendino, G.; Nilius, B. Pharmacology of vanilloid transient receptor potential cation channels. *Mol. Pharmacol.* **2009**, *75*, 1262–1279. [CrossRef] [PubMed]

39. Xu, H.; Ramsey, I.S.; Kotecha, S.A.; Moran, M.M.; Chong, J.A.; Lawson, D.; Ge, P.; Lilly, J.; Silos-Santiago, I.; Xie, Y.; et al. TRPV3 is a calcium-permeable temperature-sensitive cation channel. *Nature* **2002**, *418*, 181–186. [CrossRef] [PubMed]

40. Peier, A.M.; Moqrich, A.; Hergarden, A.C.; Reeve, A.J.; Andersson, D.A.; Story, G.M.; Earley, T.J.; Dragoni, I.; McIntyre, P.; Bevan, S.; et al. A trp channel that senses cold stimuli and menthol. *Cell* **2002**, *108*, 705–715. [CrossRef]

41. Moqrich, A.; Hwang, S.W.; Earley, T.J.; Petrus, M.J.; Murray, A.N.; Spencer, K.S.; Andahazy, M.; Story, G.M.; Patapoutian, A. Impaired thermosensation in mice lacking TRPV3, a heat and camphor sensor in the skin. *Science* **2005**, *307*, 1468–1472. [CrossRef] [PubMed]

42. Borbiro, I.; Lisztes, E.; Toth, B.I.; Czifra, G.; Olah, A.; Szollosi, A.G.; Szentandrassy, N.; Nanasi, P.P.; Peter, Z.; Paus, R.; et al. Activation of transient receptor potential vanilloid-3 inhibits human hair growth. *J. Investig. Dermatol.* **2011**, *131*, 1605–1614. [CrossRef] [PubMed]

43. Mergler, S.; Valtink, M.; Takayoshi, S.; Okada, Y.; Miyajima, M.; Saika, S.; Reinach, P.S. Temperature-sensitive transient receptor potential channels in corneal tissue layers and cells. *Ophthalmic Res.* **2014**, *52*, 151–159. [CrossRef] [PubMed]

44. Ueda, T.; Yamada, T.; Ugawa, S.; Ishida, Y.; Shimada, S. TRPV3, a thermosensitive channel is expressed in mouse distal colon epithelium. *Biochem. Biophys. Res. Commun.* **2009**, *383*, 130–134. [CrossRef] [PubMed]

45. Hamamoto, T.; Takumida, M.; Hirakawa, K.; Takeno, S.; Tatsukawa, T. Localization of transient receptor potential channel vanilloid subfamilies in the mouse larynx. *Acta Oto-Laryngol.* **2008**, *128*, 685–693. [CrossRef] [PubMed]

46. Ishibashi, T.; Takumida, M.; Akagi, N.; Hirakawa, K.; Anniko, M. Expression of transient receptor potential vanilloid (TRPV) 1, 2, 3, and 4 in mouse inner ear. *Acta Oto-Laryngol.* **2008**, *128*, 1286–1293. [CrossRef] [PubMed]

47. Denda, M.; Sokabe, T.; Fukumi-Tominaga, T.; Tominaga, M. Effects of skin surface temperature on epidermal permeability barrier homeostasis. *J. Investig. Dermatol.* **2007**, *127*, 654–659. [CrossRef] [PubMed]

48. Hu, H.Z.; Xiao, R.; Wang, C.; Gao, N.; Colton, C.K.; Wood, J.D.; Zhu, M.X. Potentiation of TRPV3 channel function by unsaturated fatty acids. *J. Cell. Physiol.* **2006**, *208*, 201–212. [CrossRef] [PubMed]

International Journal of
Molecular Sciences

MDPI

Communication

Could −79 °C Spray-Type Cryotherapy Be an Effective Monotherapy for the Treatment of Keloid?

Tae Hwan Park [1,*,†], Hyeon-Ju Cho [2,†], Jang Won Lee [1], Chan Woo Kim [1], Yosep Chong [3], Choong Hyun Chang [4] and Kyung-Soon Park [2,*]

1 Department of Plastic and Reconstructive Surgery, CHA Bundang Medical Center, CHA University, Seongnam 13496, Korea; RA22211@chamc.co.kr (J.W.L.); A176004@chamc.co.kr (C.W.K.)
2 Department of Biomedical Science, College of Life Science, CHA University, Seongnam 13488, Korea; whguswn94@chauniv.ac.kr
3 Department of Hospital Pathology, College of Medicine, The Catholic University of Korea, Seoul 06591, Korea; roja30@hanmail.net
4 Department of Plastic and Reconstructive Surgery, Kangbuk Samsung Hospital, Sungkyunkwan University School of Medicine, Seoul 03181, Korea; eppeene@hanmail.net
* Correspondence: pspark0124@cha.ac.kr (T.H.P.); kspark@cha.ac.kr (K.S.P.);
   Tel.: +82-31-881-7593 (T.H.P.); +82-31-881-7144 (K.S.P.); Fax: +82-31-780-5285 (T.H.P.); +82-31-881-7249 (K.S.P.)
† These authors contributed equally to this work.

Received: 15 September 2017; Accepted: 24 November 2017; Published: 26 November 2017

Abstract: Cryotherapy has been regarded as an effective modality for the treatment of keloids, and the spray-type device is one of the novel cryotherapeutic units. However, the biological mechanisms and therapeutic effects of this technique are incompletely studied. We evaluated the clinical efficacy of our cryotherapy protocol with molecular and pathologic evidence for the treatment of keloids. We evenly split each of ten keloid lesions into a non-treated (C−) and treated (C+) area; the C+ area was subjected to two freeze-thaw cycles of spray-type cryotherapy using −79 °C spray-type CryoPen™. This treatment was repeated after an interval of two weeks. The proliferation and migration abilities of the fibroblasts isolated from the dermis under the cryotherapy-treated or untreated keloid tissues (at least 5 mm deep) were compared and pathologic findings of the full layer were evaluated. Molecular analysis revealed that the number of dermal fibroblasts was significantly higher in C+ group as compared with C− group. The dermal fibroblasts from C+ group showed more than two-fold increase in the migration ability as compared with the fibroblasts from C− group. The expression of matrix metallopeptidase 9 was increased by more than two-fold and a significant increase in transforming growth factor beta 1 expression and Smad2/3 phosphorylation level was observed in C+ group. C+ group showed more extensive lymphoplasmacytic infiltration with thicker fibrosis and occasional "proliferating core collagen" as compared with C− group. Thus, −79 °C spray-type cryotherapy is ineffective as a monotherapy and should be used in combination with intralesional corticosteroids or botulinum toxin A for favourable outcomes in the treatment of thick keloids.

Keywords: keloid; cryotherapy; fibrosis; matrix metallopeptidase 9; transforming growth factor β 1

1. Introduction

Patients with keloids typically present with pruritus, pain, ulceration, secondary infection, restricted motion, and psychological symptoms due to cosmetic disfigurement [1]. Given their intractable nature and high recurrence rate, keloids are a great burden to the patients, physicians, scientists, and society and are associated with physical, aesthetic, and social complaints [2,3]. Numerous treatment methods, including cryotherapy, intralesional injection, laser treatment, pressure therapy, radiation, and topical treatment have been proposed for keloids [4]. Cryotherapy has been widely used for years and is regarded as an effective modality for the treatment of keloids; several techniques and devices related to cryotherapy have been developed in the clinical practice [5,6].

In 1982, Shepherd and Dawber first described cryotherapy as an optimal treatment options for keloid scars [7]. External cryotherapy, contact cryotherapy with spray technique, is an efficient and effective method for clinical use and has been performed over the past few years; however, numerous side effects such as edema, swelling, blister formation, or oozing are inevitably encountered during the wound-healing process [8].

Intralesional cryotherapy was designed in 1993 by Weshahy [9] and involves intralesional application of liquid nitrogen with a needle to destroy the core of the keloid scars. It is an effective method, wherein the scar tissue is frozen from the inside and responds well to a single use. The recurrence rate is reduced with intralesional cryotherapy and a favourable reduction in the scar volume was observed as opposed to the external cryotherapy. However, it usually requires local anaesthesia and patients often complain about severe pain and bleeding, due to its relatively invasive nature. In addition, it requires general anaesthesia in cases of large keloids, leading to increased morbidity, hospital visits, and medical cost. Furthermore, the application of intralesional cryotherapy is difficult for wide superficial keloids. Considering the recent trends of minimally invasive techniques, many physicians use technologically advanced spray-type cryotherapy such as −79 °C hand-held spray to treat warts and other skin lesions. In recent years, this spray has been used for keloid treatment, as it eliminates the need for local anaesthesia. As a consequence, it offers less pain and discomfort and reduced psychological burden as compared with the conventional cryotherapy device using −196 °C or intralesional spray. However, the underlying mechanisms and therapeutic effects of this technique are questionable. Therefore, we employed −79 °C spray-type CryoPen™ in this study.

Matrix metallopeptidase 9 (MMP9) has been implicated in the keloid pathophysiology, due to its gelatinase activity. The enzyme activity of MMP9 induces extracellular remodelling through the degradation of gelatins (collagens) and matrix-associated substrates such as aggrecan and elastin [10,11]. In addition, MMP9 converts various cytokines and chemokines into their more active forms [12–14]. Though the role of MMP9 in keloid development is incompletely studied, several reports reveal its active role in the development of keloid lesions. MMP9 is significantly overexpressed in keloid-derived fibroblasts, especially at the margins of the keloid wound [15]. Transforming growth factor beta 1 (TGF-β1)-mediated up-regulation of microRNA-21 increases the expression of MMP9 in keloid fibroblasts, wherein it promotes fibroblast proliferation and transdifferentiation [16].

In this study, we evaluated the clinical efficacy of −79 °C spray-type CryoPen™ with molecular and pathologic evidence for the treatment of keloids.

2. Results

2.1. The Schematic Illustration of Tissue Sampling for Pathologic and Molecular Study

A total of six patients with ten keloids were treated with −79 °C spray-type CryoPen™. Of these, two patients displayed multiple lesions (four and two), which were treated with cryotherapy. The mean age of the patients was 29 years (range: 17–56) and the duration of scar ranged from 0.5 to 30 years. Four keloids were of Fitzpatrick skin type III–IV, while six keloids were of skin type I–II. The locations of scars were chest wall, upper arm, pubis, axilla, shoulder, trochanter, ear lobule, and helix (Table 1). We selected two cases for the molecular and pathologic study (Case #1 and #5). The dermal keloid (3 mm thick and at least 5 mm deep) from the epidermis was used for primary cell culture and the adjacent full layer from the epidermis to reticular dermis was used for pathologic examination (Figure 1).

Table 1. Baseline patient demographics in this study.

No. of Cases (*n*)	Age (Years)	Gender	Skin Type (Fitzpatrick)	Duration of Scar (Years)	Scar Location	Cause	Sessions (*n*)	Interval (Weeks)	No. of FTCs (Freeze–Thaw Cycle) (*n*)
1	21	F	F III-IV	7	Upper arm	spontaneous	4	2	2
2	21	F	F III-IV	7	Chest wall	spontaneous	4	2	2
3	21	F	F III-IV	7	pubis	spontaneous	4	2	2
4	21	F	F III-IV	7	Axilla	spontaneous	4	2	2
5	56	F	F I-II	30	chest wall	infection	4	2	2
6	56	F	F I-II	35	Shoulder	Vaccination	4	2	2
7	26	F	F I-II	0.5	Trochanter	Surgery	4	2	2
8	33	F	F I-II	2	Lobule	Piercing	4	2	2
9	21	F	F I-II	3	Helix	Piercing	4	2	2
10	17	F	F I-II	3	Helix	Piercing	4	2	2

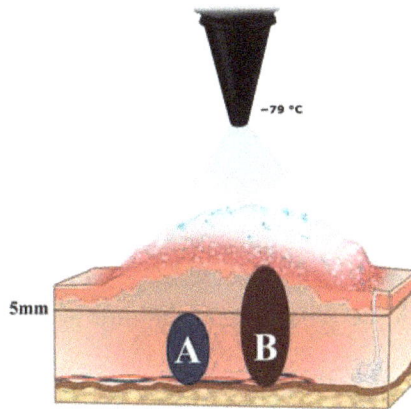

Figure 1. Schematic illustration of our tissue sampling for pathologic and molecular study. (A) The dermal keloid (3 mm thick and at least 5 mm deep) from the epidermis was used for primary cell culture, (B) while the adjacent full layer from the epidermis to reticular dermis was used for pathologic examination.

2.2. Keloids Treated with −79 °C Spray-Type CryoPen™ Rapidly Reoccurred in Clinical Trials

As shown in Figure 2A,B, the keloid lesion located at the chest wall responded well to the protocol in terms of redness and hypertrophy and the symptom improved for 2 weeks after cryotherapy. However, the lesion reverted to its original status after 2 weeks unless it was treated with cryotherapy (Figure 2C). However, the cryotherapy was ineffective for severe cases of keloid located at the upper arm and caused skin erosion at the site of cryotherapy; no volume reduction or shrinkage was reported (Figure 2D–F).

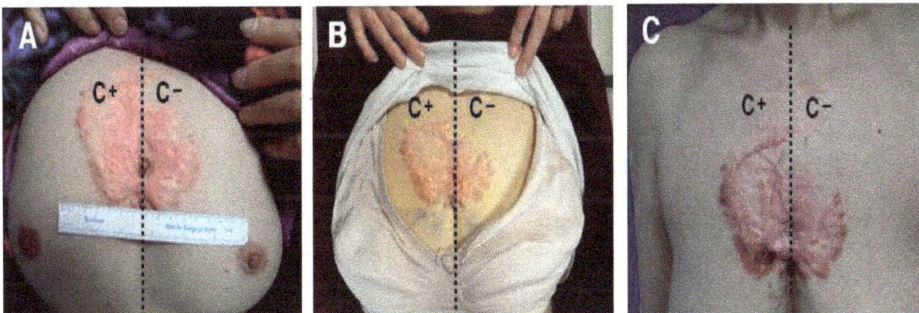

Figure 2. *Cont.*

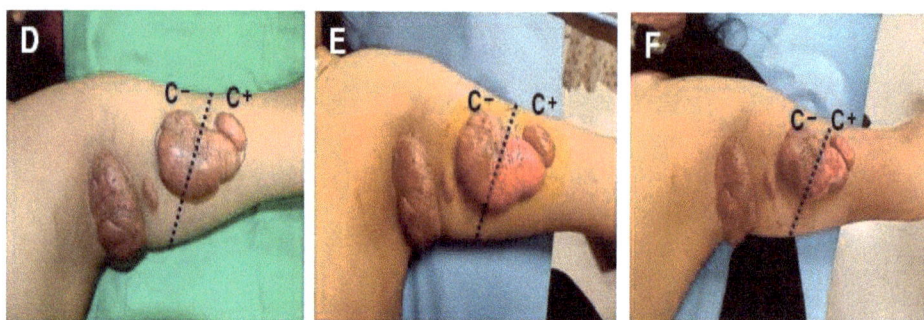

Figure 2. Clinical cases of keloids treated with −79 °C spray type CryoPen™. (**A**) A 56-year-old female patient with anterior chest keloid was presented with recurrent central ulceration, pruritus, and intermittent pain; (**B**) Split therapy was performed such that the right half was treated with cryotherapy, while the remaining left half was left untreated. The therapeutic effect was noticed 1 week following four sessions of cryotherapy in terms of improvement in the texture, softness, and redness accompanied with a slight shrinkage of the right treated half; (**C**) Three weeks after the four sessions of cryotherapy, the lesion reverted to its original status unless it was subjected to repeated cryotherapy; some skin erosion was observed in the treated half (small white arrow). Keloid excision and full thickness skin grafting from inguinal area were planned; (**D**) A 21-year-old female patient was presented with spontaneous keloids of the upper arm and axilla; (**E**) Split therapy was performed such that the lower half was treated with cryotherapy, while the upper half was left untreated. The therapeutic effect was evaluated after four sessions of cryotherapy in terms of reduction in the volume or shrinkage of keloid. Cryotherapy was ineffective and caused skin erosion at the site of cryotherapy; no volume reduction or shrinkage (small white arrow) was observed; (**F**) Post-treatment appearance of keloid at 3 weeks after final cryotherapy. The erosion was improved with conservative wound management. Keloid excision and primary closure were planned.

2.3. Keloid-Derived Fibroblasts Show Enhanced Proliferation and Migration Activity Following Treatment with −79 °C Spray-Type CryoPen™

We first compared the proliferation of dermal fibroblasts isolated from deep tissues of cryotherapy-treated or untreated keloids. As shown in Figure 3a, the number of dermal fibroblasts treated with cryotherapy was significantly higher than the untreated fibroblasts at 48 h after cell seeding, indicating that cryotherapy promotes proliferation of dermal fibroblasts. We compared the migration ability of dermal fibroblasts using the transwell assay. To exclude the effect of proliferation on the migration, the assay was performed with cells pretreated with mitomycin C, which causes cell cycle arrest. In comparison with the untreated fibroblasts, those treated with cryotherapy showed more than two-fold increase in their migration abilities (Figure 3b). Furthermore, fibroblasts subjected to cryotherapy showed significantly enhanced migration ability in the wound-healing assay (Figure 3c). Taken together, these findings indicate that the treatment of keloids with −79 °C spray-type CryoPen™ results in an increase in the proliferation and migration activity of fibroblasts.

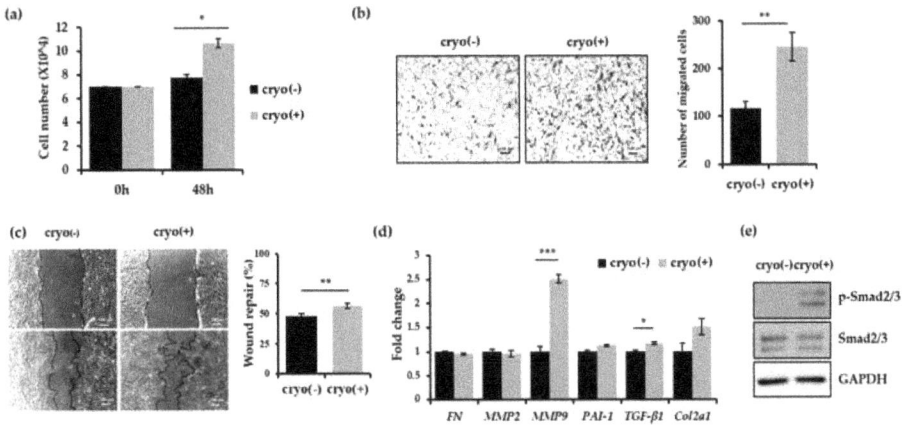

Figure 3. Cryotherapy increases the migration and proliferative activity of keloid dermal fibroblasts. (a) The comparison of the proliferative activity of the dermal fibroblasts from keloids before and after cryotherapy. Cell proliferation was promoted in the fibroblast treated with cryotherapy. The cell number was estimated at 48 h after cell seeding; (b,c) The migration activity of dermal fibroblasts was increased following cryotherapy. The migration of the cells was analysed by the transwell assay (b) and wound-healing assay (c); (d) qRT-PCR experiments were performed to evaluate the mRNA levels of fibronectin, MMP2 (Matrix metalloproteinase 2), MMP9, PAI-1 (Plasminogen activator inhibitor-1), TGF-β1, and Col2a1 (Collagen Type II Alpha 1) in fibroblasts derived from the dermis under the cryotherapy-treated (+) or untreated (−) keloids. MMP9 and TGF-β1 were significantly increased in the fibroblast treated with cryotherapy as compared with controls; (e) Immunoblot analysis of phosphorylated Smad2/3. Total Smad2/3 and GAPDH were used as the loading controls. Error bars represent the standard error from three repeated experiments. * $p < 0.05$, ** $p < 0.01$, and *** $p < 0.005$ (Student's t-test).

2.4. Expression of MMP9 and TGF-β1 was Increased in Keloids after Treatment with −79 °C Spray-Type CryoPen™

The expression of MMP9, which promotes cell migration ability by converting various cytokines and chemokines into their more active forms, 12–14 was increased by more than two-fold following cryotherapy (Figure 3d). In addition, the expression of TGF-β1 and level of Smad2/3 phosphorylation were significantly increased by cryotherapy (Figure 3d,e).

2.5. Cryotherapy-Treated Keloids Showed More Fibrosis and Extensive Lymphoplasmacytic Infiltration

The result of the molecular analysis was further confirmed by the pathologic examination. We found that keloids treated with cryotherapy showed extensive lymphoplasmacytic infiltration characterised with thicker fibrosis and occasional "proliferating core collagen" in the deeper part. On the other hand, untreated keloids showed relatively less lymphoplasmacytic infiltration and thinner, surface-parallel fibrosis without "proliferating core collagen" (Figure 4). Thus, the treatment of keloids with −79 °C spray-type CryoPen™ induces fibrotic incidence as compared with the untreated keloid tissues.

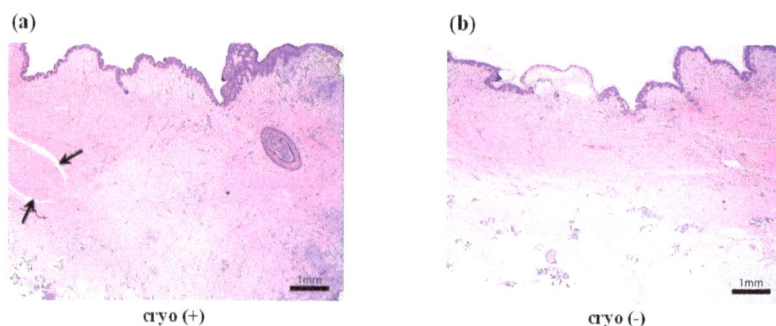

cryo (+)   cryo (-)

Figure 4. Histologic findings of keloids pretreated and untreated with cryotherapy. (**a**) Cryotherapy-treated keloid showed extensive lymphoplasmacytic infiltration characterised by thicker fibrosis and occasional "proliferating core collagen" (**black arrow**) in the deeper part (H&E stain, ×12.5); (**b**) The untreated keloid showed less inflammation and less surface-parallel fibrosis without core collagen (H&E stain, ×12.5).

3. Discussion

Keloids are a result of an overgrowth of a dense fibrous tissue that develops after trauma of skin, and cryotherapy is known as an efficient and effective method for the treatment of keloids. However, our clinical experience of treating keloid patients with −79 °C spray-type cryotherapy protocol as a monotherapy has made us doubt the efficacy of this therapy in keloid treatment.

More than 2 weeks after cryotherapy, the lesion reverted to its primary status and required additional cryotherapy (Figure 2C). In the severe case of keloids located at the upper arm, cryotherapy was ineffective and caused skin erosion at the site of cryotherapy; no volume reduction or shrinkage was observed (Figure 2D–F). These results imply that −79 °C spray-type cryotherapy may be effective only for the treatment of mild keloid and to destroy superficially located fibroblasts; the rapid recurrence of the symptom may be possibly due to the compensatory stimulation of the deep keloid tissues located under the cryotherapy-treated superficial keloid tissues.

We performed in vitro analysis of the cellular activity of the keloid-derived fibroblasts to understand the therapeutic effect of −79 °C spray-type cryotherapy. Dalkowski et al., used an experimental model for controlled cell freezing in vitro to simulate the effect of cryotherapy on keloid fibroblasts [17]. In our opinion, freezing cell may lead to mechanical destruction rather than actual physiologic effect in tissues. We also believed that superficial tissues, including epidermis and upper dermis of keloid, may be affected by the cryotherapy-mediated mechanical destruction. Hence, we harvested deep keloid tissues (at least 5 mm deep) from the epidermis for molecular analysis. Our results indicate that fibroblasts derived from the cryotherapy-treated keloids showed significantly enhanced proliferation and migration ability (Figure 3a–c). In addition, the mRNA expression of MMP9 and TGF-β1 was increased (Figure 3d) and the level of Smad2/3 phosphorylation was significantly up-regulated following cryotherapy (Figure 3d,e). Contrary to the general expectation that cryotherapy reduces the migration or proliferation ability of keloid fibroblasts through the inhibition of collagen synthesis [17], molecular analysis revealed that the proliferation and migration abilities of keloid fibroblasts from the deep tissue were significantly enhanced after cryotherapy, possibly due to the overexpression of MMP9 and activation of TGF-β signalling pathway.

Our current findings contradict the results demonstrating a significant reduction in TGF-β1 expression in the keloid tissue after cryotherapy by Awad et al., suggesting that a few sessions of cryotherapy normalised the abnormal collagen structure and reduced fibroblast proliferation by suppressing TGF-β1 expression [18]. The major discrepancy between these observations and those reported in our study arises from the quality of device used. While Awad et al., used the hand-held device (Brymill cryogenic system, CRY–AC, Ellington, CT, USA) with −196 °C liquid nitrogen, we performed our study using −79 °C liquid nitrogen. Furthermore, we used the simple spray-type cryotherapy as compared with the needle type cryotherapy used in their study.

However, Har-Shai et al. [19,20] reported successful outcome with intralesional cryosurgery treatment. Their technique resulted in major changes in collagen structure and organization including reduced the number of proliferating cells, of myofibroblasts and of mast cells. This means that intralesional cryosurgery effectively irradicates "proliferating core collagen" in keloids [21].

Taken together, we suggest that the monotherapy with −79 °C spray-type cryotherapy is ineffective for the treatment of thick keloids and should be used in the combination with intralesional injection of corticosteroids or botulinum toxin A to prevent any compensatory stimulation of deep keloid tissues. At present, we use this combination therapy as the first choice of treatment in keloid patients that are unwilling or unsuitable for any surgical treatment. However, further studies should be performed to optimise the cryotherapy protocol to provide patients and physicians with the best possible management of keloids.

4. Materials and Methods

4.1. Inclusion and Exclusion Criteria and Study Design

The current study was approved by the institutional review board of the CHA University (2017-03-051; 10 April 2017). Patients with keloids who presented to the outpatient clinic were included in the study based on the following criteria: (1) the scar was elevated and extended beyond the dimensions of the initial injury site or lesion; (2) patients were older than 18 years; (3) surgical excision was scheduled; (4) patients received no additional treatment or medication during the study and prior to surgical excision; and (5) patients signed up for the data use agreement as a basis to the clinical study. Patients were excluded from the study if they were unavailable for follow-up or wanted to stop cryotherapy treatment for any reason. Patients who had received any additional adjuvant therapy during the treatment were also excluded from the study. A total of ten keloids on six patients (all females) were included in this study and all keloids showed deep thickness. The detailed information of the cases is listed in Table 1.

4.2. Cryotherapy Protocol

All procedures were performed with −79 °C hand-held spray-type CryoPen™ (L&C BIO, Gyeonggi-do, Korea) without anaesthesia. Each keloid lesion was evenly divided into untreated area (C−) and treated area (C+). The C+ area was subjected to only two freeze-thaw cycles of spray-type cryotherapy. This treatment was repeated after an interval of 2 weeks. Eventually, the C+ lesion was treated with four sessions of cryotherapy over a period of 8 weeks prior to surgical excision. We then excised lesions completely, including a full layer of dermis until we noticed bleeding of the underneath subcutaneous tissue. The bleeding was controlled with bipolar electrocoagulator. The 3-mm thick and at least 5-mm-deep dermal keloid from the epidermis was used for primary cell culture, while the adjacent full layer from the epidermis to reticular dermis was used for pathologic examination (Figure 1). We closed wounds with an appropriate approximation using 5–0 nylon interrupted sutures. When primary closure was not possible, we covered the wound with full thickness skin graft from the inguinal area, fixed it with 5–0 black silk sutures, and a tie-over dressing was done.

4.3. Primary Cell Isolation, in Vitro Culture, and Cell Proliferation Assay

The fibroblasts were isolated from the dermis under the surgically excised and cryotherapy-treated or untreated keloids. The dermis was cut into approximately 5 mm^3 pieces. The epidermis and lipid layer were removed with 2% dispase II (Sigma, St. Louis, MO, USA) and the connective tissue was digested in 0.5 mg/mL collagenase A (Sigma) at 37 °C for 3 h using a water bath. The digested solution was filtered through 70 μm strainer (BD biosciences, San Diego, CA, USA). The cell pellets were resuspended in, and washed with, $1\times$ Dulbecco's phosphate-buffered saline (Gibco, Gaithersburg, MD, USA). The cells were cultured in Dulbecco's modified Eagle's medium (DMEM) medium (Gibco) supplemented with 10% fetal bovine serum (FBS, Gibco) and 1% penicillin/streptomycin (Gibco) at 37 °C and 5% CO_2. The medium was replaced every 2–3 days and the cells were subcultured at 70–80% confluency. All experiments were performed with cells at passage 3. For proliferation assay, cells were cultured at a density of 7.0×10^4 cells in a 60 mm^2 dish. After 48 h, cells were detached with 0.05% trypsin/ethylenediaminetetraacetic acid (EDTA; Gibco) and the cell number was estimated with a haemocytometer.

4.4. Wound-Healing and Transwell Assay

We performed the wound-healing and transwell assay to compare the cell migration activity. For the wound-healing assay, 7.0×10^4 cells were cultured in each well of 12-well plates. After 24 h, each well was treated with mitomycin C (1 mg/mL) for 2 h and scratched using yellow tips. The wounds were observed 24 h after scratching. For the transwell assay, 2.0×10^4 cells were added to the upper well of a transwell plate (Corning Inc., Corning, NY, USA). The lower portion of the well contained DMEM supplemented with 10% FBS as a chemoattractant. After incubation for 24 h, the medium and cells present in the bottom chamber were removed; the cells that migrated to the bottom chamber were fixed in 4% paraformaldehyde (Santa Cruz Biotechnology Inc., Santa Cruz, CA, USA) and stained with crystal violet for 15 min. The migrated cells were counted in random fields using ImageJ (Bethesda, MD, USA: U. S. National Institutes of Health).

4.5. Quantitative Real-Time Polymerase Chain Reaction (qRT-PCR)

Total RNA was extracted using TRIzol (Invitrogen, Carlsbad, CA, USA) and 1 μg of complementary DNA (cDNA) was synthesised using the LeGene Express 1st Strand cDNA Synthesis System (LeGene Biosciences Inc., San Diego, CA, USA) according to manufacturer's instructions. qRT-PCR analysis was performed using the synthetic cDNAs and TOPrealTM qPCR $2\times$ PreMIX (Enzynomics, Daejeon, Korea). The expression of the target genes was normalised against that of glyceraldehydes 3-phosphate dehydrogenase (GAPDH). The PCR primers are listed in Table 2.

Table 2. Quantitative real-time PCR primers.

No.	Gene	Direction	Sequences (5' to 3')
1	*GAPDH*	Forward	ACCACAGTCCATGCCATCAC
		Reverse	TCCACCACCCTGTTGCTGTA
2	*fibronectin*	Forward	CAGTGGGAGACCTCGAGAAG
		Reverse	TCCCTCGGAACATCAGAAAC
3	*TGFβ1*	Forward	GGACACCAACTATTGCTTCAG
		Reverse	TCCAGGCTCCAAATGTAGG
4	*MMP2*	Forward	ATGACAGCTGCACCACTGAG
		Reverse	ATTTGTTGCCCAGGAAAGTG
5	*MMP9*	Forward	ATTCAGGGAGACGCCCATTT
		Reverse	CTGCGT TTCCAAACCGAGTT
6	*col 2a1*	Forward	GGGAGTAATGCAAGGACCAA
		Reverse	ATCATCACCAGGCTTTCCAG

4.6. Western Blotting

Cells were harvested with phosphate-buffered saline (PBS) and lysed in the tissue lysis buffer (20 mM Tris-base [pH 7.4], 137 mM sodium chloride [NaCl], 2 mM EDTA, 1% Triton X-100, 25 mM β-glycerophosphate, 2 mM sodium pyrophosphate, 10% glycerol, 1 mM sodium orthovanadate, 1 mM benzamidine, and 1 mM phenylmethylsulfonyl fluoride). Total cell extracts were separated by sodium dodecyl sulfate polyacrylamide gel electrophoresis, transferred onto polyvinylidene fluoride membranes (BIORAD), and blotted with antibodies against Smad2/3 (Cell Signaling Technology, Beverly, MA, USA) and phospho-Smad2/3 (Cell Signaling Technology). Immunoreactivity was detected with enhanced chemiluminescence (Amersham).

4.7. Pathologic Study

Microscopic examination was performed for the cryotherapy-treated and untreated keloids using routine formalin-fixed, paraffin-embedded tissue process and hematoxylin-eosin staining. The excised keloids were fixed in 10% buffered formalin, and totally embedded in paraffin blocks after routine preparation process. Then, 7 μm-thin sections were subsequently stained by hematoxylin and eosin (H&E) for usual microscopic examination. The slides were assessed by pathologist (Yosep Chong).

5. Conclusions

Cryotherapy with the spray-type device is a relatively quick and minimally invasive technique, which may not necessitate local anaesthesia. However, we show that $-79\,°C$ hand-held spray-type cryotherapy is ineffective as a monotherapy and highlight the need for its use as a combination therapy with intralesional corticosteroids or botulinum toxin A to achieve favourable results for the treatment of keloids.

Acknowledgments: This work was supported by the Priority Research Centers Program through the National Research Foundation of Korea (NRF) (20120006679). This work was also funded by the Ministry of Education, Science, and Technology (NRF-2017-M3A9B4031169). This work was supported by the National Research Foundation of Korea (NRF) grant funded by the Korea government (MSIP; Ministry of Science, ICT & Future Planning) (no. 2017R1C1B5017180, TH Park) and was also supported by a grant of the Research Driven Hospital R&D project, funded by the CHA Bundang Medical Center (Grant no. BDCHA R&D 2017-013, TH Park).

Conflicts of Interest: The authors declare no conflict of interest.

Abbreviations

DMEM	Dulbecco's modified Eagle's medium
GAPDH	Glyceraldehyde 3-phosphate dehydrogenase
MMP9	Matrix metallopeptidase 9
PBS	Phosphate-buffered saline
qRT-PCR	Quantitative real-time polymerase chain reaction
TGF-β1	Transforming growth factor β1

References

1. Kafka, M.; Collins, V.; Kamolz, L.P.; Rappl, T.; Branski, L.K.; Wurzer, P. Evidence of invasive and noninvasive treatment modalities for hypertrophic scars: A systematic review. *Wound Repair Regen. Off. Publ. Wound Heal. Soc. Eur. Tissue Repair Soc.* **2017**, *25*, 139–144. [CrossRef] [PubMed]

2. Bijlard, E.; Timman, R.; Verduijn, G.M.; Niessen, F.B.; Van Neck, J.W.; Busschbach, J.J.; Mureau, M.A. Intralesional cryotherapy versus excision and corticosteroids or brachytherapy for keloid treatment: Study protocol for a randomised controlled trial. *Trials* **2013**, *14*, 439. [CrossRef] [PubMed]

3. Park, T.H.; Park, J.H.; Tirgan, M.H.; Halim, A.S.; Chang, C.H. Clinical implications of single-versus multiple-site keloid disorder: A retrospective study in an Asian population. *Ann. Plast. Surg.* **2015**, *74*, 248–251. [CrossRef] [PubMed]

4. Savion, Y.; Sela, M. Prefabricated pressure earring for earlobe keloids. *J. Prosthet. Dent.* **2008**, *99*, 406–407. [CrossRef]

5. Guan, H.; Zhao, Z.; He, F.; Zhou, Q.; Meng, Q.; Zhu, X.; Zheng, Z.; Hu, D.; Chen, B. The effects of different thawing temperatures on morphology and collagen metabolism of -20 degrees C dealt normal human fibroblast. *Cryobiology* **2007**, *55*, 52–59. [CrossRef] [PubMed]

6. Ogawa, R. Keloid and Hypertrophic Scars Are the Result of Chronic Inflammation in the Reticular Dermis. *Int. J. Mol. Sci.* **2017**, *18*, 606. [CrossRef] [PubMed]

7. Van Leeuwen, M.C.; Bulstra, A.E.; van der Veen, A.J.; Bloem, W.B.; van Leeuwen, P.A.; Niessen, F.B. Comparison of two devices for the treatment of keloid scars with the use of intralesional cryotherapy: An experimental study. *Cryobiology* **2015**, *71*, 146–150. [CrossRef] [PubMed]
8. Tirgan, M. Massive ear keloids: Natural history, evaluation of risk factors and recommendation for preventive measures—A retrospective case series. *F1000Research* **2016**, *5*, 2517. [CrossRef] [PubMed]
9. Weshahy, A.H. Intralesional cryosurgery. A new technique using cryoneedles. *J. Dermatol. Surg. Oncol.* **1993**, *19*, 123–126. [CrossRef] [PubMed]
10. Collier, I.E.; Wilhelm, S.M.; Eisen, A.Z.; Marmer, B.L.; Grant, G.A.; Seltzer, J.L.; Kronberger, A.H.E.C.; He, C.S.; Bauer, E.A.; Goldberg, G.I. H-ras oncogene-transformed human bronchial epithelial cells (TBE-1) secrete a single metalloprotease capable of degrading basement membrane collagen. *J. Biol. Chem.* **1988**, *263*, 6579–6587. [PubMed]
11. Senior, R.M.; Griffin, G.L.; Fliszar, C.J.; Shapiro, S.D.; Goldberg, G.I.; Welgus, H.G. Human 92- and 72-kilodalton type IV collagenases are elastases. *J. Biol. Chem.* **1991**, *266*, 7870–7875. [PubMed]
12. Opdenakker, G.; Van den Steen, P.E.; Van Damme, J. Gelatinase B: A tuner and amplifier of immune functions. *Trends Immunol.* **2001**, *22*, 571–579. [CrossRef]
13. Schonbeck, U.; Mach, F.; Libby, P. Generation of biologically active IL-1β by matrix metalloproteinases: A novel caspase-1-independent pathway of IL-1 beta processing. *J. Immunol.* **1998**, *161*, 3340–3346. [PubMed]
14. Van den Steen, P.E.; Proost, P.; Wuyts, A.; Van Damme, J.; Opdenakker, G. Neutrophil gelatinase B potentiates interleukin-8 tenfold by aminoterminal processing, whereas it degrades CTAP-III, PF-4, and GRO-alpha and leaves RANTES and MCP-2 intact. *Blood* **2000**, *96*, 2673–2681. [PubMed]
15. Li, H.; Nahas, Z.; Feng, F.; Elisseeff, J.H.; Boahene, K. Tissue engineering for in vitro analysis of matrix metalloproteinases in the pathogenesis of keloid lesions. *JAMA Facial Plast. Surg.* **2013**, *15*, 448–456. [CrossRef] [PubMed]
16. Liu, Y.; Li, Y.; Li, N.; Teng, W.; Wang, M.; Zhang, Y.; Xiao, Z. TGF-beta1 promotes scar fibroblasts proliferation and transdifferentiation via up-regulating MicroRNA-21. *Sci. Rep.* **2016**, *6*, 32231. [CrossRef] [PubMed]
17. Dalkowski, A.; Fimmel, S.; Beutler, C.; Zouboulis, C.C. Cryotherapy modifies synthetic activity and differentiation of keloidal fibroblasts in vitro. *Exp. Dermatol.* **2003**, *12*, 673–681. [CrossRef] [PubMed]
18. Awad, S.M.; Ismail, S.A.; Sayed, D.S.; Refaiy, A.E.; Makboul, R. Suppression of transforming growth factor-β1 expression in keloids after cryosurgery. *Cryobiology* **2017**, *75*, 151–153. [CrossRef] [PubMed]
19. Har-Shai, Y.; Dujovny, E.; Rohde, E.; Zouboulis, C.C. Effect of skin surface temperature on skin pigmentation during contact and intralesional cryosurgery of keloids. *J. Eur. Acad. Dermatol. Venereol. JEADV* **2007**, *21*, 191–198. [CrossRef] [PubMed]
20. Har-Shai, Y.; Mettanes, I.; Zilberstein, Y.; Genin, O.; Spector, I.; Pines, M. Keloid histopathology after intralesional cryosurgery treatment. *J. Eur. Acad. Dermatol. Venereol. JEADV* **2011**, *25*, 1027–1036. [CrossRef] [PubMed]
21. Chong, Y.; Kim, C.W.; Kim, Y.S.; Chang, C.H.; Park, T.H. Complete excision of proliferating core in auricular keloids significantly reduces local recurrence: A prospective study. *J. Dermatol.* **2017**. Available online: http://dx.doi.org/10.1111/1346-8138.14110 (accessed on 30 October 2017).

International Journal of
Molecular Sciences

MDPI

Article

Novel Application of Cultured Epithelial Autografts (CEA) with Expanded Mesh Skin Grafting Over an Artificial Dermis or Dermal Wound Bed Preparation

Sadanori Akita [1,2,*], Kenji Hayashida [2,3], Hiroshi Yoshimoto [2], Masaki Fujioka [4],
Chikako Senju [2,4], Shin Morooka [4], Gozo Nishimura [5], Nobuhiko Mukae [6], Kazuo Kobayashi [7],
Kuniaki Anraku [8], Ryuichi Murakami [9], Akiyoshi Hirano [2], Masao Oishi [2], Shintaro Ikenoya [10],
Nobuyuki Amano [9], Hiroshi Nakagawa [7] and Nagasaki University plastic surgeons group

[1] Department of Plastic Surgery, Wound Repair and Regeneration, School of Medicine, Fukuoka University, Fukuoka 814-0180, Japan
[2] Department of Plastic and Reconstructive Surgery, Nagasaki University Graduate School of Biomedical Sciences, Nagasaki 852-8523, Japan; tokimayu122710130311@gmail.com (K.H.); hy671117@nagasaki-u.ac.jp (H.Y.); csenju@nagasaki-u.ac.jp (C.S.); akiyoshi@nagasaki-u.ac.jp (A.H.); moishi999@hotmail.com (M.O.)
[3] Division of Plastic and Reconstructive Surgery, Shimane University Hospital, Shimane 693-0021, Japan
[4] Department of Plastic and Reconstructive Surgery, National Hospital Organization Nagasaki Medical Center, Nagasaki 856-8562, Japan; mfujioka@nagasaki-mc.com (M.F.); moroshin.1760mm@hotmail.co.jp (S.M.)
[5] Department of Plastic and Reconstructive Surgery, Fukuoka Tokushukai Hospital, Fukuoka 816-0864, Japan; irs.mggg@nifty.com
[6] Department of Plastic and Reconstructive Surgery, Kitakyushu General Hospital, Kitakyushu 802-8517, Japan; mojapon1220@icloud.com
[7] Department of Plastic and Reconstructive Surgery, Ehime Prefectural Central Hospital, Ehime 790-0024, Japan; kobak@silver.plala.or.jp (K.K.); hnakagawa@diary.ocn.ne.jp (H.N.)
[8] Department of Plastic and Reconstructive Surgery, Sasebo City General Hospital, Sasebo 857-0056, Japan; anraku7@yahoo.co.jp
[9] Department of Plastic and Reconstructive Surgery, Yamaguchi Prefectural Grand Medical Center, Osaki 747-8511, Japan; rychi@ymghp.jp (R.M.); amanonobu1975@gmail.com (N.A.)
[10] Department of Plastic and Reconstructive Surgery, Matsue Red Cross Hospital, Matsue 690-8506, Japan; ikesin09201225@yahoo.co.jp
* Correspondence: akitas@hf.rim.or.jp; Tel.: +81-92-866-8250

Received: 8 November 2017; Accepted: 22 December 2017; Published: 25 December 2017

Abstract: Cultured epithelial autografts (CEA) with highly expanded mesh skin grafts were used for extensive adult burns covering more than 30% of the total body surface area. A prospective study on eight patients assessed subjective and objective findings up to a 12-month follow-up. The results of wound healing for over 1:6 mesh plus CEA, gap 1:6 mesh plus CEA, and 1:3 mesh were compared at 3, 6, and 12 months using extensibility, viscoelasticity, color, and transepidermal water loss by a generalized estimating equation (GEE) or generalized linear mixed model (GLMM). No significant differences were observed among the paired treatments at any time point. At 6 and 12 months, over 1:6 mesh plus CEA achieved significantly better expert evaluation scores by the Vancouver and Manchester Scar Scales ($p < 0.01$). Extended skin grafting plus CEA minimizes donor resources and the quality of scars is equal or similar to that with conventional low extended mesh slit-thickness skin grafting such as 1:3 mesh. A longitudinal analysis of scars may further clarify the molecular changes of scar formation and pathogenesis.

Keywords: split-thickness skin grafting; cultured epithelial autografts (CEA); assessment of scar quality; generalized estimating equation (GEE); generalized linear mixed model (GLMM)

1. Introduction

Scars and scar formation have been extensively investigated on a molecular basis; however, clinical scars are of many different characteristics and there are lots of variations. Additionally, it is very difficult for experimental models to mimic clinical features, because many animals do not exhibit hypertrophic scars or keloid in natural healing patterns. Thus, careful understanding of human scars and development of novel therapeutic modalities in humans are essential. Multiple and various treatments are required for extensive burns including burn resuscitation, cardiovascular, respiratory, and renal support, nutritional support, infection control, pain control, and surgical resurfacing and reconstruction. These procedures and interdisciplinary approaches improve survival rates [1,2]. Treatments using cultured epithelial autografts (CEA) were used clinically for the first time in the 1970s and early 1980s [3]. CEA are now more widely used in the treatment of extensive burn wounds in burn care specialist centers [4,5].

Commercially processed CEA developed in Japan (J-TEC Autologous Cultured Epidermis, JACE®, Japan Tissue Engineering CO., Ltd. (J-TEC), Gamagori, Aichi, Japan) were accepted by the health insurance reimbursement policy for the treatment of severe burns covering more than 30% of the total body surface area (TBSA) [6]. After 20 years of CEA development, JACE® was produced and is a Green-type autologous cultured epidermis similar to EPICEL in the United States [7]. However, the manufacturing process for JACE® differs from that for EPICEL. In a previous study, JACE® was implanted with an artificial dermis in order to reconstruct the dermis and grafted with JACE® on meshed 6:1 split thickness autografts [8]. Although the outcomes achieved were very good and acceptable, further detailed analysis of autograft site was not described.

A few quantitative studies have been conducted on clinical changes in burn scars and longitudinal burn scar quantification according to prospective quantified clinical characteristics of patient-matching, after burn hypertrophic scar (HSc), donor site scar (D) and normal skin (N) using these instruments and each investigator measured three sites (HSc, D, N) in 46 burn survivors 3, 6, and 12 months after burns [9].

Expanded skin grafting has recently been combined with CEA for extensive burns and is beneficial for wound coverage in multicenter surveillance; however, the quality of wounds and the rationale for using CEA with split-thickness skin grafting currently remain unknown [6].

Therefore, we assessed longitudinal burn scar qualification of expanded split-thickness skin grafting using clinical and objective methods up to a 12-month follow-up.

2. Results

There was no problem in donor site of CEA wound healing in all cases.

2.1. Vancouver and Manchester Scar Scales at 6 and 12 Months

2.1.1. Vancouver Scar Scale (VSS)

At 6 and 12 months, the total value of the modified Vancouver Scar Scale (VSS) was significantly lower in over 1:6 mesh plus CEA (JACE®) than that in gap 1:6 mesh plus CEA (JACE®) or 1:3 mesh (6.6 ± 1.79, and 4.1 ± 1.35; 5.8 ± 1.41 and 5.6 ± 1.00 at 6 months; and 2.6 ± 1.04 and 5.1 ± 0.60 at 12 months, respectively, $p < 0.01$) (Table 1).

The parameters of pigmentation, pliability, and vascularity in gap 1:6 mesh plus CEA, over 1:6 mesh plus CEA, and 1:3 mesh were 2.2 ± 0.70, 1.3 ± 0.50, 2.1 ± 0.50; 2.6 ± 0.70, 2.0 ± 0.60, 2.4 ± 0.60; 1.8 ± 0.50, 0.8 ± 0.40, 1.3 ± 0.43, at 6 months and 2.3 ± 0.30, 0.7 ± 0.40, 2.2 ± 0.40; 2.0 ± 0.70, 1.3 ± 0.40, 2.1 ± 0.50; 1.3 ± 0.40, 0.5 ± 0.30, 0.8 ± 0.30 at 12 months, respectively.

2.1.2. Manchester Scar Scale (MSS)

At 6 and 12 months, the total value for the modified Manchester Scar Scale (MSS) was significantly lower in over 1:6 mesh plus CEA (JACE®) than in gap 1:6 mesh plus CEA (JACE®) or 1:3 mesh

(12.7 ± 2.20 and 8.7 ± 1.20; 10.9 ± 1.80 and 10.7 ± 1.90 at 6 months; and 7.3 ± 1.20 and 10.2 ± 1.10 at 12 months respectively, $p < 0.01$) (Table 1).

Table 1. The Vancouver Scar Scale (VSS) and the Manchester Scar Scale (MSS), at 6 and 12 months among gap 1:6 mesh plus CEA, over 1:6 mesh plus cultured epithelial autografts (CEA), and 1:3 mesh.

Graft Type	6 Months		12 Months	
	VSS (Mean ± SE)	MSS (Mean ± SE)	VSS (Mean ± SE)	MSS (Mean ± SE)
1:6 gap + CEA	6.6 (0.60)	12.7 (0.81)	5.6 (041)	10.7 (0.70)
1:6 over + CEA	4.1 (0.50) **	8.7 (0.40) **	2.6 (0.40) **	7.3 (0.41) **
1:3 skin grafting	5.8 (0.50)	10.9 (0.61)	5.1 (0.20)	10.2 (0.40)

** $p < 0.01$.

The parameter of color, matte/shiny, contour, distortion or VAS (visual analogue scale) was 2.7 ± 0.30, 1.8 ± 0.30, 2.3 ± 0.38; 1.5 ± 0.30, 1.3 ± 0.10, 1.3 ± 0.20; 1.7 ± 0.40, 1.1 ± 0.20, 1.5 ± 0.30; 2.1 ± 0.40, 1.6 ± 0.30, 2.0 ± 0.36; 4.8 ± 1.00, 3.0 ± 0.60, 3.7 ± 0.78, gap 1:6 mesh plus CEA, over 1:6 mesh plus CEA, and 1:3 mesh, at 6 months, and 2.4 ± 0.30, 1.5 ± 0.30, 2.2 ± 0.20; 1.1 ± 0.10, 1.2 ± 0.10, 1.3 ± 0.20; 1.3 ± 0.30, 1.0 ± 0.00, 1.2 ± 0.20; 1.9 ± 0.30, 1.3 ± 0.20, 1.9 ± 0.20; 4.0 ± 1.10, 2.3 ± 0.60, 3.6 ± 0.60, gap 1:6 mesh plus CEA, over 1:6 mesh plus CEA, and 1:3 mesh, at 12 months, respectively.

2.2. Longitudinal Data from a Cutometer, Mexameter, Moisture Meter, and Color Meter (Tables 2–4)

2.2.1. Cutometer

Maximal extensibility (R0) was analyzed by Model 4 and viscoelasticity (R7) by Model 2. No significant differences were observed in R0 or R7 among the over 1:6 mesh plus CEA, gap 1:6 mesh plus CEA, and 1:3 mesh. R0 increased with time, whereas R7 decreased from 3 to 12 months in all treatments.

2.2.2. Mexameter

The melanin index was analyzed by Model 4 and the hemoglobin index by Model 2. No significant differences were observed in either index among the over 1:6 mesh plus CEA, gap 1:6 mesh plus CEA, and 1:3 mesh. The melanin index increased with time in all treatments. The hemoglobin index in the over 1:6 mesh plus CEA increased, while gap 1:6 mesh plus CEA and 1:3 mesh decreased from 3 to 12 months in all treatments.

2.2.3. Moisture Meter

Transepidermal Water Loss (TEWL) was analyzed by Model 2. No significant differences were observed in TEWL among the over 1:6 mesh plus CEA, gap 1:6 mesh plus CEA, and 1:3 mesh. TEWL decreased from 3 to 12 months in all treatments.

2.2.4. Color Meter

Clarity, yellow, and red were analyzed by Models 4, 2, and 2, respectively. Clarity and yellow increased from 3 to 12 months in all treatments. Red in the gap 1:6 mesh plus CEA increased, while red in the over 1:6 mesh plus CEA and 1:3 mesh decreased in a time-dependent manner.

Table 2. Models for generalized estimating equation (GEE) and generalized linear mixed model (GLMM) analyses.

	Linearity	Intercept	Slope	Selected Statistics
Model 1	non-linear	fixed	fixed	GEE
Model 2	linear	fixed	fixed	GEE
Model 3	linear	random	random	GLMM
Model 4	linear	random	fixed	GLMM

Table 3. Value of each parameter (3, 6 and 12 months).

Parameter	Over 1:6 Mesh Plus CEA (Mean ± SE)	Gap 1:6 Mesh Plus CEA (Mean ± SE)	1:3 Mesh (Mean ± SE)
3 Months			
Maximal extensibility (R0)	67.0 (±13.52)	62.6 (±8.48)	69.7 (±5.58)
Viscoelasticity (R7)	126.6 (±18.67)	93.7 (±13.06)	108.2 (±14.87)
Melanin Index	97.9 (±12.24)	118.1 (±15.80)	123.4 (±35.54)
Hemoglobin Index	133.2 (±11.89)	155.2 (±13.68)	144.7 (±14.16)
Transepidermal Water Loss (TEWL)	177.7 (±49.34)	160.3 (±32.44)	191.9 (±50.31)
Clarity	84.9 (±4.01)	79.9 (±3.21)	85.9 (±3.23)
Red	172.2 (±21.65)	181.7 (±19.87)	179.7 (±25.21)
Yellow	72.7 (±4.30)	66.9 (±3.99)	75.9 (±5.60)
6 Months			
Maximal extensibility (R0)	70.1 (±11.06)	67.5 (±9.48)	75.3 (±6.00)
Viscoelasticity (R7)	118.0 (±16.20)	87.3 (±11.23)	95.1 (±12.53)
Melanin Index	107.1 (±16.25)	121.2 (±17.18)	124.8 (±30.55)
Hemoglobin Index	135.2 (±12.38)	150.2 (±13.73)	141.0 (±13.83)
Transepidermal Water Loss (TEWL)	158.4 (±37.52)	155.1 (±20.30)	170.5 (±40.77)
Clarity	86.8 (±3.74)	82.7 (±3.15)	87.8 (±2.82)
Red	161.6 (±20.63)	182.8 (±17.61)	166.6 (±24.01)
Yellow	77.2 (±4.98)	70.6 (±3.73)	80.4 (±5.22)
12 Months			
Maximal extensibility (R0)	76.2 (±10.66)	77.5 (±15.15)	86.4 (±8.11)
Viscoelasticity (R7)	100.7 (±15.78)	74.4 (±9.54)	68.7 (±9.08)
Melanin Index	125.6 (±26.41)	127.4 (±21.68)	127.7 (±24.02)
Hemoglobin Index	139.1 (±15.74)	140.1 (±20.33)	133.6 (±15.99)
Transepidermal Water Loss (TEWL)	120.0 (±18.52)	144.7 (±33.15)	127.7 (±25.01)
Clarity	90.6 (±3.58)	88.2 (±3.35)	91.8 (±2.38)
Red	140.3 (±20.85)	185.1 (±22.83)	140.3 (±23.32)
Yellow	86.3 (±6.61)	77.9 (±4.16)	89.2 (±5.90)

Table 4. Model selection and statistical analysis.

	Selected Model	Comparision	Statistics among Groups
Maximal extensibility (R0) by a cutometer	Model 4	Over 1:6 vs. Gap 1:6	n.s. $p = 0.7446$
		Over 1:6 vs. 1:3	n.s. $p = 0.9919$
		Gap 1:6 vs. 1:3	n.s. $p = 0.5604$
Viscoelasticity (R7) by a cutometer	Model 2	Over 1:6 vs. Gap 1:6	n.s. $p = 0.2059$
		Over 1:6 vs. 1:3	n.s. $p = 0.6296$
		Gap 1:6 vs. 1:3	n.s. $p = 0.3683$
Melanin index by a mexameter	Model 4	Over 1:6 vs. Gap 1:6	n.s. $p = 0.1653$
		Over 1:6 vs. 1:3	n.s. $p = 0.4409$
		Gap 1:6 vs. 1:3	n.s. $p = 0.8756$
Hemoglobin index by a mexameter	Model 2	Over 1:6 vs. Gap 1:6	n.s. $p = 0.1670$
		Over 1:6 vs. 1:3	n.s. $p = 0.3930$
		Gap 1:6 vs. 1:3	n.s. $p = 0.6004$

Table 4. *Cont.*

	Selected Model	Comparision	Statistics among Groups
Transepidermal Water Loss (TEWL) by a moisture meter	Model 2	Over 1:6 vs. Gap 1:6 Over 1:6 vs. 1:3 Gap 1:6 vs. 1:3	n.s. $p = 0.6926$ n.s. $p = 0.8510$ n.s. $p = 0.5432$
Clarity by a color meter	Model 4	Over 1:6 vs. Gap 1:6 Over 1:6 vs. 1:3 Gap 1:6 vs. 1:3	n.s. $p = 0.3013$ n.s. $p = 0.8910$ n.s. $p = 0.1992$
Red by a color meter	Model 2	Over 1:6 vs. Gap 1:6 Over 1:6 vs. 1:3 Gap 1:6 vs. 1:3	n.s. $p = 0.9463$ n.s. $p = 0.7801$ n.s. $p = 0.7376$
Yellow by color meter	Model 2	Over 1:6 vs. Gap 1:6 Over 1:6 vs. 1:3 Gap 1:6 vs. 1:3	n.s. $p = 0.4162$ n.s. $p = 0.6559$ n.s. $p = 0.3040$

Over 1:6, over 1:6 mesh plus CEA; gap 1:6, gap 1:6 mesh plus CEA, 1:3, 1:3 mesh; n.s.: not significant.

3. Discussion

CEA for the wound coverage of extensive burns are useful for life-saving [2–4], and recent advances in bioengineered matrices have not only contributed to wound bed preparation, but also facilitated the expansion of coverage [5]. In Japan, the commercially available and reimbursed CEA was reported and followed Green-type CEA [7], and their indications and usefulness were discussed [6]. When a burn wound bed is prepared with an artificial dermis or with dermis remained wound beds, highly expanded (1:6 ratio) skin grafting with CEA may achieve similar healing to that with usual ratio skin grafting, such as a 1:3 mesh [8]. Too expanded mesh skin grafting such 1:6 may result in the irregularity or discoloration due to lack of skin component and thus scar formation, especially in the gap. However, the quality of wound healing, scars, and quality of life of patients, including the range of motion, texture, color, or function of reconstruction sites, have not yet been investigated. In the present study, the criteria, particularly of total burn surface areas, strictly followed the reimbursement category of TBSA of 30% or more, resulting in fewer cases being enrolled and precise each patient's body surface is calculated and the surface area used with CEA was also achieved. In Japan, due to significant decreases in heavy industry production or maybe effective safety systems in industry or the environment, the number of patients with extensive burns has decreased. More than 90% of burn victims have total burn surface areas of less than 30% (the registry system of the Japanese Society for Burn Injuries), and, thus, this multi-center study only examined eight patients over a three-year period. Although one patient died due to hepatic cancer after eight months, the remaining seven patients were analyzed for scar and wound healing qualities using objective analyses with a cutometer, mexameter, moisture meter, and color meter in addition to expert evaluations with VSS and MSS at 3, 6, and 12 months. Except severe scars, the cutometer, mexameter and color reliability were acceptable. Concurrent validity correlations with the modified VSS (mVSS) were significant except for the comparison of the mVSS pliability subscale and the cutometer maximum deformation measure comparison in severe scar [10]. Longitudinal wound evaluation by a mixed model regression analysis demonstrated redder at 3 and 6 months but normalized by 12 months [9]. We adopted GEE and GLMM in objective data analyses. When data for each subject was considered to be correlated, and correlated structure is incorporated into GEE analysis, variation for each subject can be analyzed by a linear mixed model, and this is generalized as GLMM [11–13].

All data showed no significant differences when pairing among the over 1:6 mesh plus CEA, gap 1:6 mesh plus CEA or 1:3 mesh at all time points. This may explain why an increased ratio of 1:6 mesh plus CEA can bring about equal or similar results to conventional 1:3 meshes. With extensive burns, the donor site is limited and, thus, a smaller donor surface with greater expansion may reduce morbidity and be more economically resourceful [14]. Expert evaluations at 6 and 12 months demonstrated significantly lower VSS and MSS, suggesting some interactions between expanded skin grafts and CEA that improved burn scar qualities.

A prospective randomized multicenter intra-patient comparative study showed significantly increased epithelization, patient and observer scores, and some objective analysis by Wilcoxon's non-parametric analysis at 3 and 12 months [15].

The usefulness of combining highly expanded mesh split skin grafting with CEA may be supported by studies involving larger subject numbers and appropriate statistical analyses. Furthermore, inter-racial and cultural backgrounds may strongly influence the outcomes achieved. These detailed clinical findings may clarify the longitudinal molecular pathogenesis and healing mechanisms.

4. Materials and Methods

Between 10 September 2012 and 31 March 2015, a prospective multicenter clinical study on longitudinal extensive burn scar assessments was approved by the Internal Review Board (#12090390, 10 September 2012) in full accordance with the Declaration of Helsinki. After a verbal explanation to each patient and their family, written informed consent was obtained at enrollment to this study. Patients with severe extensive burns greater than 30% of the total burn surface area of second degree or third degree burns, age older than 20 years, and survival at least 6 months post-operatively were included. Ten patients were initially recruited; however, due to limitations and decreases in the number of severe burn patients, 8 were ultimately enrolled and analyzed.

Patients were aged between 46 and 70 years, with a mean of 52.4 years, and comprised 7 males and one female. The total burn surface area was between 40% and 50%, with a mean of 44.9 ± 4.49%, the prognostic burn index was between 56 and 114, with a mean of 88.1 ± 17.02. Two patients did not receive dermal template temporal coverage until CEA, JACE® was developed, 5 were treated with Integra® (Century Medical, Tokyo, Japan), and one was treated with Terudermis (Olympus Terumo Biomaterials, Tokyo, Japan) (Tables 5 and 6). The CEA sized 80 cm^2 and each patient body surface was calculated by the Japanese body surface equation [16]. The CEA was applied 4.8 to 12.1% of patient's body surface (Table 6).

Table 5. Criteria for patients to participate in the present study. TBSA: total body surface area.

Inclusion Criteria
20 years of age or older
Acute full-thickness burn wounds that require widely meshed skin grafting
Minimal TBSA of 30% with full thickness wounds
Minimal study wound area of 100 cm^2
Maximal study wound area of 300 cm^2
Informed consent
Exclusion criteria
Immunocompromised patients or immunosuppressed physical conditions
Non-compliance by the patient, judged by medical experts
Active infected wounds
Known drug allergy

Int. J. Mol. Sci. **2018**, *19*, 57

Table 6. Patient demographics. PBI: prognostic burn index. wks: weeks.

Case	Age	Sex	Burn Depth and Percent	TBSA	PBI	How to Reconstruct "Dermis"	Body Surface Area (m²)	Frequency of Grafts	Number of Sheets	Total CEA Covered BSA (%)	Percent Epithelialization at Four Weeks (%)	Prognosis
1	56	M	DDB 5% DB 40%	45	98	Terudermis	1.441	2	20 + 20	22.2	100	survive
2	59	M	DDB 30% DB 10%	40	84	none	1.6705	1	7	3.4	100	survive
3	31	M	DDB 30% DB 10%	40	56	none	1.8623	1	20	8.6	100	survive
4	41	M	DDB 10% DB 40%	50	86	Integra®	1.9984	1	32 + 12	17.1	50	survive
5	63	M	DDB 20% DB 25%	45	98	Integra®	1.5958	2	20 + 20	20.1	100	survive
6	53	F	DDB 35% DB 5%	40	78	Integra®	1.42	1	22	12.4	95	survive
7	70	M	DDB 10% DB 39%	49	114	Integra®	1.7985	2	24 + 11	15.6	50	deseased at 8 months
8	46	M	DDB 5% DB 45%	50	91	Integra®	1.98	2	30 + 30	24.2	95	survive

One patient died 8 months post-operatively due to hepatic cancer, while the remaining 7 patients were followed up until 12 months.

Intermediate split-thickness skin autografts, 15/1000-inch, were harvested with an electric dermatome and mesher (Zimmer Biomet, Dover, OH, USA) and expanded to a ratio of 1:3 or 1:6. The 1:3 mesh is frequently used clinically and heals in approximately 4 days, whereas the 1:6 mesh takes more than 17 days to heal [13], which is covered by CEA (JACE®) to simulate wound healing by factors from CEA or through an interaction between CEA and mesh skin grafting (Figure 1). It takes 3 to 4 weeks before CEA is ready for clinical application and harvesting of the "donor" is 3–4 cm in length and 1–2 cm in width in full thickness skin distant from the burns.

Figure 1. Schematic cross-section of 1:3 mesh without CEA (JACE®), (**A**) and 1:6 mesh plus CEA (JACE®) coverage, (**B**).

Patients underwent assessments 6 and 12 months after healing using VSS, which evaluates height, pliability, vascularity, and pigmentation on a scale of 0 to 13 [17], and MSS [18] and VSS were evaluated in a blind manner among 12 Nagasaki University plastic surgeons, who are all board certified, and independently by photographs. In VSS, "height" and "texture" in MSS were deleted because they cannot be assessed in a 2-dimensional analysis.

The following parameters were repeatedly and longitudinally evaluated in patients 3, 6, and 12 months after healing.

4.1. Moisture Meter

A compact and portable moisture meter was used (ASA-M2; Asahi Biomed, Yokohama, Japan). It only weighed 250 g (200 g for the power supply and 50 g for the hand piece) and had the ability to detect TEWL using an effective contact coefficient [19], water quantity, and the thickness of the stratum corneum skin layer by formulating the value of the effective contact coefficient, evaluated by electrolytes in the stratum corneum layer. The principle of this machine is to record and analyze conduction susceptibility using a low-frequency (160 Hz) alternate currency and to detect conduction admittance using a high frequency (143 kHz) alternate current.

The proposed formula is as follows:

$$\text{Skin conductance } (\mu c) = \text{Effective contact coefficient } (\%) \times \text{Quantity of water } (\mu S).$$

Low and high frequency electric voltages were added to effectively enable these formula factors. The round probe of the hand piece was 5 mm in diameter and detection was set to 5 s after probe contact with the participant in order to stabilize the electrodes and skin condition. The measurement of each contact point was always perpendicular to the participant, thereby avoiding unnecessary pressure or loading; it was repeated 5 times. The mean value of 3 adjacent points at least 10 mm apart and 20 mm from the edge of intact skin was assessed following the manufacturer's instructions.

Measurements were performed by 2 of the authors who are very familiar with this system. All data were immediately transferred to a personal computer for further analyses.

In written informed consent, there was a description of data collection for all patients, and no complications occurred as a result of moisture meter measurements.

4.2. Color Meter

A color meter was used to assess scar clarity (L): "a" is red in color when the value is positive and green when it is negative, and "b" is yellow in color when the value is positive and blue when it is negative. The hand-held color meter weighed 420 g, including batteries, for the main body of the system and 110 g for the hand-piece probe and color analyzer (NF-333; Nippon Denshoku, Osaka, Japan). The light source was a multicolored LED. All data were easily transferred to Microsoft Excel 2013 files on a laptop computer via a data connector, and the differentials of each polarized color criterion parameter (L, a, and b) were standardized with the surrounding intact skin. The delta ratio of each parameter was then compared and statistically analyzed. The measurement of each point was always perpendicular to the scar and was repeated 5 times immediately after touching the scar surface. The mean value of 3 adjacent points at least 8 mm apart and 12 mm from the edge of intact skin was assessed in a single room. The accuracy of this system is traceable to the standard of the National Institute of Standards and Technology, Gaithersburg, MD. The accuracy of this function is based on the choice of optical filters, which are assessed by an optimization criterion using a combination of methodologies from differential geometry with a statistical error analysis. The magnitude of errors associated with the optimal filters was previously reported to be typically half that for typical RGB filters in a 3-parameter model of human skin coloration [20]. A relatively simple and easy-to-use skin chromatometer was employed to assess temporal changes after skin grafting, with multiple relevant factors such as age, the type of skin grafting, anatomical differences at the donor site or recipient site, and the Fitzpatrick skin type [21].

4.3. Cutometer

The cutometer MPA 580 (Courage + Khazaka Electronic GmbH, Cologne, Germany) was used to evaluate skin elasticity parameters 1 year after complete wound healing. The cutometer has the ability to measure the vertical deformation of skin by suctioning into a round probe that is 6 mm in diameter. A vacuum load of 500 mbar was used over the skin (or scar) surface for 1 s, followed by normal pressure for 1 s. Each measurement was repeated 3 times, and the mean value of 4 adjacent points at least 6 mm apart and 12 mm from intact skin was assessed. As discussed and reported previously, 2 parameters of the cutometer were used in the present study. Uf (depicted as R0) represents the maximal skin (or scar) extension of the deformation at the end of the vacuum period. Ur/Uf (R7) represents the ratio of retraction (Ur) to maximal extension (Uf) and reflects the elasticity of the measured skin [22,23]. These measurements were performed by 3 authors at 25 °C and 50% humidity with air conditioning to standardize patient conditions at the completion of wound healing and after 3 and 6 months.

4.4. Mexameter

The mexameter (MX18, Courage & Khazaka Electronic GmbH, Köln, Germany) quantifies scar erythema and melanin based on the tissue's narrow wavelength light absorption. The probe has 16 light-emitting diodes that send 3 defined wavelengths of light (568, 660, and 880 nm). A receiver then measures the light reflected by the skin. Since the quantity of emitted light is known, the absorption rate of defined wavelengths may be ascertained, which are selectively absorbed by melanin (660 nm) pigments or hemoglobin (568 nm). The measurement area is 5 mm in diameter, although the total surface contacted by the probe is 2 cm in diameter. Similar to the cutometer, the central portion of the mexameter is spring-mounted to maintain constant pressure on the skin. In each measurement, the probe was held perpendicular to the skin. It lightly touched the skin surface, without the outer ring making contact, activating the light emitter. The reflected light was measured by the receiver and the erythema and melanin index (range 1–1000) was immediately displayed on the console; therefore, the probe only remains in contact with the skin for several seconds [10].

Evaluator indices of VSS and MSS as well as objective measurements from the moisture meter, color meter, cutometer, and mexameter were obtained in the gap 1:6 mesh plus CEA (JACE®), over 1:6 mesh plus CEA (JACE®), and 1:3 mesh (Figure 2).

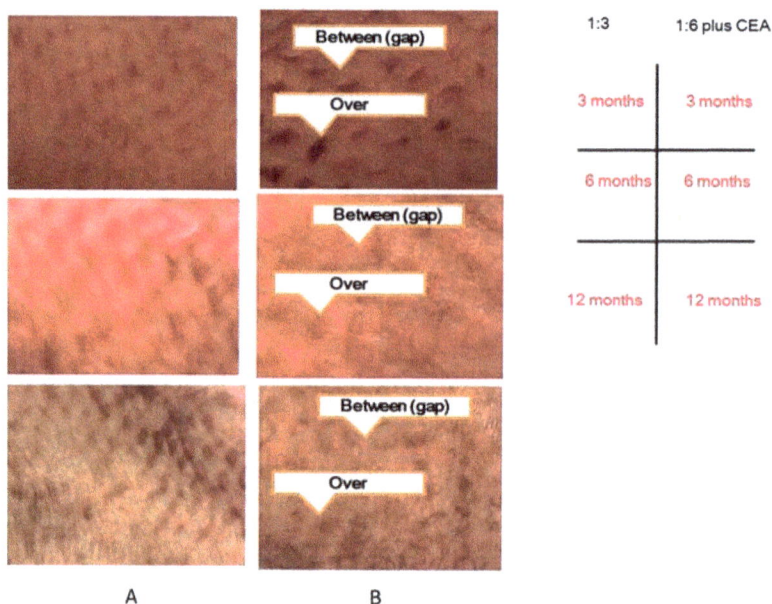

Figure 2. Measurement points of the gap 1:6 gap mesh plus CEA or over 1:6 mesh plus CEA (**A**) and 1:3 mesh (**B**).

4.5. Statistical Analysis

Regarding VSS and MSS, results are expressed as the mean ± Standard errors (SE). Data among groups were evaluated using an independent *t*-test. All tests were 2-tailed, and the significance of differences was defined as $p < 0.05$. All analyses were performed using IBM SPSS, Statistics, and version 21 (Japan IBM, Tokyo, Japan).

Since repetitive and longitudinal data were obtained, GEE or GLMM was used in statistical analyses for the moisture meter, color meter, cutometer, and mexameter. When data for each subject was considered to be correlated, and correlated structure is incorporated into GEE analysis, variation for each subject can be analyzed by a linear mixed model, and this is generalized as GLMM.

There were 4 models (Table 2) and data from each time point were expressed as the mean ± SE (Table 3). Data analyses were performed with SAS version 9.2 (SAS Institute Inc., Cary, NC, USA) (Table 4).

5. Conclusions

When a burn wound bed is prepared with an artificial dermis or with dermis remained wound beds, highly expanded (1:6 ratio) skin grafting with CEA may achieve similar healing to that with usual ratio skin grafting, such as a 1:3 mesh. At 6 and 12 months, over 1:6 mesh plus CEA achieved significantly better expert evaluation scores by the Vancouver and Manchester Scar Scales ($p < 0.01$). Extended skin grafting plus CEA minimizes donor resources and the quality of scars is equal or similar to that with conventional low extended mesh slit-thickness skin grafting such as 1:3 mesh.

Acknowledgments: Nagasaki University plastic surgeons' group consist of plastic surgeons in Nagasaki University, National Hospital Organization Nagasaki Medical Center, Fukuoka Tokushukai Medical Center, Kitakyushu General Hospital, Yamaguchi Prefectural Grand Medical Center, Ehime Prefectural Central Hospital, Sasebo City General Hospital and Matsue Red Cross Hospital participated in the project, Japan Tissue Engineering Co., Ltd. kindly supported and provided shipping cost of measuring devices between hospitals and opportunity of expert meetings. (12090390, Nagasaki University). The authors appreciate for Professor Katsumi Tanaka, department of plastic and reconstructive surgery, Nagasaki University encouragement in the project.

Author Contributions: Sadanori Akita conceived and designed the study; Sadanori Akita, Kenji Hayashida, Hiroshi Yoshimoto, Masaki Fujioka, Chikako Senju, Shin Morooka, Gozo Nishimura, Nobuhiko Mukae, Kazuo Kobayashi, Kuniaki Anraku, Ryuichi Murakami, Akiyoshi Hirano, Masao Oishi, Shintaro Ikenoya, Nobuyuki Amano and Hiroshi Nakagawa performed the study; Sadanori Akita and Kuniaki Anraku analyzed the data; Sadanori Akita wrote the paper.

Conflicts of Interest: The authors declare no conflict of interest.

Abbreviations

CEA	Cultured Epithelial Autografts
GEE	generalized estimating equation
GLMM	generalized linear mixed model
JACE®	Commercially processed CEA (J-TEC Autologous Cultured Epidermis, JACE®, Japan Tissue Engineering CO., Ltd. (J-TEC)

References

1. Thompson, P.; Herndon, D.N.; Abston, S.; Rutan, T. Effect of early excision on patients with major thermal injury. *J. Trauma* **1987**, *27*, 205–207. [CrossRef] [PubMed]
2. White, C.E.; Renz, E.M. Advances in surgical care: Management of severe burn injury. *Crit. Care Med.* **2008**, *36* (Suppl. S7), S318–S324. [CrossRef] [PubMed]
3. Hefton, J.M.; Madden, M.R.; Finkelstein, J.L.; Shires, G.T. Grafting of burn patients with allografts of cultured epidermal cells. *Lancet* **1983**, *2*, 428–430. [CrossRef]
4. Cirodde, A.; Leclerc, T.; Jault, P.; Duhamel, P.; Lataillade, J.J.; Bargues, L. Cultured epithelial autografts in massive burns: A single-center retrospective study with 63 patients. *Burns* **2011**, *37*, 964–972. [CrossRef] [PubMed]
5. Fang, T.; Lineaweaver, W.C.; Sailes, F.C.; Kisner, C.; Zhang, F. Clinical application of cultured epithelial autografts on acellular dermal matrices in the treatment of extended burn injuries. *Ann. Plast. Surg.* **2014**, *73*, 509–515. [CrossRef] [PubMed]
6. Matsumura, H.; Matsushima, A.; Ueyama, M.; Kumagai, N. Application of the cultured epidermal autograft "JACE®" for treatment of severe burns: Results of a 6-year multicenter surveillance in Japan. *Burns* **2016**, *42*, 769–776. [CrossRef] [PubMed]
7. Green, H.; Kehinde, O.; Thomas, J. Growth of cultured human epidermal cells into multiple epithelia suitable for grafting. *Proc. Natl. Acad. Sci. USA* **1979**, *76*, 5665–5668. [CrossRef] [PubMed]
8. Hayashi, M.; Muramatsu, H.; Nakano, M.; Ito, H.; Inoie, M.; Tomizuka, Y.; Inoue, M.; Yoshimoto, S. Experience of using cultured epithelial autografts for the extensive burn wounds in eight patients. *Ann. Plast. Surg.* **2014**, *73*, 25–29. [CrossRef] [PubMed]
9. Nedelec, B.; Correa, J.A.; de Oliveira, A.; LaSalle, L.; Perrault, I. Longitudinal burn scar quantification. *Burns* **2014**, *40*, 1504–1512. [CrossRef] [PubMed]
10. Nedelec, B.; Correa, J.A.; Rachelska, G.; Armour, A.; LaSalle, L. Quantitative measurement of hypertrophic scar: Interrater reliability and concurrent validity. *J. Burn Care Res.* **2008**, *29*, 501–511. [CrossRef] [PubMed]
11. Saris, D.B.; Vanlauwe, J.; Victor, J.; Almqvist, K.F.; Verdonk, R.; Bellemans, J.; Luyten, F.P. Treatment of symptomatic cartilage defects of the knee. *Am. J. Sports Med.* **2009**, *37* (Suppl. S1), 10S–19S. [CrossRef] [PubMed]

12. Lattka, E.; Rzehak, P.; Szabó, É.; Jakobik, V.; Weck, M.; Weyermann, M.; Grallert, H.; Rothenbacher, D.; Heinrich, J.; Brenner, H.; et al. Genetic variants in the *FADS* gene cluster are associated with arachidonic acid concentrations of human breast milk at 1.5 and 6 mo postpartum and influence the course of milk dodecanoic, tetracosenoic, and trans-9-octadecenoic acid concentrations over the duration of lactation. *Am. J. Clin. Nutr.* **2011**, *93*, 382–391. [PubMed]

13. Jacka, F.N.; Cherbuin, N.; Anstey, K.J.; Sachdev, P.; Butterworth, P. Western diet is associated with a smaller hippocampus: A longitudinal investigation. *BMC Med.* **2015**, *13*, 215. [CrossRef] [PubMed]

14. Vandeput, J.; Nelissen, M.; Tanner, J.C.; Boswick, J. A review of skin meshers. *Burns* **1995**, *21*, 364–370. [CrossRef]

15. Gardien, K.L.; Marck, R.E.; Bloemen, M.C.; Waaijman, T.; Gibbs, S.; Ulrich, M.M.; Middelkoop, E.; Dutch Outback Study Group. Outcome of Burns Treated With Autologous Cultured Proliferating Epidermal Cells: A Prospective Randomized Multicenter Intrapatient Comparative Trial. *Cell Transplant.* **2016**, *25*, 437–448. [CrossRef] [PubMed]

16. Fujimoto, S.; Watanabe, T.; Sakamoto, A.; Yukawa, K.; Morimoto, K. Studies on the Physical Surface Area of Japanese. *Jpn. J. Hyg.* **1968**, *23*, 443–456. [CrossRef]

17. Sullivan, T.; Smith, J.; Kermode, J.; McIver, E.; Courtemanche, D.J. Rating the burn scar. *J. Burn Care Rehabil.* **1990**, *11*, 256–260. [CrossRef] [PubMed]

18. Beausang, E.; Floyd, H.; Dunn, K.W.; Orton, C.I.; Ferguson, M.W. A new quantitative scale for clinical scar assessment. *Plast. Reconstr. Surg.* **1998**, *102*, 1954–1956. [CrossRef] [PubMed]

19. Tanaka, K.; Akita, S.; Yoshimoto, H.; Houbara, S.; Hirano, A. Lipid-colloid dressing shows improved reepithelialization, pain relief, and corneal barrier function in split-thickness skin-graft donor wound healing. *Int. J. Low Extrem. Wounds* **2014**, *13*, 220–225. [CrossRef] [PubMed]

20. Preece, S.J.; Claridge, E. Spectral filter optimization for the recovery of parameters which describe human skin. *IEEE Trans. Pattern Anal. Mach. Intell.* **2004**, *26*, 913–922. [CrossRef] [PubMed]

21. Kim, J.S.; Park, S.W.; Choi, T.H.; Kim, N.G.; Lee, K.S.; Kim, J.R.; Lee, S.I.; Kang, D.; Han, K.H.; Son, D.G.; et al. The evaluation of relevant factors influencing skin graft changes in color over time. *Dermatol. Surg.* **2008**, *34*, 32–39. [CrossRef] [PubMed]

22. Morita, A.; Kobayashi, K.; Isomura, I.; Tsuji, T.; Krutmann, J. Ultraviolet A1 (340–400 nm) phototherapy for scleroderma in systemic sclerosis. *J. Am. Acad. Dermatol.* **2000**, *43*, 670–674. [CrossRef] [PubMed]

23. Draaijers, L.J.; Botman, Y.A.; Tempelman, F.R.; Kreis, R.W.; Middelkoop, E.; van Zuijlen, P.P. Skin elasticity meter or subjective evaluation in scars: A reliability assessment. *Burns* **2004**, *30*, 109–114. [CrossRef] [PubMed]

International Journal of
Molecular Sciences

MDPI

Article

High-Mobility Group Box 1 Mediates Fibroblast Activity via RAGE-MAPK and NF-κB Signaling in Keloid Scar Formation

Jihee Kim [1], Jong-Chul Park [2], Mi Hee Lee [2], Chae Eun Yang [3], Ju Hee Lee [1,*] and Won Jai Lee [3,*]

[1]   Department of Dermatology & Cutaneous Biology Research Institute, Yonsei University College of Medicine, Seoul 03722, Korea; mygirljihee@yuhs.ac
[2]   Department of Medical Engineering, Yonsei University College of Medicine, Seoul 03722, Korea; parkjc@yuhs.ac (J.-C.P.); LEEMH1541@yuhs.ac (M.H.L.)
[3]   Department of Plastic and Reconstructive Surgery, Institute for Human Tissue Restoration, Yonsei University College of Medicine, Seoul 03722, Korea; CHENIYA@yuhs.ac
*   Correspondence: juhee@yuhs.ac (J.H.L.); pswjlee@yuhs.ac (W.J.L.); Tel.: +82-2-2228-2080 (J.H.L.); +82-2-2228-2215 (W.J.L.); Fax: +82-2-393-6947 (J.H.L. & W.J.L.)

Received: 11 December 2017; Accepted: 22 December 2017; Published: 28 December 2017

Abstract: Emerging studies have revealed the involvement of high-mobility group box 1 (HMGB1) in systemic fibrotic diseases, yet its role in the cutaneous scarring process has not yet been investigated. We hypothesized that HMGB1 may promote fibroblast activity to cause abnormal cutaneous scarring. In vitro wound healing assay with normal and keloid fibroblasts demonstrated that HMGB1 administration promoted the migration of both fibroblasts with increased speed and a greater traveling distance. Treatment of the HMGB1 inhibitor glycyrrhizic acid (GA) showed an opposing effect on both activities. To analyze the downstream mechanism, the protein levels of extracellular signal-regulated kinase (ERK) 1/2, protein kinase B (AKT), and nuclear factor kappa-light-chain-enhancer of activated B cells (NF-κB) were measured by western blot analysis. HMGB1 increased the expression levels of ERK1/2, AKT, and NF-κB compared to the control, which was suppressed by GA. HMGB1 promoted both normal and keloid fibroblasts migration to a degree equivalent to that achieved with TGF-β. We concluded that HMGB1 activates fibroblasts via the receptor for advanced glycation end product (RAGE)—mitogen-activated protein kinases (MAPK) and NF-κB interaction signaling pathways. Further knowledge of the relationship of HMGB1 with skin fibrosis may lead to a promising clinical approach to manage abnormal scarring.

Keywords: keloid; hypertrophic scar; fibroblast; HMGB1

---

1. Introduction

Keloids, or hypertrophic scars, are formed by the excessive accumulation of extracellular matrix (ECM) substrates caused by abnormal wound healing processes. Recently, numerous hypotheses regarding the cause of keloids have been proposed, including aberrant cellular responses to mechanical strains, upregulation of growth factors and inflammatory cytokines, the epithelial-mesenchymal transition (EMT), and regulation of cellular apoptosis [1–4]. Abnormal fibroblast activity is a histopathological hallmark for keloid pathogenesis, as it results in excessive synthesis of ECM components, especially collagen. Among the various signaling molecules involved in keloid pathogenesis, transforming growth factor-beta (TGF-β) is a crucial mediator. In keloid tissues and fibroblasts, proteins involved in TGF-β signal transduction are overexpressed. TGF-β promotes signaling via the Smad and Wnt-dependent pathways to enhance cell proliferation, resulting in excessive ECM production and deposition [5,6].

High-mobility group box 1 (HMGB1) is a member of the HMGB protein family, which was originally identified as a nuclear non-histone DNA-binding protein that acts as a critical co-factor of somatic cell transcriptional regulation [7]. It plays roles in a wide variety of processes, including inflammation, immune responses, apoptosis, and responses to injury [8,9]. As an intracellular transcription factor, HMGB1 modulates cellular transcription, recombination, replication, and repair [10]. Extracellular HMGB1 is released and stimulated by necrotic, damaged, or inflammatory cells after tissue injury and acts as a potent inflammatory cytokine. Extracellular HMGB1 also functions as a chemo-attractant; these functions are mediated by interactions with cell-surface receptors such as Toll-like receptors (TLRs) and the receptor for advanced glycation end product (RAGE). Exogenous HMGB1 also pairs with the chemokine C-X-C motif chemokine 12 (CXCL12), which binds to the CXCR4 chemokine receptor [11]. These receptors activate multiple intracellular signaling pathways, including mitogen-activated protein kinases (MAPKs), extracellular signal-regulated kinase (ERK) 1/2, and phosphatidylinositol-4,5-bisphosphate 3-kinase (PI3K)/protein kinase B (AKT), which are known to be related to wound healing processes. In keloid pathogenesis, activation of the PI3K/AKT signaling pathway is known to be responsible for stimulating collagen synthesis.

The release of HMGB1 after tissue injury and the subsequent onset of fibrosis has been reported in multiple organs. Emerging studies have revealed that HMGB1 is strongly associated with fibrotic diseases in the liver, renal, lung, and myocardial tissue, while the inhibition of HMGB1-related signaling pathways was shown to prevent fibrosis in experimental animal models [12–15]. During wound healing, HMGB1 is thought to influence the healing process as a pro-fibrotic element. Recent studies have demonstrated that HMGB1 binds to cell-surface receptors such as RAGE and TLR2/4 in keratinocytes and fibroblasts [16]. When administered to early embryonic murine skin, which normally heals without scarring, HMGB1 induced scarring and fibrosis [17]. However, the roles of HMGB1 in wound healing and associated cutaneous scarring are currently unclear.

Although the role of HMGB1 in wound healing has been investigated, no studies to date have evaluated its role in cutaneous scarring and keloid development. We hypothesized that HMGB1 may promote fibroblast activity, ultimately causing abnormal cutaneous scarring. Therefore, in this study, we investigated the motility and migration activity of fibroblasts in response to HMGB1 and its inhibitor glycyrrhizic acid (GA). Additionally, to study the underlying molecular mechanism upon HMGB1 activity, we analyzed the expression of RAGE and MAPK signaling pathway molecules.

2. Results

2.1. HMGB1 Administration Promoted Fibroblast Migration

Enhanced migration and invasion of normal fibroblasts are critical factors for the development of keloid diseases. Therefore, we investigated whether HMGB1 would affect the behavior of normal fibroblasts in vitro using a wound healing assay model. TGF-β was administered to compare its effect with HMGB1. The initial wound width of untreated normal fibroblasts was approximately 500 μm; after 36 h, migrating cells almost completely covered the denuded area. However, the wound area on the 50 ng/mL and 100 ng/mL HMGB1-treated and 10 ng/mL TGF-β-treated groups showed rapid closure compared to the untreated control group. Furthermore, more than half of the wound area in the untreated control group remained uncovered after 24-h incubation (Figure 1).

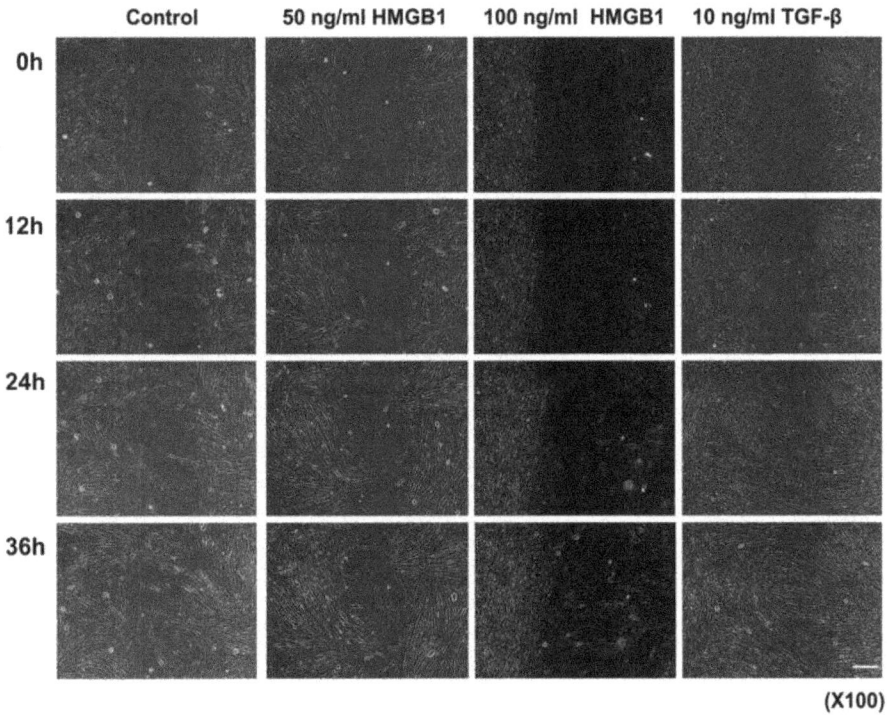

Figure 1. In vitro wound healing assay and cell-tracking system for cell migration. Scale bar: 50 μm. Recovery of the denuded space was completed after 36 h in normal fibroblasts. However, the wound area on the 50 ng/mL and 100 ng/mL High-mobility group box 1 (HMGB1)-treated and 10 ng/mL TGF-β-treated groups showed more rapid closure than the untreated control group.

2.2. HMGB1 Administration Increased the Migration Speed and Distance Traveled in Normal Fibroblasts

Cells from the control, HMGB1-treated (50 and 100 ng/mL), and TGF-β-treated (10 ng/mL) groups were selected from both sides of the wound, and the cell migration speed, distance, and directionality were analyzed (Figure 2A).

The average migration velocities of the 50 ng/mL and 100 ng/mL HMGB1-treated normal fibroblast were 26.31 ± 4.00 μm/h and 29.06 ± 4.33 μm/h, respectively. This was significantly higher than the speed of the untreated normal fibroblasts (19 ± 4.25 μm/h; $p < 0.05$; Figure 2B). The average migration speed of the TGF-β-treated group was higher than that of the control group (30.51 ± 2.78 μm/h). However, there was no significant difference between the HMGB1-treated group and the TGF-β-treated group (Figure 2B).

The total distances traveled by the 50 ng/mL and 100 ng/mL HMGB1-treated normal fibroblasts were 307.60 ± 47.45 μm and 338.70 ± 50.90 μm, respectively, which was significantly higher than that of the untreated group (220.60 ± 49.26 μm; $p < 0.05$; Figure 2C). Furthermore, the distance traveled by the 10 ng/mL TGF-β-treated group was greater than that of the untreated normal fibroblasts (355.70 ± 33.35 μm; $p < 0.05$), but there was no significant difference compared to the HMGB1-treated groups. The total distance traveled by the 50 ng/mL (297.80 ± 61.12 μm) and 100 ng/mL (285.30 ± 28.45 μm) HMGB1-treated groups was greater than that of the untreated group (219.20 ± 30.06 μm; $p < 0.05$; Figure 2C). In addition, the distance traveled by the 10 ng/mL TGF-β-treated group (323.20 ± 42.71 μm) was greater than that of the untreated group (Figure 2C).

Next, we investigated the directionality values of normal fibroblast cells in response to HMGB1 treatment. The directionality value indicates the potential of the cell to migrate closer to the midline when initiating vertically from the wound edge. The results for the control, 50 ng/mL, 100 ng/mL HMGB1-treated, and 10 ng/mL TGF-β-treated normal fibroblasts were 0.65 ± 0.08, 0.52 ± 0.06, 0.76 ± 0.09, and 0.69 ± 0.05, respectively. As shown in Figure 2D, only the 100 ng/mL HMGB1-treated group showed a significant increase in directionality compared to the control ($p < 0.05$). These results indicate that fibroblasts showed greater directionality toward the wound center, displayed higher migration velocities, and traveled longer distances after HMGB1 treatment. Similar results were noted in TGF-β-treated cells.

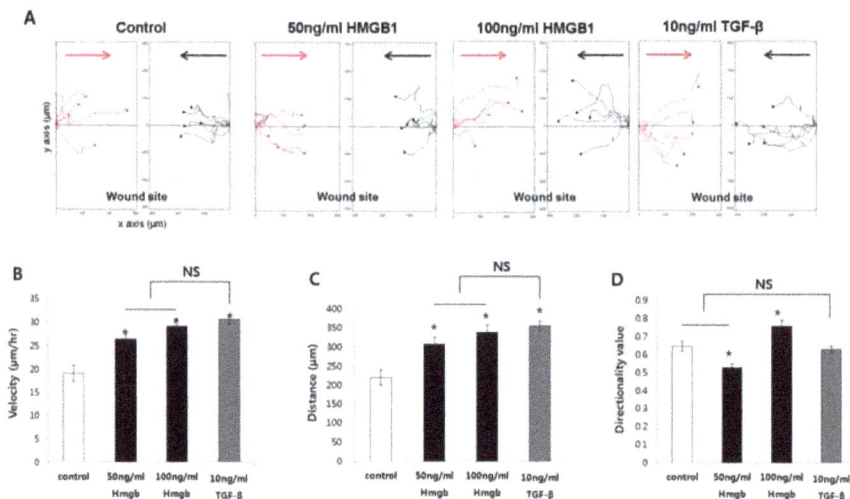

Figure 2. The cell tracking system for fibroblast migration was analyzed using cells selected from both sides of the wound, cells were tracked toward the site of wound insert after cultivation for 24 h; black, cells toward the midline from the left edge; red, cells toward the midline from the right edge (**A**); The average migration speed (**B**) and average accumulated distance (**C**) traveled after treatment with HMGB1 or TGF-β; (**D**) Directionality values in response to HMGB1 or TGF-β treatment. All results are shown as mean $\pm$ SD (* $p < 0.05$).

2.3. Treatment with Glycyrrhizic Acid (GA), an HMGB1 Inhibitor, Decreased the Migration Speed and Distance Traveled in Normal Fibroblasts

The above results demonstrated that HMGB1 induces an increase in the cell migration speed and distance of normal fibroblasts, as in TGF-β-treated cells. We added GA, a known inhibitor of HMGB1, to evaluate its effect on HMGB1-induced fibroblast migration. In all three groups treated with 200 μM GA, fibroblast migration was considerably reduced. Most of the area remained uncovered after 36-h incubation, although cells at the margin showed forward movement (Figure 3A).

The average migration speed of the 100 ng/mL HMGB1- and 10 ng/mL TGF-β-treated normal fibroblasts was 26.11 ± 3.27 μm/h and 29.99 ± 5.18 μm/h, respectively. The TGF-β-treated group showed significant increase compared to that of untreated normal fibroblasts (23.76 ± 2.92 μm/h; $p < 0.033$). Treatment with 100 μM GA did not significantly affect the fibroblast migration speed or distance. However, treatment with 200 μM GA induced a significant decrease in migration speed in both the HMGB1- and TGF-β-treated groups (12.84 ± 3.37 μm/h and 12.18 ± 2.23 μm/h, respectively; $p < 0.001$, Figure 3B).

The total distances traveled by the 100 ng/mL HMGB1- and 10 ng/mL TGF-β-treated normal fibroblasts were 312.41 ± 39.01 μm and 357.52 ± 61.78 μm, respectively, greater than that of the untreated group (283.23 ± 34.86 μm; $p < 0.05$). After treatment with 200 μM GA, the distance was significantly reduced in both the HMGB1- and TGF-β-treated groups (153.06 ± 40.21 μm and 145.19 ± 26.56 μm, respectively; $p < 0.001$, Figure 3C).

(X100)

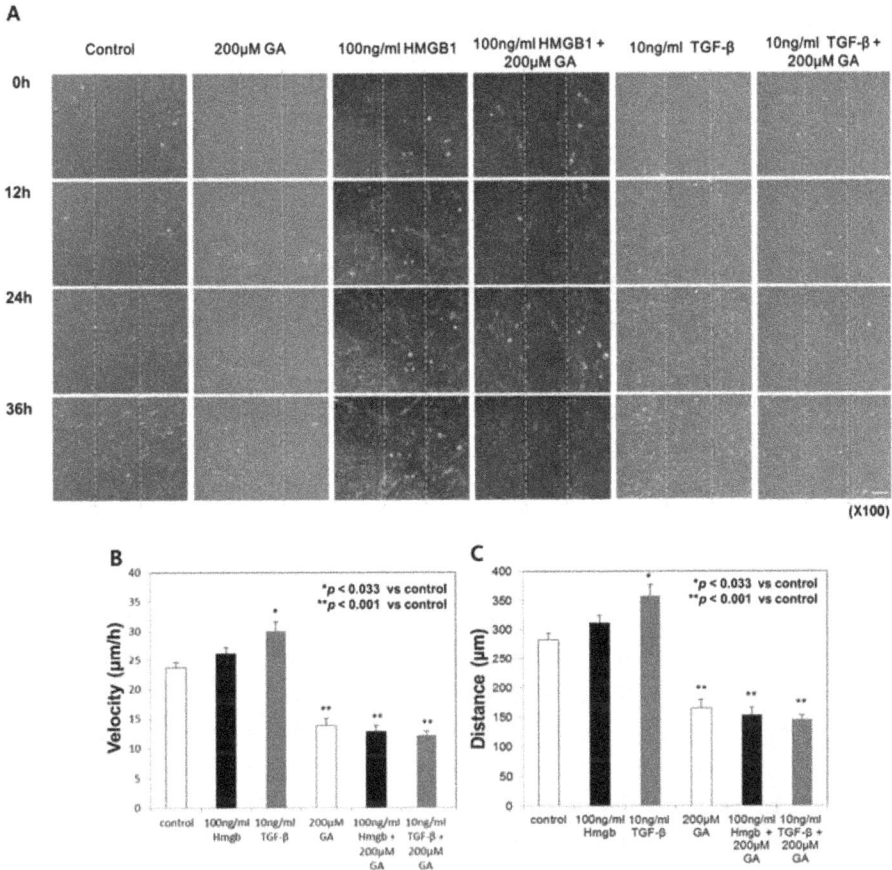

Figure 3. Fibroblast migration was delayed by glycyrrhizic acid (GA) treatment. Scale bar: 50 μm (**A**) GA induced a decrease in the average migration speed (**B**) and the average distance (**C**) traveled of normal fibroblasts. All results are shown as mean ± SD (* $p < 0.033$, ** $p < 0.001$).

2.4. GA Treatment Decreased the Migration Speed and Distance Traveled in Keloid Fibroblasts

In vitro migration assays were performed with keloid fibroblasts. Cell migration was initiated after 12 h, and the denuded area was almost completely covered by migrating cells after 36 h. When GA was administered, wound recovery was inhibited in a dose-dependent manner. After treatment with 200 μM GA, more than half of the area remained uncovered, even after 24-h incubation (Figure 4A,B).

The average migration speed of 200 μM GA-treated keloid fibroblasts was 9.11 ± 2.78 μm/h, which was significantly lower than that of untreated keloid fibroblasts (18.91 ± 5.56 μm/h; $p < 0.01$, Figure 4C). The average total distance traveled by untreated keloid fibroblasts was 225.37 ± 66.28 μm. After treatment with significantly decreased following treatment with 200 μM GA ($p < 0.001$, Figure 4D). The directionality values for the untreated and 200 μM GA-treated groups were 0.68 ± 0.21 and 0.68 ± 0.11, respectively, indicating no significant change.

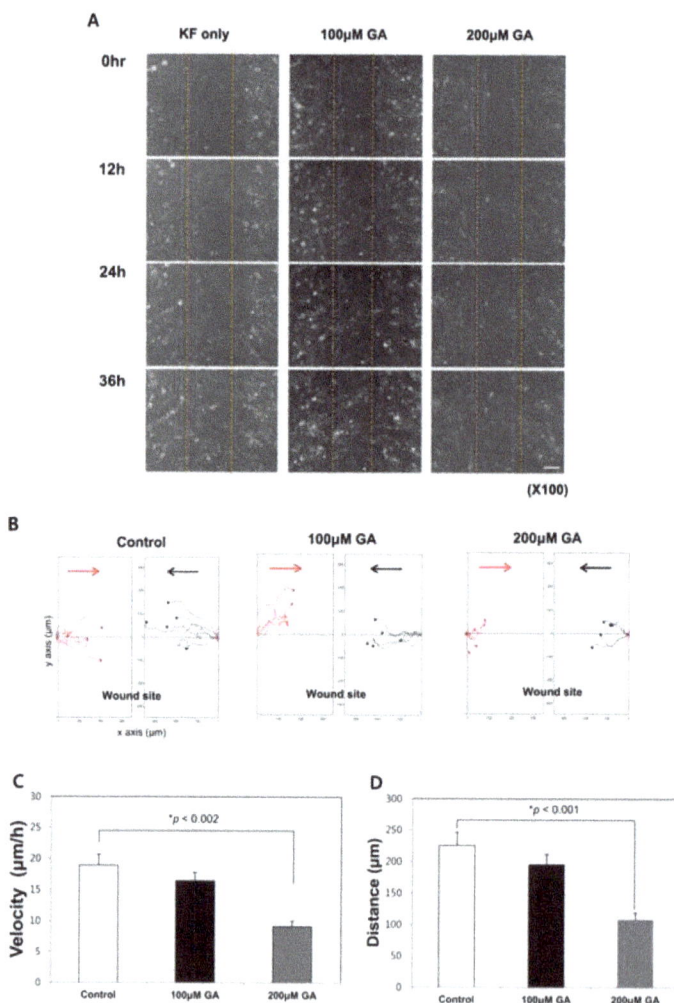

Figure 4. Glycyrrhizic acid (GA) inhibited keloid fibroblast activity. Scale bar: 50 μm. (**A**) The in vitro wound healing assay showed decreased keloid fibroblast migration after GA treatment; (**B**) The cell tracking system for fibroblast migration was analyzed using cells selected from both sides of the wound, keloid fibroblasts were tracked toward the site of wound insert after cultivation for 24 h; black, cells toward the midline from the left edge; red, cells toward the midline from the right edge. GA-induced decreases in the average migration speed (**C**) and the average distance (**D**) traveled of keloid fibroblasts. All results are shown as mean ± SD (* $p < 0.05$).

2.5. HMGB1-Induced ERK1/2, AKT, and NF-κB Protein Expression was Suppressed by GA Treatment in Human Normal Fibroblasts

When HMGB1 is released by damaged or necrotic cells, NF-κB activation is required to promote cell migration. When HMGB1 binds to TLR2/4, the activated TLR triggers activation of the PI3K/AKT signaling cascade. Therefore, we assessed whether HMGB1-induced cell migration may involve the intracellular signaling pathways of ERK1/2, AKT and NF-κB, and whether GA treatment would inhibit these pathways.

The protein levels of ERK1/2, AKT, and NF-κB were measured by western blot analysis. HMGB1 (100 ng/mL) treatment significantly increased the expression levels of ERK1/2, AKT, and NF-κB compared to those of the control group ($p < 0.001$; $p < 0.01$; $p < 0.05$; Figure 5A). In addition, the HMGB1-induced expression of all three proteins was significantly decreased by treatment with 200 μM GA (Figure 5B). Furthermore, challenging HMGB1 on HDF also increased ERK1/2 and AKT phosphorylation and its inhibition attenuated the activation of ERK1/2, AKT signaling molecules (Figure 6).

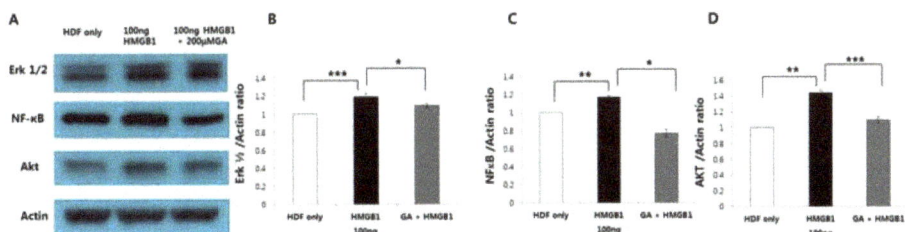

Figure 5. HMGB1 (100 ng/mL) increased the expression levels of ERK1/2, AKT, and NF-κB. (**A**) HMGB1-induced expression of internal signaling cascade molecules such as ERK1/2 (**B**); NF-κB (**C**); and AKT (**D**) was significantly decreased by treatment with 200 μM GA (*** $p < 0.001$; ** $p < 0.01$; * $p < 0.05$).

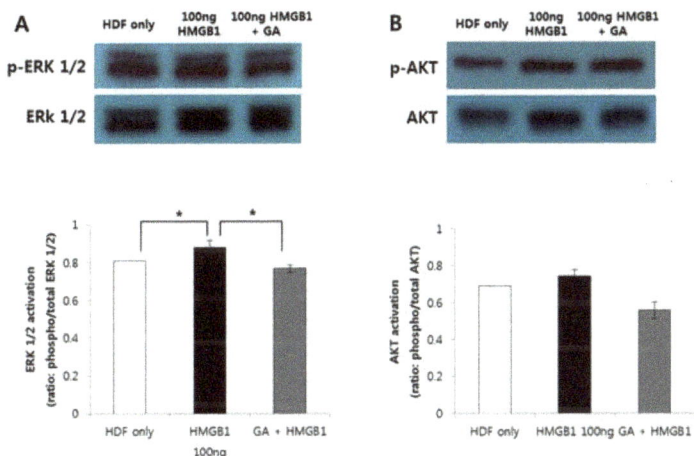

Figure 6. HMGB1 (100 ng/mL) increased the expression levels of phosphorylated form of ERK1/2 (p-ERK) and AKT (p-AKT). Assessment of both phosphorylated and total ERK1/2 (**A**) and AKT (**B**) was done to detect the activation of signaling cascade. The graph shows the ratio between the optical density of the bands of the phosphorylated form and the bands of the corresponding total protein. HMGB1-induced increase of p-ERK1/2 and p-AKT was decreased by treatment with 200 μM GA and the change was statistically significant in p-ERK1/2 (* $p < 0.05$).

3. Discussion

HMGB1 is normally found in the nucleus, and functions as an intranuclear architectural DNA-binding protein that is associated with transcription factors. When cell damage occurs, HMGB1 is translocated to the cytoplasm and released by the cell to act as a multifunctional cytokine with roles in infection, organ dysfunction, inflammation, and immune responses [17,18]. Recent studies of wound healing have demonstrated that HMGB1 binds to cell-surface receptors and acts as a promoter of wound closure through its chemotactic effect on skin fibroblasts and keratinocytes [16,17]. HMGB1 levels were found to be decreased in diabetic human and mouse skin, which may account for the altered wound healing in diabetic patients [16].

Fibroblast dysfunction and associated ECM deposition is a histopathological hallmark in keloid pathogenesis. Unlike normal fibroblasts, keloid fibroblasts possess tumor-like properties, showing excessive proliferation and invasion of surrounding tissues [19,20]. Enhanced migration and invasion of normal fibroblasts are critical factors for the development of keloids [21].

The present study was designed to evaluate the role of HMGB1 in the development of keloids by conducting an in vitro assay to uncover its mechanism of action in normal and keloid fibroblasts. The results of this study demonstrated that HMGB1 alters the behavior of both normal and keloid fibroblasts. Treatment with HMGB1 promoted an increase in the migration of both normal and keloid fibroblasts to a degree equivalent to that achieved by treatment with TGF-β. Subsequent in vitro wound healing assays demonstrated an increase in the migration speed and distance traveled in both cell lines. Furthermore, HMGB1-induced increases in the motility of normal and keloid fibroblasts were significantly inhibited by treatment with GA.

There are several mechanisms by which HMGB1 promotes the fibroblast proliferation associated with keloid development. In keloid tissues, fibroblast proliferation is influenced by epithelial-mesenchymal interactions between the surrounding keratinocytes and fibroblasts [22]. In a previous study, we demonstrated that the EMT-like transition via Wnt3a activation substantially contributes to collagen accumulation during the development of keloids [23]. Emerging experimental evidence indicates that HMGB1 also plays an important role in EMT. HMGB1-induced EMT has been identified in colorectal and gastric cancers, and is activated by the RAGE/NF-κB pathways [24,25]. In estrogen-mediated wound healing, HMGB1 has been shown to greatly accelerate keratinocyte migration, and its knockdown blocked estrogen-induced keratinocyte migration [26]. Cardiac fibroblasts actively secrete HMGB1 in response to mechanical stress or inflammation, which leads to cardiac collagen deposition via the PKCβ/ERK1/2 signaling pathway [27,28]. In human airway epithelial wound closure, HMGB1 has been shown to induce TLR4- and RAGE-mediated wound closure, ECM protein and receptor expression, and intracellular signaling. In particular, the similarities between HMGB1 and TGF-β in wound closure in human bronchial models suggest that changes associated with EMT may occur as part of the repair process upon exposure to HMGB1 [13,29].

To evaluate the presence of HMGB1-mediated EMT in normal fibroblasts, we investigated the expression of downstream molecules involved in HMGB1 signaling. Among its multiple cellular receptors, HMGB1 has been shown to interact with TLR2/4 and RAGE in injury and inflammation models to trigger subsequent inflammatory signaling [30–34]. Specifically, HMGB1 induces an increase in the expression levels of ERK1/2, AKT, and NF-κB. Moreover, these increases were significantly decreased in the presence of an HMGB1 inhibitor. Considering that ERK1/2 is a downstream molecule of RAGE, and that AKT is secreted via TLR2/4 activation and the PI3K cascade, our results support the current hypothesis that extracellular HMGB1 acts on normal fibroblasts by binding to both RAGE and TLR2/4. Activation of the ERK and PI3K/AKT pathways is associated with the proliferation of keloid fibroblasts, along with keratinocytes, and eventually leads to excessive collagen accumulation [35]. In keloid tissue, simultaneous activation of the ERK and PI3K/AKT pathways is important for the production of collagen and other ECM components [36].

TGF-β/Smad signaling has long been considered a pivotal fibrogenic factor in abnormal wound healing. Subsequently, anti-fibrotic strategies based on the blockade or elimination of TGF-β signaling emerged as an important pharmacological target for treating keloids [37–39]. Although there are currently no effective therapeutic interventions for keloids, small-molecule inhibitors targeting HMGB1 or its receptors based on silencing HMGB1, RAGE, or TLRs have proven successful in reducing the severity of symptoms in various experimental models of fibrotic diseases [13,14,25,40,41].

Glycyrrhizin and its metabolites have been reported to suppress tissue fibrosis by targeting TGF-β and other fibrosis-related pathological signaling molecules [42,43]. In the present study, administration of the HMGB1 inhibitor GA resulted in decreased migration and motility in HMGB1- and TGF-β treated normal fibroblasts. In addition, treatment with GA caused decreased migratory property of normal fibroblasts. It is well demonstrated that inflammation during the wound healing period induces transition of normal fibroblast to keloid fibroblasts [44]. GA is known to inhibit HMGB1 activity by inhibiting its pro-inflammatory activity [45]. When applied to keloid fibroblasts, cellular activity represented by migration speed and distance was significantly impeded. This effect of GA can be attributed to its inhibitory effects on both the RAGE and TLR2/4-activated signaling of HMGB1. In a previous study, we demonstrated that HMGB1 induces the expression of EMT-associated proteins in normal fibroblasts; thus, we hypothesized that inhibiting HMGB1 activity or its related signals could be beneficial for treating or even preventing keloids. Theoretically, inhibition of HMGB1 can suppress EMT and the activation of associated pro-fibrotic cytokines. However, because HMGB1 interacts with multiple signaling systems and undergoes dynamic post-translation modifications, it is important to optimize strategies for blocking its abnormal activation.

Currently available studies have already demonstrated that HMGB1 promotes wound healing. Our study further demonstrates that HMGB1 may induce the fibroproliferative properties associated with excessive collagen accumulation, as observed in keloid tissue. Unlike previous studies, which showed increased or decreased expression levels of HMGB1 under certain conditions, our study focused on the molecular signaling pathways associated with keloid development. We suggest that after skin injury, extracellular HMGB1 induces RAGE- and TLR2/4-mediated fibroblast activation. This may provoke the activation of a TGF-β-like profibrotic effect via the ERK1/2 and PI3K/AKT pathways, which would eventually lead to EMT-like changes, causing abnormal ECM accumulation in keloid tissues.

4. Materials and Methods

4.1. Keloid Tissues, Keloid-Derived Fibroblasts, and Human Dermal Fibroblast Cell Culture

Keloid tissues were collected from patients with active-stage keloid (n = 5) after obtaining informed consent according to the protocol approved by the Yonsei University College of Medicine Institutional Review Board (IRB No. 4-2015-0228, 29 JUL 2015) All experiments involving humans were performed in accordance with the Declaration of Helsinki. Human normal dermal fibroblasts were obtained from the American Type Culture Collection (Manassas, VA, USA). Separated cells were cultured in Dulbecco's modified Eagle's medium (Gibco, Grand Island, NY, USA) supplemented with 10% heat-inactivated fetal bovine serum, penicillin (30 U/mL), streptomycin (300 μg/mL), and actinomycin. The culture medium was replaced every 2–3 days.

4.2. In Vitro Wound Healing Assay and Cell Tracking System for Cell Migration

An in vitro wound healing model was established using silicon culture inserts (Ibidi, Munich, Germany) with two individual wells for cell seeding. Each insert was placed in a culture dish; 1×10^4 normal or keloid fibroblasts were plated in each well and grown to form a confluent and homogeneous layer. Twenty-four hours after cell seeding, the culture insert was removed to form an approximately 500-μm-wide cell-free "wound" area for observation. Cells were treated with 50 or 100 ng/mL HMGB1 and 10 ng/mL TGF-β in media. Wound healing by cell migration over time was measured under a light microscope (IX-70, Olympus, Tokyo, Japan).

To track the cells, they were treated with HMGB1 and TGF-β in media, cultured in a mini-incubator (Live Cell Instrument, Seoul, Korea), visualized by light microscopy, and cell images were recorded every 5 min for 36 h by a charge-coupled device camera (Electric Biomedical Co., Ltd., Osaka, Japan) attached to an inverted microscope (Olympus). Captured images were analyzed using Image J software, version 1.48J. Image analysis was carried out by manual tracking, as well as the chemotaxis and migration tool plug-in V1.01 (Ibidi, Planegg, Germany). Cells were selected from both sides of the wound. We obtained the datasets of XY coordinates by using manual tracking, then these datasets were imported into chemotaxis plug-in. These datasets were then imported into the chemotaxis and migration tool plug-in, which computed the cell migration speed and directionality and plotted the cell migration pathway.

The migration speed was calculated as an accumulated distance of the cell divided by time. The migration distance of the cell was defined as the straight-line distance along the y axis between the start position and the end position of cell divided by accumulated distance. For each experiment, cells were randomly selected along each edge of the wound. Cells undergoing division, death or migration outside the field of the view were excluded from the analysis.

We added GA, a known inhibitor of HMGB1, to evaluate its effect on HMGB1-induced fibroblast migration. GA (200 μM) was administered to the 100 ng/mL HMGB1- and 10 ng/mL TGF-β-treated normal fibroblasts to analyze its effects. Additionally, 100 μM and 200 μM GA was treated to keloid fibroblasts.

4.3. Western Blotting Analysis

Normal and keloid fibroblasts were grown to 70% confluence in 100×20-mm cell culture dishes. Cultured keloid fibroblasts were exposed to HMGB1 (100 ng/mL) for 48 h. The protein (20 μg) was subjected to 10% sodium dodecyl sulfate-polyacrylamide gel electrophoresis and electrophoretically transferred onto a polyvinylidene fluoride membrane (Millipore, Billerica, MA, USA). The membranes were blocked with blocking buffer for 1 h and incubated with primary antibodies against AKT (1:1000, rabbit polyclonal; Cell Signaling Technology, Danvers, MA, USA), p-AKT (Cell Signaling Technology), ERK 1/2 (Cell Signaling Technology), p-ERK 1/2 (Cell Signaling Technology) and actin (1:5000, mouse monoclonal; Sigma-Aldrich, St. Louis, MO, USA) before incubating overnight at 4 °C. Secondary antibodies against horseradish peroxidase-conjugated rabbit antibody (1:2000, Santa Cruz Biotechnology, Santa Cruz, CA, USA) and mouse antibody (1:2000, Santa Cruz Biotechnology) were then added and the membrane was incubated for another 2 h at room temperature. After incubation with secondary antibodies, the membrane blot was developed using an electrochemiluminescence blotting system (Amersham Pharmacia Biotech, Piscataway, NJ, USA) according to the manufacturer's instructions; the densities of the bands on the developed film were analyzed using Image J software. Protein expression levels were normalized to those of actin. Relative quantitation is expressed as fold-induction compared to control conditions (normal fibroblast group).

4.4. Statistical Analysis

All data are presented as mean ± standard error of the mean (SEM) and were analyzed using a paired t-test or one-way analysis of variance; $p < 0.05$ was considered statistically significant. SPSS version 19.0 (SPSS Inc., Chicago, IL, USA) was used for all statistical analyses.

5. Conclusions

Our study demonstrated that HMGB1 induces the activation of fibroblasts via activating the RAGE-MAPK and NF-κB interaction signaling pathways. Further study is needed to elucidate the mechanisms of ECM accumulation with exogenous HMGB1 administration and its action on EMT-mediated canonical signaling in wound repair. We expect that further knowledge of the relationship of skin fibrosis to HMGB1 may help to develop a new clinical approach to treat and prevent keloids and hypertrophic scars.

Acknowledgments: This work was supported by grants from a National Research Foundation of Korea (NRF) grant funded by the Korean government (MEST) (No. 2014051295, to Won Jai Lee).

Author Contributions: Jihee Kim conceived and performed the experiments, collected and analyzed the data, and drafted the manuscript. Jong-Chul Park and Mi Hee Lee performed the experiments and analyzed the data. Chae Eun Yang participated in analyzing the data. Ju Hee Lee and Won Jai Lee conceived and designed the project,

Int. J. Mol. Sci. **2018**, *19*, 76

oversaw the collection of results and data interpretation, wrote the manuscript, and share responsibility for the final content. All authors approved the final manuscript.

Conflicts of Interest: The authors declare no conflict of interest. The funding sponsors had no role in the design of the study; in the collection, analyses, or interpretation of data; in the writing of the manuscript, and in the decision to publish the results.

Abbreviations

AKT	Protein kinase B
CXCL12	C-X-C motif chemokine 12
CXCR4	Chemokine Receptor type 4
ECM	Extracellular matrix
EMT	Epithelial-mesenchymal transition
ERK	Extracellular signal-regulated kinases
GA	Glycyrrhizic acid
HMGB1	High-mobility group box 1
MAPK	Mitogen-activated protein kinase
NF-κB	Nuclear factor kappa-light-chain-enhancer of activated B cells
PBS	Phosphate-buffered saline
PI3K	Phosphatidylinositol-4,5-bisphosphate 3-kinase
RAGE	Receptor for advanced glycation end product
TGF-β	Transforming growth factor-β
TLR	Toll-like receptor

References

1. Huang, C.; Akaishi, S.; Ogawa, R. Mechanosignaling pathways in cutaneous scarring. *Arch. Dermatol. Res.* **2012**, *304*, 589–597. [CrossRef] [PubMed]
2. Huang, C.; Akaishi, S.; Hyakusoku, H.; Ogawa, R. Are keloid and hypertrophic scar different forms of the same disorder? A fibroproliferative skin disorder hypothesis based on keloid findings. *Int. Wound J.* **2014**, *11*, 517–522. [CrossRef] [PubMed]
3. Song, N.; Wu, X.; Gao, Z.; Zhou, G.; Zhang, W.J.; Liu, W. Enhanced expression of membrane transporter and drug resistance in keloid fibroblasts. *Hum. Pathol.* **2012**, *43*, 2024–2032. [CrossRef] [PubMed]
4. Qu, M.; Song, N.; Chai, G.; Wu, X.; Liu, W. Pathological niche environment transforms dermal stem cells to keloid stem cells: A hypothesis of keloid formation and development. *Med. Hypotheses* **2013**, *81*, 807–812. [CrossRef] [PubMed]
5. Gauglitz, G.G.; Korting, H.C.; Pavicic, T.; Ruzicka, T.; Jeschke, M.G. Hypertrophic scarring and keloids: Pathomechanisms and current and emerging treatment strategies. *Mol. Med.* **2011**, *17*, 113–125. [CrossRef] [PubMed]
6. Tuan, T.L.; Nichter, L.S. The molecular basis of keloid and hypertrophic scar formation. *Mol. Med. Today* **1998**, *4*, 19–24. [CrossRef]
7. Bianchi, M.E.; Beltrame, M.; Paonessa, G. Specific recognition of cruciform DNA by nuclear protein HMG1. *Science* **1989**, *243 Pt 1*, 1056–1059. [CrossRef] [PubMed]
8. Ellerman, J.E.; Brown, C.K.; de Vera, M.; Zeh, H.J.; Billiar, T.; Rubartelli, A.; Lotze, M.T. Masquerader: High mobility group box-1 and cancer. *Clin. Cancer Res.* **2007**, *13*, 2836–2848. [CrossRef] [PubMed]
9. Ranzato, E.; Martinotti, S.; Pedrazzi, M.; Patrone, M. High mobility group box protein-1 in wound repair. *Cells* **2012**, *1*, 699–710. [CrossRef] [PubMed]
10. Yanai, H.; Ban, T.; Wang, Z.; Choi, M.K.; Kawamura, T.; Negishi, H.; Nakasato, M.; Lu, Y.; Hangai, S.; Koshiba, R.; et al. HMGB proteins function as universal sentinels for nucleic-acid-mediated innate immune responses. *Nature* **2009**, *462*, 99–103. [CrossRef] [PubMed]
11. Yanai, H.; Taniguchi, T. Nucleic acid sensing and beyond: Virtues and vices of high-mobility group box 1. *J. Intern. Med.* **2014**, *276*, 444–453. [CrossRef] [PubMed]

12. Ho, Y.Y.; Lagares, D.; Tager, A.M.; Kapoor, M. Fibrosis—A lethal component of systemic sclerosis. *Nat. Rev. Rheumatol.* **2014**, *10*, 390–402. [CrossRef] [PubMed]
13. Li, L.C.; Gao, J.; Li, J. Emerging role of HMGB1 in fibrotic diseases. *J. Cell. Mol. Med.* **2014**, *18*, 2331–2339. [CrossRef] [PubMed]
14. Lynch, J.; Nolan, S.; Slattery, C.; Feighery, R.; Ryan, M.P.; McMorrow, T. High-mobility group box protein 1: A novel mediator of inflammatory-induced renal epithelial-mesenchymal transition. *Am. J. Nephrol.* **2010**, *32*, 590–602. [CrossRef] [PubMed]
15. Wang, F.P.; Li, L.; Li, J.; Wang, J.Y.; Wang, L.Y.; Jiang, W. High mobility group box-1 promotes the proliferation and migration of hepatic stellate cells via TLR4-dependent signal pathways of PI3K/Akt and JNK. *PLoS ONE* **2013**, *8*, e64373. [CrossRef] [PubMed]
16. Straino, S.; Di Carlo, A.; Mangoni, A.; De Mori, R.; Guerra, L.; Maurelli, R.; Panacchia, L.; Di Giacomo, F.; Palumbo, R.; Di Campli, C.; et al. High-mobility group box 1 protein in human and murine skin: Involvement in wound healing. *J. Investig. Dermatol.* **2008**, *128*, 1545–1553. [CrossRef] [PubMed]
17. Dardenne, A.D.; Wulff, B.C.; Wilgus, T.A. The alarmin HMGB-1 influences healing outcomes in fetal skin wounds. *Wound Repair Regen.* **2013**, *21*, 282–291. [CrossRef] [PubMed]
18. Erlandsson Harris, H.; Andersson, U. Mini-review: The nuclear protein HMGB1 as a proinflammatory mediator. *Eur. J. Immunol.* **2004**, *34*, 1503–1512. [CrossRef] [PubMed]
19. Schafer, M.; Werner, S. Cancer as an overhealing wound: An old hypothesis revisited. *Nat. Rev. Mol. Cell Biol.* **2008**, *9*, 628–638. [CrossRef] [PubMed]
20. Vincent, A.S.; Phan, T.T.; Mukhopadhyay, A.; Lim, H.Y.; Halliwell, B.; Wong, K.P. Human skin keloid fibroblasts display bioenergetics of cancer cells. *J. Investig. Dermatol.* **2008**, *128*, 702–709. [CrossRef] [PubMed]
21. Yun, I.S.; Lee, M.H.; Rah, D.K.; Lew, D.H.; Park, J.C.; Lee, W.J. Heat Shock Protein 90 Inhibitor (17-AAG) Induces Apoptosis and Decreases Cell Migration/Motility of Keloid Fibroblasts. *Plast. Reconstr. Surg.* **2015**, *136*, 44e–53e. [CrossRef] [PubMed]
22. Yan, C.; Grimm, W.A.; Garner, W.L.; Qin, L.; Travis, T.; Tan, N.; Han, Y.P. Epithelial to mesenchymal transition in human skin wound healing is induced by tumor necrosis factor-alpha through bone morphogenic protein-2. *Am. J. Pathol.* **2010**, *176*, 2247–2258. [CrossRef] [PubMed]
23. Lee, W.J.; Park, J.H.; Shin, J.U.; Noh, H.; Lew, D.H.; Yang, W.I.; Yun, C.O.; Lee, K.H.; Lee, J.H. Endothelial-to-mesenchymal transition induced by Wnt 3a in keloid pathogenesis. *Wound Repair Regen.* **2015**, *23*, 435–442. [CrossRef] [PubMed]
24. Chung, H.W.; Jang, S.; Kim, H.; Lim, J.B. Combined targeting of high-mobility group box-1 and interleukin-8 to control micrometastasis potential in gastric cancer. *Int. J. Cancer* **2015**, *137*, 1598–1609. [CrossRef] [PubMed]
25. Zhu, L.; Li, X.; Chen, Y.; Fang, J.; Ge, Z. High-mobility group box 1: A novel inducer of the epithelial-mesenchymal transition in colorectal carcinoma. *Cancer Lett.* **2015**, *357*, 527–534. [CrossRef] [PubMed]
26. Shin, J.U.; Noh, J.Y.; Lee, J.H.; Lee, W.J.; Yoo, J.S.; Kim, J.Y.; Kim, H.; Jung, I.; Jin, S.; Lee, K.H. In vivo relative quantitative proteomics reveals HMGB1 as a downstream mediator of oestrogen-stimulated keratinocyte migration. *Exp. Dermatol.* **2015**, *24*, 478–480. [CrossRef] [PubMed]
27. Su, Z.; Yin, J.; Wang, T.; Sun, Y.; Ni, P.; Ma, R.; Zhu, H.; Zheng, D.; Shen, H.; Xu, W.; et al. Up-regulated HMGB1 in EAM directly led to collagen deposition by a PKCbeta/Erk1/2-dependent pathway: Cardiac fibroblast/myofibroblast might be another source of HMGB1. *J. Cell. Mol. Med.* **2014**, *18*, 1740–1751. [CrossRef] [PubMed]
28. Kohno, T.; Anzai, T.; Naito, K.; Miyasho, T.; Okamoto, M.; Yokota, H.; Yamada, S.; Maekawa, Y.; Takahashi, T.; Yoshikawa, T.; et al. Role of high-mobility group box 1 protein in post-infarction healing process and left ventricular remodelling. *Cardiovasc. Res.* **2009**, *81*, 565–573. [CrossRef] [PubMed]
29. Ojo, O.O.; Ryu, M.H.; Jha, A.; Unruh, H.; Halayko, A.J. High-mobility group box 1 promotes extracellular matrix synthesis and wound repair in human bronchial epithelial cells. *Am. J. Physiol. Lung Cell. Mol. Physiol.* **2015**, *309*, L1354–L1366. [CrossRef] [PubMed]
30. Yu, M.; Wang, H.; Ding, A.; Golenbock, D.T.; Latz, E.; Czura, C.J.; Fenton, M.J.; Tracey, K.J.; Yang, H. HMGB1 signals through toll-like receptor (TLR) 4 and TLR2. *Shock* **2006**, *26*, 174–179. [CrossRef] [PubMed]

31. Tsung, A.; Sahai, R.; Tanaka, H.; Nakao, A.; Fink, M.P.; Lotze, M.T.; Yang, H.; Li, J.; Tracey, K.J.; Geller, D.A.; et al. The nuclear factor HMGB1 mediates hepatic injury after murine liver ischemia-reperfusion. *J. Exp. Med.* **2005**, *201*, 1135–1143. [CrossRef]

32. Park, J.S.; Svetkauskaite, D.; He, Q.; Kim, J.Y.; Strassheim, D.; Ishizaka, A.; Abraham, E. Involvement of toll-like receptors 2 and 4 in cellular activation by high mobility group box 1 protein. *J. Biol. Chem.* **2004**, *279*, 7370–7377. [CrossRef]

33. Kokkola, R.; Andersson, A.; Mullins, G.; Ostberg, T.; Treutiger, C.J.; Arnold, B.; Nawroth, P.; Andersson, U.; Harris, R.A.; Harris, H.E. RAGE is the major receptor for the proinflammatory activity of HMGB1 in rodent macrophages. *Scand. J. Immunol.* **2005**, *61*, 1–9. [CrossRef]

34. Ibrahim, Z.A.; Armour, C.L.; Phipps, S.; Sukkar, M.B. RAGE and TLRs: Relatives, friends or neighbours? *Mol. Immunol.* **2013**, *56*, 739–744. [CrossRef]

35. Zhang, Z.; Nie, F.; Kang, C.; Chen, B.; Qin, Z.; Ma, J.; Ma, Y.; Zhao, X. Increased periostin expression affects the proliferation, collagen synthesis, migration and invasion of keloid fibroblasts under hypoxic conditions. *Int. J. Mol. Med.* **2014**, *34*, 253–261. [CrossRef]

36. Lim, I.J.; Phan, T.T.; Tan, E.K.; Nguyen, T.T.; Tran, E.; Longaker, M.T.; Song, C.; Lee, S.T.; Huynh, H.T. Synchronous activation of ERK and phosphatidylinositol 3-kinase pathways is required for collagen and extracellular matrix production in keloids. *J. Biol. Chem.* **2003**, *278*, 40851–40858. [CrossRef] [PubMed]

37. Washio, H.; Fukuda, N.; Matsuda, H.; Nagase, H.; Watanabe, T.; Matsumoto, Y.; Terui, T. Transcriptional inhibition of hypertrophic scars by a gene silencer, pyrrole-imidazole polyamide, targeting the TGF-beta1 promoter. *J. Investig. Dermatol.* **2011**, *131*, 1987–1995. [CrossRef]

38. Seifert, O.; Mrowietz, U. Keloid scarring: Bench and bedside. *Arch. Dermatol. Res.* **2009**, *301*, 259–272. [CrossRef] [PubMed]

39. Liu, W.; Wang, D.R.; Cao, Y.L. TGF-beta: A fibrotic factor in wound scarring and a potential target for anti-scarring gene therapy. *Curr. Gene Ther.* **2004**, *4*, 123–136. [CrossRef] [PubMed]

40. Wang, W.K.; Wang, B.; Lu, Q.H.; Zhang, W.; Qin, W.D.; Liu, X.J.; Liu, X.Q.; An, F.S.; Zhang, Y.; Zhang, M.X. Inhibition of high-mobility group box 1 improves myocardial fibrosis and dysfunction in diabetic cardiomyopathy. *Int. J. Cardiol.* **2014**, *172*, 202–212. [CrossRef]

41. He, M.; Kubo, H.; Ishizawa, K.; Hegab, A.E.; Yamamoto, Y.; Yamamoto, H.; Yamaya, M. The role of the receptor for advanced glycation end-products in lung fibrosis. *Am. J. Physiol. Lung Cell. Mol. Physiol.* **2007**, *293*, L1427–L1436. [CrossRef] [PubMed]

42. Gao, L.; Tang, H.; He, H.; Liu, J.; Mao, J.; Ji, H.; Lin, H.; Wu, T. Glycyrrhizic acid alleviates bleomycin-induced pulmonary fibrosis in rats. *Front. Pharmacol.* **2015**, *6*, 215. [CrossRef] [PubMed]

43. Moro, T.; Shimoyama, Y.; Kushida, M.; Hong, Y.Y.; Nakao, S.; Higashiyama, R.; Sugioka, Y.; Inoue, H.; Okazaki, I.; Inagaki, Y. Glycyrrhizin and its metabolite inhibit Smad3-mediated type I collagen gene transcription and suppress experimental murine liver fibrosis. *Life Sci.* **2008**, *83*, 531–539. [CrossRef] [PubMed]

44. Ogawa, R. Keloid and Hypertrophic Scars Are the Result of Chronic Inflammation in the Reticular Dermis. *Int. J. Mol. Sci.* **2017**, *18*, 606. [CrossRef] [PubMed]

45. Mollica, L.; De Marchis, F.; Spitaleri, A.; Dallacosta, C.; Pennacchini, D.; Zamai, M.; Agresti, A.; Trisciuoglio, L.; Musco, G.; Bianchi, M.E. Glycyrrhizin binds to high-mobility group box 1 protein and inhibits its cytokine activities. *Chem. Biol.* **2007**, *14*, 431–441. [CrossRef] [PubMed]

International Journal of
Molecular Sciences

MDPI

Article

Extracorporeal Shock Wave Therapy Alters the Expression of Fibrosis-Related Molecules in Fibroblast Derived from Human Hypertrophic Scar

Hui Song Cui [1,†], A Ram Hong [1,†], June-Bum Kim [2] , Joo Hyang Yu [1], Yoon Soo Cho [3], So Young Joo [3,*] and Cheong Hoon Seo [3,*]

[1] Burn Institute, Hangang Sacred Heart Hospital, College of Medicine, Hallym University, Seoul 07247, Korea; bioeast@hanmail.net (H.S.C.); bsnunki@naver.com (A.R.H.); koko6394@naver.com (J.H.Y.)
[2] Department of Pediatrics, Hallym University Hangang Sacred Heart Hospital, Seoul 07247, Korea; hoppdoctor@hanmail.net
[3] Department of Rehabilitation Medicine, Hangang Sacred Heart Hospital, College of Medicine, Hallym University, Seoul 07247, Korea; hamays@hanmail.net
* Correspondence: anyany98@naver.com (S.Y.J.); chseomd@gmail.com (C.H.S.); Tel.: +82-2-2639-5730 (S.Y.J. & C.H.S.); Fax: +82-2-2635-7820 (S.Y.J. & C.H.S.)
† These authors contributed equally to this work.

Received: 31 October 2017; Accepted: 26 December 2017; Published: 2 January 2018

Abstract: Extracorporeal shock wave therapy (ESWT) considerably improves the appearance and symptoms of post-burn hypertrophic scars (HTS). However, the mechanism underlying the observed beneficial effects is not well understood. The objective of this study was to elucidate the mechanism underlying changes in cellular and molecular biology that is induced by ESWT of fibroblasts derived from scar tissue (HTSFs). We cultured primary dermal fibroblasts derived from human HTS and exposed these cells to 1000 impulses of 0.03, 0.1, and 0.3 mJ/mm². At 24 h and 72 h after treatment, real-time PCR and western blotting were used to detect mRNA and protein expression, respectively, and cell viability and mobility were assessed. While HTSF viability was not affected, migration was decreased by ESWT. Transforming growth factor beta 1 (TGF-β1) expression was reduced and alpha smooth muscle actin (α-SMA), collagen-I, fibronectin, and twist-1 were reduced significantly after ESWT. Expression of E-cadherin was increased, while that of N-cadherin was reduced. Expression of inhibitor of DNA binding 1 and 2 was increased. In conclusion, suppressed epithelial-mesenchymal transition might be responsible for the anti-scarring effect of ESWT, and has potential as a therapeutic target in the management of post-burn scars.

Keywords: extracorporeal shock wave therapy; burn hypertrophic scar; hypertrophic scar-derived fibroblast; epithelial-mesenchymal transition; inhibitor of DNA binding protein

1. Introduction

Post-burn hypertrophic scars (HTSs) are a most common complication of burn injury and result from excessive scar tissue formation in prolonged aberrant wound healing, and are characterized by hyalinized collagen bundles. They can functionally (arthrentasis), symptomatically (pruritis, pain), and aesthetically impact the patient's quality of life. Depending on the depth of the wound, hypertrophic scar (HTS) occurs in up to 91% of burn injuries [1]. The pathophysiology of HTS formation involves an overactive proliferative phase in wound healing [2]. A variety of cells (macrophages, fibroblasts, keratinocytes), cytokines, and growth factors participate this process. Activated T cells, macrophages, and Langerhans cells disrupt normal wound healing and tissue remodeling, and contribute to abnormal extracellular matrix (ECM) accumulation with an increased cellular activity [3,4].

Int. J. Mol. Sci. **2018**, *19*, 124

In particular, transforming growth factor beta 1 (TGF-β1) has potent stimulatory effects on ECM protein synthesis, and fibroblast proliferation and differentiation [3]. A previous clinical study revealed that the serum levels of TGF-β in patients with burn injuries are approximately twice of those in control patients [5]. Furthermore, TGF-β1 and 2 receptors are overexpressed in fibroblasts of HTS tissues [6].

There is clear evidence of elevated fibrotic markers, such as collagen type I, alpha smooth muscle actin (α-SMA), vimentin, fibroblast specific protein 1 (FSP-1), and N-cadherin, and a loss of E-cadherin in human HTS, but not in fibroblasts derived from scar tissue (HTSFs) [7]. The mRNA and protein expression of TGF-β1 is significantly increased in HTSF as compared to normal cells [8]; and the numbers of total fibroblasts and αSMA-positive myofibroblasts are significantly higher in HTS than in normotrophic scar or normal skin [9]. Myofibroblasts are fibroblasts that are activated by TGF-β and other growth factors, and are an important source of cells for the synthesis and secretion of ECM components, which are critical for pathological HTS formation [2].

At least three subpopulations have been identified in the dermis: superficial fibroblasts, reticular fibroblasts, and fibroblasts associated with hair follicles; of these, reticular cells originating from the deep dermis contribute to HTS formation [10,11]. At the same time, reticular cells show specific characteristics, including larger cell size, slower proliferation in culture, higher TGF-β1 and collagen production, and higher α-SMA expression when compared with fibroblasts from other dermal layers [10]. However, studies on the pathological characteristics of HTSFs are lacking.

Extracorporeal shockwave therapy (ESWT) is a non-invasive physiotherapy that was first used in the lithotripsy of kidney stones [12]. With its development, it has been gradually applied in the treatment of musculoskeletal diseases, such as plantar fasciitis and chronic lateral epicondylitis (tennis elbow), for which it is approved by the US Food and Drug Administration [13]. Several experimental studies demonstrated that ESWT induces nitric oxide (NO) production and inhibition of nuclear factor kappa B (NF-κB) activation in an in vitro model [14], and significantly induces the expression of 84 angiogenic genes in normal mice or mice with non-healing diabetic ulcers [15]. Further, ESWT induces angiogenic and proliferative growth factors, such as nitric oxide synthase (eNOS), vascular endothelial growth factor (VEGF), and proliferating cell nuclear antigen (PCNA), at joints to stimulate the formation of new capillaries and muscularized vessels, thus improving blood supply and tendon regeneration in animal models [16,17]. Furthermore, clinical trials have shown encouraging, significant correlations of ESWT, with an improved healing rate and complete epithelialization, depending on wound size [18]. One clinical study in patients with post-burn HTSs showed that ESWT softened scars and improved their appearance [19]. Very recently, we reported that ESWT significantly reduced scar pain and pruritus in burn patients during rehabilitation [20,21].

Here, we examined molecular changes induced in HTSFs by ESWT, to reveal the mechanism underlying the beneficial effect of ESWT.

2. Results

2.1. Characterization of HTSFs

To determine the character of HTSFs, we conducted western blot analysis. epithelial-mesenchymal transition (EMT) markers were significantly more strongly expressed by HTSFs than by human normal fibroblasts (HNFs) of the same passage. Protein expression of TGF-β1, α-SMA, collagen-I, collagen-III, fibronectin, vimentin, FSP-1, E-cadherin, and twist1 was significantly higher in HTSFs than in HNFs (Figure 1). In contrast, N-cadherin expression was significantly lower in HTSFs than in HNFs (Figure 1). These results indicated that EMT is closely involved in the pathology of HTS formation.

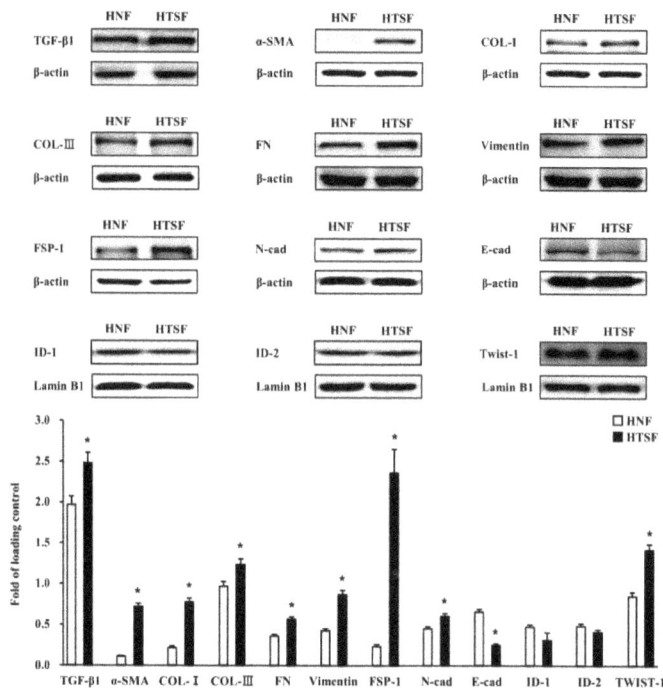

Figure 1. The characteristics of fibroblasts derived from scar tissue (HTSFs). Matched human normal fibroblasts (HNF) and HTSF were cultured from four patients with post burn hypertrophic scar tissue. Protein expression of transforming growth factor beta 1 (TGF-β1), alpha smooth muscle actin (α-SMA), COL-I (collagen type I), COL-III (collagen type III), FN (fibronectin), Vimentin, fibroblast specific protein 1 (FSP-1), Twist-1 and N-cad (N-cadherin) was significantly higher in HTSFs compared with HNF from skin dermis. The protein expression of E-cad (E-cadherin), inhibitor of DNA binding protein 1 (ID-1) and inhibitor of DNA binding protein 2 (ID-2) were lower in HTSFs when compared with HNF from skin dermis. That expression of those proteins was measured by western blotting against specific antibody. The intensity of band was normalized with that of loading control, β-actin or lamin B1, respectively; HNF, Human normal skin derived fibroblast; HTSF, human hypertrophic scar derived fibroblast. * $p < 0.05$ vs. the corresponding HNF.

2.2. Effects of ESWT on HTSF Viability

We investigated the viability of HTSFs after ESWT with 1000 impulses/cm^2 at 0.03, 0.1, and 0.3 mJ/mm^2 of energy flux densities at 24 h after plating using a viability assay (Figure 2A). ESWT had no effect on the viability of HTSFs when compared to non-treated cells. ESWT-HTSFs showed a normal growth pattern, with no effect of ESWT on growth rate (Figure 2B). The mRNA levels of glyceraldehyde-3-phosphate dehydrogenase (GAPDH) and β-actin were not affected by ESWT. Furthermore, mRNA expression of bcl-2-associated X protein (bax) and B-cell lymphoma 2 (bcl-2), apoptotic-related factors, was not affected at 24 h and 72 h after ESWT in HTSF (Figures S1 and S2).

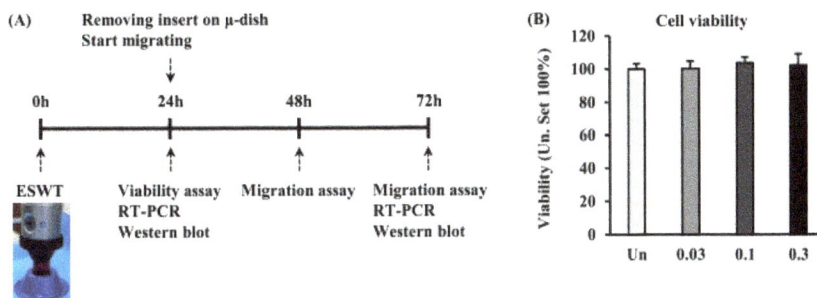

Figure 2. Experimental schematic diagrams and viability of HTSFs. The dermis was separated from human hypertrophic scar tissue by dispase, and then digested with collagenase type IV. HTSF was released, collected, suspended in medium, and continue cultured. After detachment, HTSFs were suspended in to a 17 mL conical tube. ESWT is performed with 1000 impulse/cm^2 at 0.03, 0.1, and 0.3 mJ/mm^2 of energy flux densities. Then, HTSFs were seeded on 96 well cell culture plates for viability assays, μ-dish for migration assays, and T75 culture plates for RT-PCR and western blot (0 h). After 24 h, the viability of HTSF was measured. Once removes insert of μ-dish, the HTSF begins to move, and then analyzed after migration 24 and 48 h (48 and 72 h after ESWT). Real time polymerase chain reaction (RT-PCR) and western blot were performed 24 h and 72 h after plating, respectively (**A**). ESWT no influence on viability of HTSFs (**B**). Cell viability was determined using an Cell Titer-Glo® Luminescent cell viability assay kit 24 h after ESWT. Each group was assayed in sextuplicate, and the experiments were performed at three times independently. HTSF viability was expressed as a percentage value of untreated cells. Un: untreated cells.

2.3. Effects of ESWT on TGF-β1, α-SMA and Vimentin Expression in HTSFs

As TGF-β1 plays a critical role as a potent EMT inducer in the pathogenesis of HTS formation, we first investigated whether the anti-scarring effect of ESWT was partially mediated via suppression of TGF-β1 expression. The TGF-β1 mRNA level was significantly reduced 24 h after ESWT with 1000 impulses/cm^2 at 0.03, 0.1, and 0.3 mJ/mm^2, when compared with non-treated cells ($p < 0.05$) (Figure 3A), while no further changes were observed at 72 h after ESWT. TGF-β1 protein expression was also significantly decreased 24 h and 72 h after ESWT with 1000 impulses/cm^2 at all tested energy flux densities, as compared to non-treated cells ($p < 0.05$) (Figure 3C). Next, we investigated the expression of α-SMA, an EMT marker directly regulated by TGF-β1. α-SMA mRNA expression was reduced at 24 h and 72 h after ESWT with 1000 impulses/cm^2 at 0.03, 0.1, and 0.3 mJ/mm^2, compared with non-treated cells ($p < 0.05$ or 0.01) (Figure 3B). Similar to TGF-β1 protein expression, α-SMA protein expression was significantly decreased at 24 h and 72 h after all ESWT regimens, as compared with non-treated cells ($p < 0.05$) (Figure 3D). Finally, we measured the expression of vimentin, another EMT marker. Vimentin expression was also significantly decreased at 24 h and 72 h after ESWT, when compared with non-treated cells ($p < 0.05$) (Figure S3). On the other hand, non-treated cells were larger and expressed higher levels of vimentin in the cell body. After ESWT, depending on the energy flux density, the HTSFs exhibited spindle- or stellate-shaped fibroblast-like appearance (Figure S3A,B).

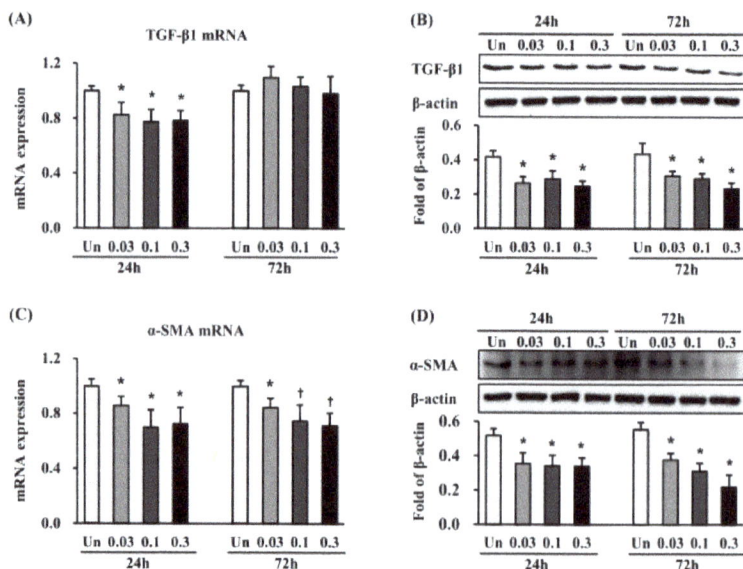

Figure 3. Extracorporeal shockwave therapy (ESWT) decreases the expression of TGF-β1 and alpha smooth muscle actin (α–SMA) in HTSFs. HTSF was cultured from four patients with post burn hypertrophic scar tissue. The mRNA expression of TGF-β1 (**A**) and α–SMA (**C**) were measured 24 h and 72 h after ESWT using a Light Cycler real-time PCR system. Each sample was assayed in duplicate, and experiments were performed least three times independently. The mRNA expression was normalized as ratio = $2^{-\Delta\Delta Ct}$, and data are the mean ± S.E. * $p < 0.05$ and † $p < 0.01$ vs. the corresponding untreated control group. Protein expression of TGF-β1 (**B**) and α–SMA (**D**) were measured with western blot analysis 24 and 72 h after ESWT, respectively. The protein expression was normalized with β-actin, respectively; and data are the mean ± S.E. * $p < 0.05$ vs. the corresponding untreated control group. Un: Untreated cells.

2.4. Effects of ESWT on Expression of ECM-Related Proteins in HTSFs

We investigated the expression of Collagen1a1 and fibronectin, both markers of EMT, at 24 h and 72 h after ESWT. Collagen1a1 mRNA expression was significantly decreased at 24 h after ESWT with 1000 impulses/cm², at 0.03 and 0.1 mJ/mm² and at 72 h after all the ESWT regimens, compared with non-treated cells ($p < 0.05$ or 0.01) (Figure 4A). Accordingly, collagen-I protein expression was significantly reduced at 24 h and 72 h after ESWT ($p < 0.05$) (Figure 4C). Fibronectin mRNA expression was significantly decreased at 24 h after ESWT with 1000 impulses/cm², at 0.03, 0.1 and 0.3 mJ/mm², and at 72 h after ESWT with 1000 impulses/cm², at 0.03, and 0.1 mJ/mm², when compared with non-treated cells ($p < 0.05$ or 0.01) (Figure 4B). Surprisingly, fibronectin protein was significantly increased 24 h after ESWT with 1000 impulses/cm² at 0.03, 0.1, and 0.3 mJ/mm² ($p < 0.05$) (Figure 4D), but significantly reduced at 72 h after ESWT under the same conditions ($p < 0.05$) (Figure 4D).

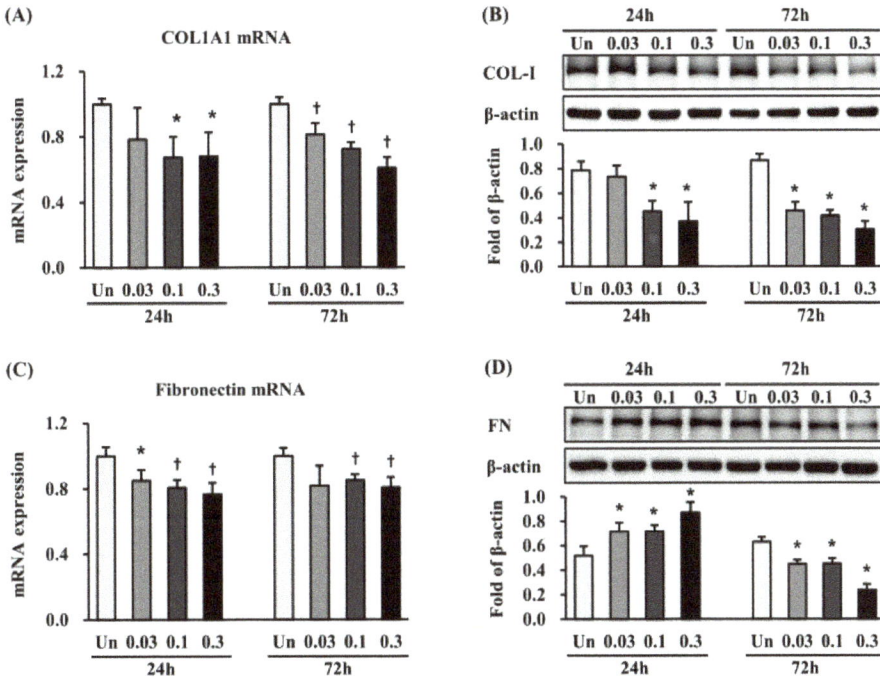

Figure 4. ESWT decreases the expression of extracellular matrix protein in HTSFs. HTSF was cultured from four patients with post burn hypertrophic scar tissue. The mRNA expression of collagen-I (**A**) and fibronectin (**B**) were measured 24 and 72 h after ESWT using a Light Cycler real-time PCR system. Each sample was assayed in duplicate, and experiments were performed least three times independently. The mRNA expression was normalized as ratio = $2^{-\Delta\Delta Ct}$, and data are the mean ± S.E. * $p < 0.05$ and † $p < 0.01$ vs. the corresponding untreated control group. Protein expression of collagen-I (**C**) and fibronectin (**D**) were measured with western blot analysis 24 and 72 h after ESWT, respectively. The protein expression was normalized with β-actin, respectively; and data are the mean ± SE. * $p < 0.05$ vs. the corresponding untreated control group. Un: Untreated cells.

2.5. Effects of ESWT on Expression of N- and E-Cadherin in HTSFs

We measured the expression of N-cadherin and E-cadherin, which are cell-surface markers of EMT. mRNA expression of N-cadherin was significantly decreased at 24 h and 72 h after ESWT in all the regimens ($p < 0.05$ or 0.01) (Figure 5A), while that of E-cadherin was significantly increased 24 h and 72 h after ESWT with 1000 impulse/cm^2 at 0.1 and 0.3 mJ/mm^2 energy flux densities, respectively ($p < 0.01$) (Figure 5B). Similar to mRNA expression, at 24 and 72 h after ESWT in all the regimens, N-cadherin was significantly decreased ($p < 0.05$) (Figure 5C), while E-cadherin was significantly increased 24 h and 72 h after treated with 1000 impulse/cm^2 at 0.1 and 0.3 mJ/mm^2, respectively ($p < 0.05$) (Figure 5D).

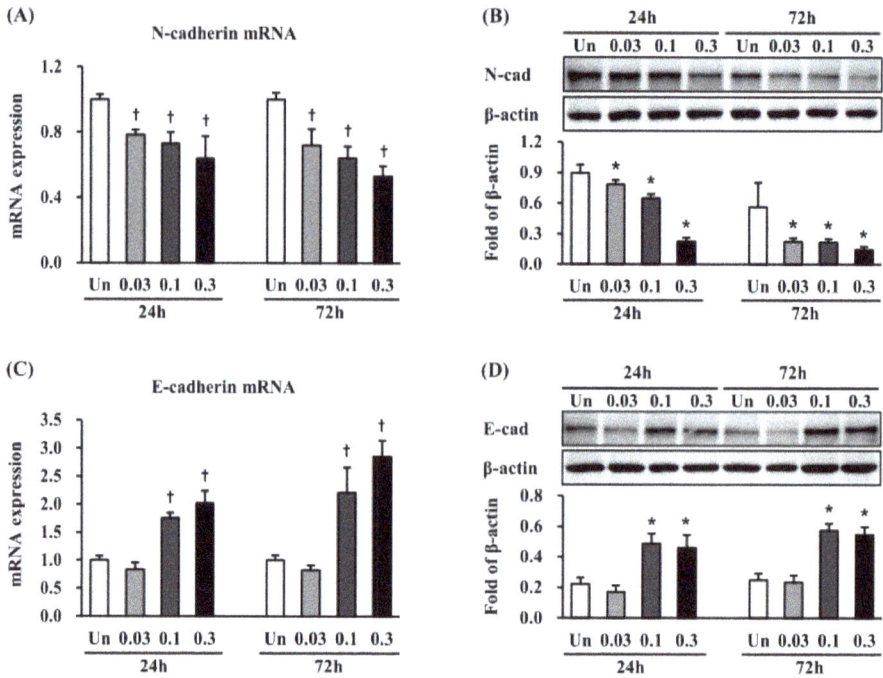

Figure 5. Effects of ESWT on the expression of N-cadherin and E-cadherin in HTSFs. HTSFs were cultured from four patients with post burn hypertrophic scar tissue. The ESWT decreased the mRNA expression of N-cadherin (**A**), and increased the mRNA expression of E-cadherin (**B**). The mRNA expression was measured 24 h and 72 h after ESWT using a Light Cycler real-time PCR system. Each sample was assayed in duplicate, and experiments were performed least three times independently. The mRNA expression was normalized as ratio = $2^{-\Delta\Delta Ct}$, and data are the mean $\pm$ S.E. * $p < 0.05$ and † $p < 0.01$ vs. the corresponding untreated control group. ESWT decreased the protein expression of N-cadherin (**C**), and increased protein expression of E-cadherin (**D**). Protein expression of N-cadherin and E-cadherin were measured with western blot analysis 24 h and 72 h after ESWT, respectively. The protein expression was normalized with β-actin, respectively, and data are the mean $\pm$ S.E. * $p < 0.05$ vs. the corresponding untreated control group. Un: Untreated cells.

2.6. Effects of ESWT on Transcription Factor Expression in HTSFs

We investigated whether ESWT regulates the expression of the transcription factors inhibitor of DNA binding 1 (ID-1) and inhibitor of DNA binding 2 (ID-2), both of which are known to be involved in anti-fibrotic effects [22,23]. ID-1 protein expression was significantly induced at 24 h after ESWT with 1000 impulses/cm^2 at 0.03 mJ/mm^2 and at 72 h after 1000 impulses/cm^2 at 0.03, 0.1, and 0.3 mJ/mm^2 ($p < 0.05$) (Figure 6A). In contrast, ID-2 protein expression was significantly induced at 24 h after ESWT with 1000 impulses/cm^2 at 0.03, 0.1 and 0.3 mJ/mm^2, and at 72 h after 1000 impulses/cm^2 at 0.03 mJ/mm^2 ($p < 0.05$) (Figure 6B). Furthermore, twist-1 protein expression was significantly reduced at 24 and 72 h after ESWT, respectively ($p < 0.05$) (Figure 6C). These results suggested that expression of transcription factors ID-1, ID-2, and twist-1 might be implicated in the anti-scarring effect of ESWT.

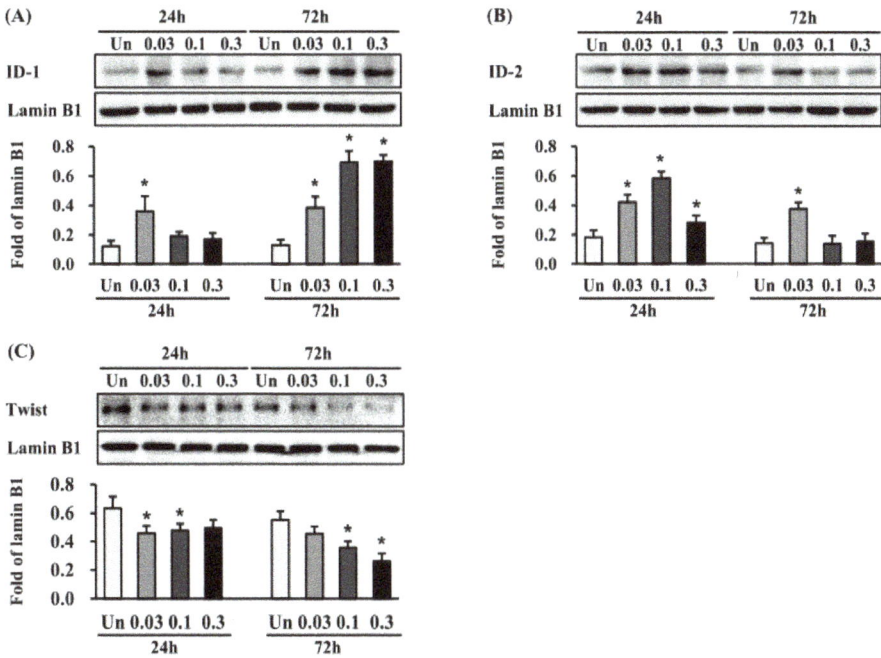

Figure 6. ESWT regulates the expression of ID-1, ID-2, and twist-1 in HTSFs. HTSF was cultured from four patients with post burn hypertrophic scar tissue. The Protein expression of ID-1 (**A**), ID-2 (**B**) and (**C**) were measured with western blot analysis 24 h and 72 h after ESWT, respectively. The protein expression was normalized with lamin B1, and data are the mean $\pm$ S.E. $*p < 0.05$ vs. the corresponding untreated control group. Un: Untreated cells.

2.7. Effects of ESWT on HTSF Migration

For migration assays, HTSFs were seeded in an insert culture system after ESWT, because when ESWT was applied to cells cultured on a dish, the cells detached from the dish. HTSF migration was significantly reduced 24 h (48 h after ESWT) and 48 h (72 h after ESWT) after transfer to the insert upon ESWT with 1000 impulses/cm^2 at 0.03 and 0.1 mJ/mm^2, when compared with non-treated cells ($p < 0.05$) (Figure 7). Even more strongly reduced migration was observed when the energy flux density was 0.3 mJ/mm^2 ($p < 0.01$) (Figure 6).

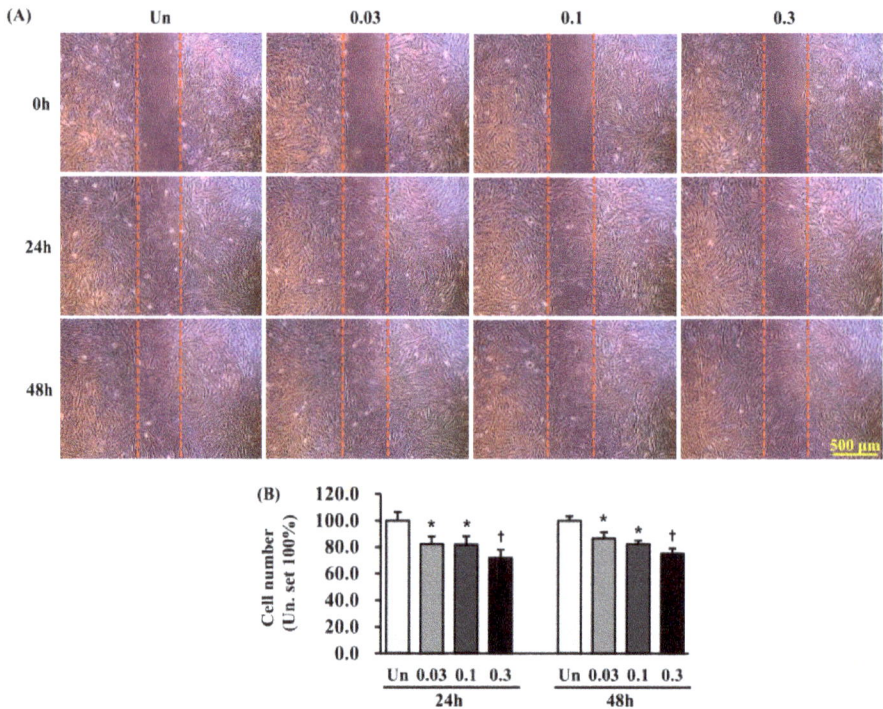

Figure 7. ESWT decreases migration of HTSFs. HTSFs were cultured from four patients with post burn hypertrophic scar tissue. (**A**) The HTSF cells seeded in a culture insert in a 35 mm μ-dish after ESWT, and then after 24 h, the culture insert was removed, made a cell-free gap, allow cells to migrate for 24 h and 48 h. The images were photographed under a light microscopy (scale bar, 500 μm). (**B**) Quantitative analysis of the migration assay was expressed as a percentage relative to untreated cells. The untreated cells were used as control, set to 100%. Data are the mean ± S.E. * $p < 0.05$ and † $p < 0.01$ vs. the corresponding untreated control group. Un: Untreated cells.

3. Discussion

Previous studies have well documented the clinical effectiveness of ESWT in burn scars [19–21]. However, the molecular mechanism remained poorly understood. In the present study, we investigated changes in cell biological behaviors to elucidate the therapeutic mechanism of ESWT underlying its anti-scarring effects. The experimental results suggested that unchanged viability, reduced migration, and the suppressed expression of typical EMT makers in HTFS might be involved in the ESWT anti-scar effect.

Recent evidence suggests that EMT, the reverse of MET, is a process that is essential to wound healing that plays a role in fibrogenesis during HTS formation [6,24,25]. EMT is a process by which epithelial cells lose their epithelial cell characteristics and develop properties typical of mesenchymal cells [26]. During EMT, epithelial cells lose cell–cell connections and downregulate epithelial markers, such as E-cadherin, while they upregulate mesenchymal markers, such as α-SMA, N-cadherin, and fibronectin, and display increased migratory activity [25].

We used HTSFs, which are considered "abnormal" cells, as a research model because fibroblasts have extremely heterogeneous multifunctional potency, and as deep dermal fibroblasts have been suggested to play an important role in HTS formation in deep dermis burn injury [11]. Furthermore, HTSFs have an EMT-like phenotype as compared with matched normal fibroblasts. For example,

HTSFs reportedly have a myofibroblast-like character [27]; high levels of TGF-β1 and its receptors; elevated expression of growth factors and inflammatory cytokines, such as CTGF, IL-6, and IL-8; and, increased ECM components as fibronectin and collagen [28]. We observed similar characters in our study, suggesting that HTSFs are a suitable model for studying HTS pathology in vitro. However, interestingly, in our results, the HTSF expressed various mesenchymal properties, such as increased expression of EMT markers, including TGF-β1, α-SMA, collagen-I, collagen-III, fibronectin, vimentin, FSP-1, E-cadherin, and twist1, and decreased expression of N-cadherin. This suggests that the characteristics of HTSF should be further studied.

In the current literature on the use of ESWT, not only were different parameters used in each study, but there is also no agreement on how to obtain maximum potential. In in vitro studies, ESWT has been used with 500–1500 pulses at 0.03 mJ/mm^2 in human umbilical vein endothelial cells [14], 1000 impulses at 0.14 mJ/mm^2 in human tenocytes [29], and 0.06–0.50 mJ/mm^2 in human osteoblasts [30]. In in vivo studies, ESWT has been used with 1000 impulses at 0.18 mJ/mm^2 on tendon bone in dogs [16], 500 impulses at 0.12 mJ/mm^2 on tendon bone in rabbits [17], and 200 impulses at 0.1 mJ/mm^2 on wound in rabbit [15]. In clinical applications, there are reports of 100 impulses at 0.05–0.15 mJ/mm^2 in burn pruritus and pain [20,21], 500 impulses at 0.13 mJ/mm^2 in burned hands [31], 500 impulses at 0.15 mJ/mm^2 in full-thickness burns [32], and 100 impulses at 0.037 mJ/mm^2 on burn scars [19]. In our study, we applied 1000 impulses/cm^2 at 0.03, 0.1, and 0.3 mJ/mm^2 to HTSFs to determine the molecular therapeutic mechanism, and convincing results were obtained under these experimental conditions. Thus, these experimental conditions can serve as a reference for animal or clinical studies.

Previous studies indicated that HTSFs exhibit resistance to apoptosis, which is one of the causes of scar formation [11,33]. Our results showed an unchanged viability of HTSFs at 24 h after ESWT with 1000 impulses/cm^2 at 0.03, 0.1, and 0.3 mJ/mm^2 energy flux densities. In addition to GAPDH and β-actin, the mRNA expression of bax and bcl-2, apoptosis-related factors, was not affected in HTSF by ESWT (Supplementary Figures S1 and S2), suggesting that the anti-scarring effect of ESWT is not related to cell death.

In the wound-healing process, fibroblasts that migrate from the wound edge into the wound core participate in re-epithelialization and granulation tissue formation through proliferation. In the remodeling phase, fibroblasts differentiate into myofibroblasts, as characterized by high α-SMA expression. These myofibroblasts synthesize and deposit ECM components, which are responsible for granulation tissue contraction and the development of mature scar tissue. At the same time, matrix metalloproteinases (MMPs) and collagenase are involved in functional degradation and remodeling of the ECM [34]. It has been suggested that MMPs induced by interleukin 13 (IL-13) convert inactive TGF-β1 into its active form [35]. In an unbalanced inflammatory reaction or under severe inflammation, over production TGF-β1 induces over activation and over proliferation of fibroblasts, which in turn, leads to excess ECM deposition, eventually resulting in HTS formation [34]. This study showed that the inhibits expression of TGF-β1, α-SMA, and vimentin, both of which are cytoskeletal markers of EMT directly regulated by TGF-β1, are likely involved in the anti-scarring therapeutic effect. TGF-β1 is a potent EMT inducer that plays a critical role in wound healing and pathological development of HTS. In addition, our results show that ESWT distinctly decreased ECM components, including collagen-I and fibronectin. These results indicate that the induction of MET and suppression of EMT might be involved in the anti-scarring effect of ESWT.

Cell migration starts at embryonic development and occurs throughout the life cycle. Generally, for cells to migrate, they must adhere to ECM through interacting with or binding to cell adhesion molecules, such as selectins, integrins, and cadherins. It has become increasingly evident that cadherin is important in cell migration. In particular, expression levels of N- and E-cadherin are closely involved in EMT [36], where the loss of E-cadherin is correlated with an upregulation of N-cadherin [37]. Increased myofibroblast invasion or migration depends on N-cadherin in cancer or wound healing [38]. Moreover, E-cadherin has been shown to be essential for cell migration in epithelial wound healing [39]. Given the correlation between cadherin expression and cell migration, we evaluated whether the changes in the cadherin expression are accompanied by changes in the migration of HTSFs. qRT-PCR and western blotting results showed that at 24 h and 72 h after ESWT, E- and N-cadherin expression increased and decreased, respectively, under all the treatment conditions, except when 0.03 mJ/mm² energy flux density was used. Moreover, we observed that HTSF migration decreased at 24 and 72 h after ESWT. The increased E-cadherin expression and decreased migration, both MET-related processes, also suggests a relevant association between MET and the anti-fibrosis effect of ESWT.

The ID protein family has four family members (ID1-4), which can bind to basic helix–loop–helix (bHLH) transcription factors to form a heterodimer that inhibits transcription factor–DNA binding [40,41]. Thus, the proteins control cell differentiation and proliferation, and can act as oncogene or tumor suppressor [41]. ID proteins can interact with Myo D protein, which is a myogenic bHLH transcription factor, to negatively regulate myogenic differentiation [40]. Previous studies have demonstrated that TGF-β1-induced differentiation of fibroblasts into myofibroblasts is partially dependent on Myo D. Moreover, shRNA-mediated knockdown of Myo D expression in myofibroblasts resulted in the reversion of TGF-β1-induced responses [42]. Particularly, an in vitro study indicated that overexpression of ID-1 suppresses TGF-β1-induced Smad 2/3 signaling, thus decreasing collagen expression in human dermal fibroblasts [22]. ID-2 protein also has broader anti-fibrotic effects; ID2 transgenic mice show resistance to bleomycin-induced pulmonary fibrosis partially through down-regulation twist. Although, mice with alveolar epithelial cell deletion of Twist developed fibrosis after bleomycin [23]. Furthermore, a report indicated soluble ECM peptide downregulates the mRNA expression of ID1 and ID2 on epithelial cells, MCF-10A, while it induced TGF-β signaling [43]. In our study, ESWT significantly upregulated the protein expression of ID-1 and ID-2, while downregulating the protein expression of twist-1. Our results, supported by the above findings, indicate that the anti-scarring effect or induction MET action of ESWT is probably via the regulation of ID and twist-1 protein expression, although we did not detect the expression of the other ID subtype proteins, ID3 and ID4, which is a limitation of the present study. Moreover, additional models, such as three-dimensional (3D) culture models, should provide more physiologically relevant information. While we showed that the transcription factor twist-1 was increased in HTSF, the involvement of other transcription factors in the response of HTSF to ESWT remains unclear. In addition, more importantly, further study is required in keratinocytes from normal skin or HTS, and animal models.

4. Materials and Methods

4.1. Primary Cell Culture

HNFs used in this study were derived from skin biopsy, while HTSFs were isolated from burn-injured HTS tissues derived from surgical procedures, and the HNFs and HTSFs were matched from four patients. The scars ranged in age from one to two years. The study was approved by the Hallym University Hangang Sacred Heart Hospital Institutional Review Board (2 July 2014, registration number 2014-062). Briefly, skin and scar tissues were cut into small pieces, soaked in dispase II (Gibco, Waltham, MA, USA) solution, and maintained at 4 °C overnight. The next day, the epidermis was separated from the dermis, and the dermis was digested with collagenase type IV solution (500 U/mL) at 37 °C for 30 min (Gibco, Waltham, MA, USA). The samples were inactivated with complete medium (DMEM) containing 10% fetal bovine serum (FBS) and 1% antibiotic-antimycotic containing penicillin, streptomycin, and amphotericin B (Gibco, Waltham, MA, USA), filtered, and centrifuged at $300 \times g$ for 5 min. The pellet was resuspended in complete medium, followed by culture at 37 °C in 5% CO_2. HTSFs at passage 2 were used for all of the experiments [44].

4.2. ESWT

HTSFs were serum-starved for 24 h with 0.1% FBS and were then removed from the cell culture flask using cell detachment solution, Accutase® (Thermo Fisher Scientific, Waltham, MA, USA). The cells were suspended 17 mL of starvation medium in 4.5-cm-long conical tubes (Thermo Fisher

Scientific, Waltham, MA, USA) at 1.0×10^5/mL. ESWT was conducted using a Duolith SD-1® device (StorzMedical, Tägerwilen, Switzerland) with an electromagnetic cylindrical coil source of focused shock wave (Figure 1). Cells were treated with 1000 impulses/cm^2 at 0.03, 0.1, and 0.3 mJ/mm^2 of energy flux density, with a frequency of 4 Hz. After ESWT, the HTSFs were maintained for 24 h or 72 h in starvation medium (0.1% FBS, 1% antibiotic-antimycotic) in 96-well cell culture plates, μ-dishes, and T75 culture plates to continue cultivation for further experiments.

4.3. Cell Viability Assay

HTSF viability was assessed using the CellTiter-Glo® Luminescent cell viability assay kit (Promega, Madison, WI, USA). After ESWT, cells were seeded at 1.0×10^4/well on cell culture plates (Corning, New York, NY, USA). After 24 h or 72 h of cultivation, 100 μL of Cell Titer-Glo® reagent was added to the medium, mixed well, and incubated for 10 min at room temperature. Luminescence was recorded using a microplate reader (Brea, CA, USA). Viability was calculated as follows: viability (%) = (sample luminescence − background luminescence)/(control sample luminescence − background luminescence) × 100.

4.4. HTSF Migration Assay

Cell migration was analyzed, as described previously [44], using a culture insert in a 35-mm μ-dish (Ibidi, GmbH, Planegg, Germany), according to the manufacturer's instructions. After ESWT, cells were seeded at 2.0×10^4/insert. After 24 h, the culture insert was removed, thus generating a cell-free gap of 500 ± 50 μm. To eliminate the impact of cell proliferation during migration, mitomycin C (5 μg/mL) (Sigma, St. Louis, MO, USA) was added to the cultures. Cells were imaged at 24 and 48 h under a light microscope (IX 70, Olympus, Tokyo, Japan), and the number of cells that had migrated into the gap was analyzed with Image J software (NIH, Bethesda, MD, USA). ESWT cells were compared with untreated HTSFs, as a control, for which migration was set to 100%. Each analysis was performed in triplicate.

4.5. qRT-PCR

HTSFs were collected 24 or 72 h after re-cultivation. Total RNA was isolated using a ReliaPrep™ RNA Miniprep Systems (Promega), according to the manufacturer's instructions. RNA concentration was measured using a nano-drop spectrophotometer (BioTek, Winooski, VT, USA), and 2 μg of RNA was used to generate cDNA with a high-capacity cDNA reverse transcription kit (Thermo Fisher Scientific, Waltham, MA, USA). qPCR was performed on a LightCycler 480 system (Roche, Basel, Switzerland) using 50 ng of cDNA, 0.5 μM primers (Table S1), and a PCR premix (Takara, Siga, Japan). The reaction conditions were as follows: initial denaturation at 95 °C for 10 min, amplification by 40 cycles of 95 °C for 10 s, 60 °C for 30 s, and extension at 72 °C for 20 s. Target gene mRNA levels of were normalized to the level of GAPDH using the $2^{-\triangle\triangle Ct}$ method [45]. Each qPCR was performed in duplicate with cDNA from at least three different HTSF cultures.

4.6. Western Blotting

HTSFs were harvested 24 or 72 h after re-cultivation. The cells were washed with ice-cold phosphate-buffered saline (PBS), resuspended in ice-cold RIPA buffer (Biosesang, Seongnam, Korea) containing a complete phosphatase inhibitor (Roche, Basel, Switzerland) and protease inhibitor cocktail (Sigma, St. Louis, MO, USA), and agitated for 30 min at 4 °C. The samples were centrifuged for 20 min (15,000× g, 4 °C), and the protein concentrations of the supernatants were determined with a BCA kit (Thermo Fisher Scientific, Waltham, MA, USA). The samples were mixed with 5× sample buffer and were heated at 70 °C for 10 min. Then, they were (30 μg protein/well) electrophoresed in 7.5% or 15% sodium dodecyl sulfate polyacrylamide gel (SDS-PAGE) gel and electro-transferred onto polyvinylidene difluoride (PVDF) membranes, pore size 40 or 20 μm, respectively (Merck Millipore, Billerica, MA, USA). The membranes were blocked with 5% bovine serum albumin (BSA) for 1 h at room temperature and then incubated for 16 h with polyclonal rabbit anti-TGFβ1 antibody (1:500, Santa Cruz Biotechnology, CA, USA), polyclonal mouse anti-αSMA antibody (1:500, Abcam, Cambridge, UK), monoclonal rabbit anti-fibronectin (1:2000, Abcam, Cambridge, UK), polyclonal rabbit anti-collagen-I; antibody (1:100, Abcam, Cambridge, UK), monoclonal rabbit anti-collagen-III (1:2000, Abcam, Cambridge, UK), monoclonal rabbit anti-vimentin (1:1000, Abcam, Cambridge, UK), polyclonal rabbit anti-FSP-1 antibody (1:1000, Merck Millipore, Billerica, MA, USA), monoclonal anti E-cadherin (1:1000, Cell Signaling, Danvers, MA, USA), monoclonal mouse anti N-cadherin (1:1000, Thermo Fisher Scientific, Waltham, MA, USA), monoclonal mouse anti-ID1 antibody (1:200, Santa Cruz Biotechnology, Santa Cruz, CA, USA), monoclonal mouse anti-ID2 antibody (1:200, Santa Cruz Biotechnology, Santa Cruz, CA, USA), and anti-β-actin (1:5000, Cell Signaling, Danvers, MA, USA). The membranes were washed three times (10 min/wash) with tris-buffered saline-tween 20 (TBST) buffer, and then incubated with peroxidase-conjugated anti-rabbit IgG (1:5000, Merck Millipore, Billerica, MA, USA) for 2 h at room temperature. They were then washed three times (10 min/wash) and were developed with an ECL detection kit (Thermo Fisher Scientific, Waltham, MA, USA). Images were obtained using a chemiluminescence imaging system (WSE-6100; Atto, Tokyo, Japan). Band densities were determined with CS Analyzer4 software (Atto, Tokyo, Japan), and normalized to β-actin density.

4.7. Statistical Analysis

All results are presented as the mean ± SEM. The Mann-Whitney U test was used for comparisons between two groups. Statistical analyses were conducted with PASW statistics 18 (SPSS Inc., Chicago, IL, USA), and $p < 0.05$ was considered significant.

5. Conclusions

In present study, we demonstrated that ESWT suppresses EMT in HTSFs by inhibiting the expression of TGF-β1, a potent EMT inducer, as well as that of α-SMA, collagen-I, fibronectin, and N-cadherin, and by upregulating E-cadherin. Moreover, ESWT inhibits the migratory ability of HTSF. These molecular changes contribute to the anti-fibrotic effects of ESWT on HTSFs. The findings explain well the beneficial effects of ESWT observed in the clinic.

Supplementary Materials: The following are available online at www.mdpi.com/1422-0067/19/1/124/s1, Figure S1: mRNA expression of β-actin and GAPDH in HTSFs after ESWT, Figure S2: mRNA expression of apoptotic related factors in HTSFs after ESWT, Figure S3: Immunocytochemical analysis of vimentin expression in HTSFs after ESWT, Table S1: Real-time PCR primer sequences.

Acknowledgments: This research was supported by the Basic Science Research Program through the National Research Foundation of Korea (NRF) funded by the Ministry of Education (NRF-2014R1A1A4A01007956, 2017R1D1A1A02018478, 2017R1D1A1B03029731), Hallym University Research Fund, and the Korea Health Technology R&D Project through the Korea Health Industry Development Institute (KHIDI), funded by the Ministry of Health & Welfare, Republic of Korea (HI15C1486).

Author Contributions: Hui Song Cui performed most of the experiments, analyzed data, wrote the manuscript, and contributed to the study concept and design; A Ram Hong contributed to the study concept, and design the project; Joo Hyang Yu performed the experiments; June-Bum Kim analyzed data and developed the study concept; and So Young Joo and Cheong Hoon Seo performed most of the experiments, analyzed data, wrote the manuscript, developed the study concept, and supervised the project. All authors read and approved the manuscript.

Conflicts of Interest: The authors declare no conflict of interest.

References

1. Gauglitz, G.G.; Korting, H.C.; Pavicic, T.; Ruzicka, T.; Jeschke, M.G. Hypertrophic scarring and keloids: Pathomechanisms and current and emerging treatment strategies. *Mol. Med.* **2011**, *17*, 113–125. [CrossRef] [PubMed]
2. Aarabi, S.; Longaker, M.T.; Gurtner, G.C. Hypertrophic scar formation following burns and trauma: New approaches to treatment. *PLoS Med.* **2007**, *4*, e234. [CrossRef] [PubMed]

3. Armour, A.; Scott, P.G.; Tredget, E.E. Cellular and molecular pathology of HTS: Basis for treatment. *Wound Repair Regen.* **2007**, *15* (Suppl. 1), S6–S17. [CrossRef] [PubMed]
4. Wynn, T.A. Fibrotic disease and the T(H)1/T(H)2 paradigm. *Nat. Rev. Immunol.* **2004**, *4*, 583–594. [CrossRef] [PubMed]
5. Tredget, E.; Shankowsky, H.; Pannu, R.; Nedelec, B.; Iwashina, T.; Ghahary, A.; Taerum, T.; Scott, P. Transforming growth factor-β in thermally injured patients with hypertrophic scars: Effects of interferon α2b. *Plast. Reconstr. Surg.* **1998**, *102*, 1317–1328. [CrossRef] [PubMed]
6. Schmid, P.; Itin, P.; Cherry, G.; Bi, C.; Cox, D.A. Enhanced expression of transforming growth factor-β type I and type II receptors in wound granulation tissue and hypertrophic scar. *Am. J. Pathol.* **1998**, *152*, 485–493. [PubMed]
7. Yan, C.; Grimm, W.A.; Garner, W.L.; Qin, L.; Travis, T.; Tan, N.; Han, Y.P. Epithelial to mesenchymal transition in human skin wound healing is induced by tumor necrosis factor-α through bone morphogenic protein-2. *Am. J. Pathol.* **2010**, *176*, 2247–2258. [CrossRef] [PubMed]
8. Wang, R.; Ghahary, A.; Shen, Q.; Scott, P.G.; Roy, K.; Tredget, E.E. Hypertrophic scar tissues and fibroblasts produce more transforming growth factor-β1 mRNA and protein than normal skin and cells. *Wound Repair. Regen.* **2000**, *8*, 128–137. [CrossRef] [PubMed]
9. Nedelec, B.; Shankowsky, H.; Scott, P.G.; Ghahary, A.; Tredget, E.E. Myofibroblasts and apoptosis in human hypertrophic scars: The effect of interferon-α2b. *Surgery* **2001**, *130*, 798–808. [CrossRef] [PubMed]
10. Garner, W.L.; Karmiol, S.; Rodriguez, J.L.; Smith, D.J., Jr.; Phan, S.H. Phenotypic differences in cytokine responsiveness of hypertrophic scar versus normal dermal fibroblasts. *J. Investig. Dermatol.* **1993**, *101*, 875–879. [CrossRef] [PubMed]
11. Honardoust, D.; Ding, J.; Varkey, M.; Shankowsky, H.A.; Tredget, E.E. Deep dermal fibroblasts refractory to migration and decorin-induced apoptosis contribute to hypertrophic scarring. *J. Burn Care Res.* **2012**, *33*, 668–677. [CrossRef] [PubMed]
12. Notarnicola, A.; Moretti, B. The biological effects of extracorporeal shock wave therapy (ESWT) on tendon tissue. *Muscles Ligaments Tendons J.* **2012**, *2*, 33–37. [PubMed]
13. Wang, C.J. Extracorporeal shockwave therapy in musculoskeletal disorders. *J. Orthop. Surg. Res.* **2012**, *20*, 11. [CrossRef] [PubMed]
14. Mariotto, S.; Cavalieri, E.; Amelio, E.; Ciampa, A.R.; de Prati, A.C.; Marlinghaus, E.; Russo, S.; Suzuki, H. Extracorporeal shock waves: From lithotripsy to anti-inflammatory action by NO production. *Nitric Oxide* **2005**, *12*, 89–96. [CrossRef] [PubMed]
15. Zins, S.R.; Amare, M.F.; Tadaki, D.K.; Elster, E.A.; Davis, T.A. Comparative analysis of angiogenic gene expression in normal and impaired wound healing in diabetic mice: Effects of extracorporeal shock wave therapy. *Angiogenesis* **2010**, *13*, 293–304. [CrossRef] [PubMed]
16. Wang, C.J.; Huang, H.Y.; Pai, C.H. Shock wave enhances neovascularization at the tendon-bone junction. *J. Foot Ankle Surg.* **2002**, *41*, 16–22. [CrossRef]
17. Wang, C.J.; Yang, K.D.; Wang, F.S.; Huang, C.C.; Yang, L.J. Shock wave induces neovascularization at the tendon-bone junction. A study in rabbits. *J. Orthop. Res.* **2003**, *21*, 984–989. [CrossRef]
18. Schaden, W.; Thiele, R.; Kolpl, C.; Pusch, M.; Nissan, A.; Attinger, C.E.; Maniscalco-Theberge, M.E.; Peoples, G.E.; Elster, E.A.; Stojadinovic, A. Shock wave therapy for acute and chronic soft tissue wounds: A feasibility study. *J. Surg. Res.* **2007**, *143*, 1–12. [CrossRef] [PubMed]
19. Fioramonti, P.; Cigna, E.; Onesti, M.G.; Fino, P.; Fallico, N.; Scuderi, N. Extracorporeal shock wave therapy for the management of burn scars. *Dermatol. Surg.* **2012**, *38*, 778–782. [CrossRef] [PubMed]
20. Cho, Y.S.; Joo, S.Y.; Cui, H.; Cho, S.R.; Yim, H.; Seo, C.H. Effect of extracorporeal shock wave therapy on scar pain in burn patients: A prospective, randomized, single-blind, placebo-controlled study. *Medicine* **2016**, *95*, e4575. [CrossRef] [PubMed]
21. Joo, S.Y.; Cho, Y.S.; Seo, C.H. The clinical utility of extracorporeal shock wave therapy for burn pruritus: A prospective, randomized, single-blind study. *Burns* **2017**. [CrossRef] [PubMed]
22. Je, Y.J.; Choi, D.K.; Sohn, K.C.; Kim, H.R.; Im, M.; Lee, Y.; Lee, J.H.; Kim, C.D.; Seo, Y.J. Inhibitory role of Id1 on TGF-β-induced collagen expression in human dermal fibroblasts. *Biochem. Biophys. Res. Commun.* **2014**, *444*, 81–85. [CrossRef] [PubMed]

23. Yang, J.; Velikoff, M.; Agarwal, M.; Disayabutr, S.; Wolters, P.J.; Kim, K.K. Overexpression of inhibitor of DNA-binding 2 attenuates pulmonary fibrosis through regulation of c-Abl and Twist. *Am. J. Pathol.* **2015**, *185*, 1001–1011. [CrossRef] [PubMed]

24. Barriere, G.; Fici, P.; Gallerani, G.; Fabbri, F.; Rigaud, M. Epithelial Mesenchymal Transition: A double-edged sword. *Clin. Transl. Med.* **2015**, *14*, 4. [CrossRef] [PubMed]

25. Weber, C.E.; Li, N.Y.; Wai, P.Y.; Kuo, P.C. Epithelial-mesenchymal transition, TGF-β, and osteopontin in wound healing and tissue remodeling after injury. *J. Burn. Care Res.* **2012**, *33*, 311–318. [CrossRef] [PubMed]

26. Kalluri, R.; Neilson, E.G. Epithelial-mesenchymal transition and its implications for fibrosis. *J. Clin. Investig.* **2003**, *112*, 1776–1784. [CrossRef] [PubMed]

27. Song, R.; Bian, H.N.; Lai, W.; Chen, H.D.; Zhao, K.S. Normal skin and hypertrophic scar fibroblasts differentially regulate collagen and fibronectin expression as well as mitochondrial membrane potential in response to basic fibroblast growth factor. *Braz. J. Med. Biol. Res.* **2011**, *44*, 402–410. [CrossRef] [PubMed]

28. Penn, J.W.; Grobbelaar, A.O.; Rolfe, K.J. The role of the TGF-β family in wound healing, burns and scarring: A review. *Int. J. Burns Trauma* **2012**, *2*, 18–28. [PubMed]

29. Vetrano, M.; d'Alessandro, F.; Torrisi, M.R.; Ferretti, A.; Vulpiani, M.C.; Visco, V. Extracorporeal shock wave therapy promotes cell proliferation and collagen synthesis of primary cultured human tenocytes. *Knee Surg. Sports Traumatol. Arthrosc.* **2011**, *19*, 2159–2168. [CrossRef] [PubMed]

30. Hofmann, A.; Ritz, U.; Hessmann, M.H.; Alini, M.; Rommens, P.M.; Rompe, J.D. Extracorporeal shock wave-mediated changes in proliferation, differentiation, and gene expression of human osteoblasts. *J. Trauma* **2008**, *65*, 1402–1410. [CrossRef] [PubMed]

31. Saggini, R.; Saggini, A.; Spagnoli, A.M.; Dodaj, I.; Cigna, E.; Maruccia, M.; Soda, G.; Bellomo, R.G.; Scuderi, N. Extracorporeal Shock Wave Therapy: An Emerging Treatment Modality for Retracting Scars of the Hands. *Ultrasound Med. Biol.* **2016**, *42*, 185–195. [CrossRef] [PubMed]

32. Arnó, A.; García, O.; Hernán, I.; Sancho, J.; Acosta, A.; Barret, J.P. Extracorporeal shock waves, a new non-surgical method to treat severe burns. *Burns* **2010**, *36*, 844–849. [CrossRef] [PubMed]

33. Linge, C.; Richardson, J.; Vigor, C.; Clayton, E.; Hardas, B.; Rolfe, K. Hypertrophic scar cells fail to undergo a form of apoptosis specific to contractile collagen-the role of tissue transglutaminase. *J. Investig. Dermatol.* **2005**, *125*, 72–82. [CrossRef] [PubMed]

34. Albanna, M.; Homes, J.H., IV. *Skin Tissue Engineering and Regenerative Medicine*, 1st ed.; Academic Press: London, UK, 2016; pp. 23–32. ISBN 9780128016541.

35. Lee, C.G.; Homer, R.J.; Zhu, Z.; Lanone, S.; Wang, X.; Koteliansky, V.; Shipley, J.M.; Gotwals, P.; Noble, P.; Chen, Q.; et al. Interleukin-13 induces tissue fibrosis by selectively stimulating and activating transforming growth factor β1. *J. Exp. Med.* **2001**, *194*, 809–821. [CrossRef] [PubMed]

36. Zeisberg, M.; Neilson, E.G. Biomarkers for epithelial-mesenchymal transitions. *J. Clin. Investig.* **2009**, *119*, 1429–1437. [CrossRef] [PubMed]

37. Derycke, L.D.; Bracke, M.E. N-Cadherin in the spotlight of cell–cell adhesion, differentiation, embryogenesis, invasion and signalling. *Int. J. Dev. Biol.* **2004**, *48*, 463–476. [CrossRef] [PubMed]

38. Bhowmick, N.A.; Ghiassi, M.; Bakin, A.; Aakre, M.; Lundquist, C.A.; Engel, M.E.; Arteaga, C.L.; Moses, H.L. Transforming growth factor-β1 mediates epithelial to mesenchymal transdifferentiation through a RhoA-dependent mechanism. *Mol. Biol. Cell* **2001**, *12*, 27–36. [CrossRef] [PubMed]

39. Hwang, S.; Zimmerman, N.P.; Agle, K.A.; Turner, J.R.; Kumar, S.N.; Dwinell, M.B. E-Cadherin is critical for collective sheet migration and is regulated by the chemokine CXCL12 protein during restitution. *J. Biol. Chem.* **2012**, *287*, 22227–22240. [CrossRef] [PubMed]

40. Benezra, R.; Davis, R.L.; Lockshon, D.; Turner, D.L.; Weintraub, H. The protein Id: A negative regulator of helix-loop-helix DNA binding proteins. *Cell* **1990**, *6*, 49–59. [CrossRef]

41. Lasorella, A.; Benezra, R.; Iavarone, A. The ID proteins: Master regulators of cancer stem cells and tumour aggressiveness. *Nat. Rev. Cancer* **2014**, *14*, 77–91. [CrossRef] [PubMed]

42. Hecker, L.; Jagirdar, R.; Jin, T.; Thannickal, V.J. Reversible differentiation of myofibroblasts by MyoD. *Exp. Cell Res.* **2011**, *317*, 1914–1921. [CrossRef] [PubMed]

43. Garamszegi, N.; Garamszegi, S.P.; Samavarchi-Tehrani, P.; Walford, E.; Schneiderbauer, M.M.; Wrana, J.L.; Scully, S.P. Extracellular matrix-induced transforming growth factor-β receptor signaling dynamics. *Oncogene* **2010**, *29*, 2368–2380. [CrossRef] [PubMed]

44. Cui, H.S.; Joo, S.Y.; Lee, D.H.; Yu, J.H.; Jeong, J.H.; Kim, J.B.; Seo, C.H. Low temperature plasma induces angiogenic growth factor via up-regulating hypoxia-inducible factor 1α in human dermal fibroblasts. *Arch. Biochem. Biophys.* **2017**, *630*, 9–17. [CrossRef] [PubMed]
45. Livak, K.J.; Schmittgen, T.D. Analysis of relative gene expression data using real-time quantitative PCR and the $2^{-\Delta\Delta Ct}$ Method. *Methods* **2001**, *25*, 402–408. [CrossRef] [PubMed]

International Journal of
Molecular Sciences

MDPI

Article

HtrA1 Is Specifically Up-Regulated in Active Keloid Lesions and Stimulates Keloid Development

Satoko Yamawaki [1], Motoko Naitoh [2,*], Hiroshi Kubota [3], Rino Aya [2], Yasuhiro Katayama [2], Toshihiro Ishiko [4], Taku Tamura [3] , Katsuhiro Yoshikawa [5], Tatsuki Enoshiri [2], Mika Ikeda [6] and Shigehiko Suzuki [2]

[1] Department of Plastic and Reconstructive Surgery, Japanese Red Cross Fukui Hospital, 2-4-1, Tsukimi, Fukui-City, Fukui 918-8501, Japan; satokoy@kuhp.kyoto-u.ac.jp
[2] Department of Plastic and Reconstructive Surgery, Graduate School of Medicine, Kyoto University, 54 Kawahara-cho, Sakyo-ku, Kyoto 606-8507, Japan; rinok@kuhp.kyoto-u.ac.jp (R.A.); hemim@kuhp.kyoto-u.ac.jp (Y.K.); enotatsu@kuhp.kyoto-u.ac.jp (T.E.); ssuzuk@kuhp.kyoto-u.ac.jp (S.S.)
[3] Department of Life Science, Faculty of Engineering Science, Akita University, 1-1 Tegata Gakuenmachi, Akita 010-8502, Japan; hkubota@gipc.akita-u.ac.jp (H.K.); taku@gipc.akita-u.ac.jp (T.T.)
[4] Department of Plastic and Reconstructive Surgery, Japanese Red Cross Otsu Hospital, 1-1-35, Nagara, Otsu City, Shiga 520-8511, Japan; ishiko@otsu.jrc.or.jp
[5] Department of Plastic and Reconstructive Surgery, Shiga Medical Center for Adults, 5-4-30, Moriyama, Moriyama City, Shiga 524-8524, Japan; khiro@kuhp.kyoto-u.ac.jp
[6] Department of Plastic and Reconstructive Surgery, Kobe City Medical Center General Hospital, 2-1-1, Minatojima minami-machi, Cyuou-ku, Kobe City, Hyogo 650-0047, Japan; mikaring@crux.ocn.ne.jp
* Correspondence: mnaitoh@kuhp.kyoto-u.ac.jp; Tel.: +81-75-751-3613

Received: 3 March 2018; Accepted: 16 April 2018; Published: 24 April 2018

Abstract: Keloids occur after failure of the wound healing process; inflammation persists, and various treatments are ineffective. Keloid pathogenesis is still unclear. We have previously analysed the gene expression profiles in keloid tissue and found that HtrA1 was markedly up-regulated in the keloid lesions. HtrA1 is a serine protease suggested to play a role in the pathogenesis of various diseases, including age-related macular degeneration and osteoarthritis, by modulating extracellular matrix or cell surface proteins. We analysed HtrA1 localization and its role in keloid pathogenesis. Thirty keloid patients and twelve unrelated patients were enrolled for in situ hybridization, immunohistochemical, western blot, and cell proliferation analyses. Fibroblast-like cells expressed more HtrA1 in active keloid lesions than in surrounding lesions. The proportion of HtrA1-positive cells in keloids was significantly higher than that in normal skin, and HtrA1 protein was up-regulated relative to normal skin. Silencing *HtrA1* gene expression significantly suppressed cell proliferation. HtrA1 was highly expressed in keloid tissues, and the suppression of the *HtrA1* gene inhibited the proliferation of keloid-derived fibroblasts. HtrA1 may promote keloid development by accelerating cell proliferation and remodelling keloid-specific extracellular matrix or cell surface molecules. HtrA1 is suggested to have an important role in keloid pathogenesis.

Keywords: keloids; fibroproliferative disorder; HtrA1; inflammation

1. Introduction

Keloids are a dermal fibrotic disease characterized by abnormal accumulation of extracellular matrix (ECM) and fibroproliferation in the dermis [1,2]. They appear as raised, red, and inflexible scar tissue that develops during the wound-healing process, even from tiny wounds including vaccination and insect bites. Keloid lesions expand over the boundaries of the initial injury site, and the lesions continue to develop and become larger [3,4]. The many treatments for keloids include steroid injections, steroid tape, and surgery with postoperative irradiation. The cure rate following surgery and postoperative radiation varies widely from 28~89% [3,5–8] and depends on the individual. Clarifying keloid pathogenesis could improve the treatment outcome.

Previously, we studied the molecular mechanism of keloid pathogenesis using cDNA microarray and Northern blot analysis to compare gene expression patterns in keloid lesions and normal skin [9]. HtrA1, a member of the HtrA family of serine protease and a mammalian homolog of *Escherichia coli* HtrA (DegP), was markedly upregulated in the keloid lesions. As human HtrA1 has multiple domains, including protease, IGFBP, and PDZ domains, HtrA1 has been expected to be a multifunctional protein. Several cellular and molecular studies suggested that HtrA1 plays a key role in regulating various cellular processes via the cleavage and/or binding of pivotal factors that participate in cell proliferation, migration, and cell fate [10–13] HtrA1 has been suggested to be closely associated with the pathology of various diseases, including osteoarthritis, age-related macular degeneration (AMD), familial cerebral small vessel disease (CARASIL), and malignant tumours. HtrA1 was also suggested to stimulate progression of arthritis through degrading cartilage matrix in osteoarthritis [14]. Recently, the increased expression of human HtrA1 in the mouse retinal pigment epithelium (RPE) was shown to induce vasculogenesis and degeneration of the elastic lamina and tunica media of the vessels, similar to that observed in AMD patients [15,16]. These observations imply that HtrA1 plays a role in the pathogenesis of various diseases by modulating proteins in the ECM or cell surface. Although controversial, HtrA1 has been proposed as a key molecule in osteogenesis and chondrogenesis [14,17,18]. HtrA1 expression is induced during hypertrophic change in chondrocytes, with the up-regulation of the type X collagen marker in keloid lesions [9,18]. HtrA1 is closely concerned with normal osteogenesis and in pathogenesis of arthritis [14]. In arthritis, synovial fibroblasts identified as a major source of HtrA1 degrading cartilage matrix, such as fibronectin and aggrecan, which are abundant in keloid lesions [9,14,18].

Based on the foregoing data, in this study, we focused on HtrA1. We examined the expression and localization of HtrA1 in keloid tissues, using in situ hybridization and immunohistochemical studies. HtrA1 was strongly up-regulated at both the mRNA and protein levels in the hypercellular and active keloid lesions. Silencing *HtrA1* gene expression in keloid fibroblasts significantly inhibited cell proliferation, and additional recombinant HtrA1 stimulated keloid fibroblast proliferation. We propose that HtrA1 may be a pivotal molecule in keloid pathogenesis, and our discussion centres on the possible roles of HtrA1 in the molecular mechanism of keloid development.

2. Results

2.1. In Situ Hybridization of HtrA1 mRNA in Keloid Lesions and Normal Skin

To confirm the up-regulation of the mRNA level for HtrA1, we previously observed using microarray and Northern blot analyses, and to determine the localization of *HtrA1* mRNA in keloid lesions, in situ hybridization was performed using skin samples from six keloid patients. In one specimen (No. 27 in Table 1), in situ hybridization was performed on several parts of lesions which differed in keloid activity. The expression of the *HtrA1* gene was clearly detected in the fibroblasts in the hypercellular and actively growing area of keloid lesions (Figure 1a, Supplementary Figure S1a,c,e), but not in unaffected skin (Figure 1b). In the sections hybridized with sense probe, no signal was observed (Supplementary Figure S1b,d,f), demonstrating specific staining by the antisense probe. All keloid sections were hard and elevated in the keloid lesions. In these regions, the antisense probe provided strong signals (Figure 1a, Supplementary Figure S1a,c,e). Clinical findings and the results of in situ hybridization of sample 27, which was an abdominal keloid after laparoscopic surgery for removal of uterine myoma, as depicted in Figure 2. Keloid activity was in the order of a, b and c. Higher activity in the affected portion of the lesion was associated with greater cell proliferation and greater up-regulation of HtrA1 (Figure 2). *HtrA1* mRNA was strongly up-regulated, and expression of HtrA1 was more pronounced in keloid lesions.

Table 1. Samples used in this study.

Tissue Source	No.	Age	Sex	Region	HtrA1-Positve Cells (%) [1]	Assays [2]
Keloid	1	75	M	shoulder	18.83	A
	2	51	F	chest	12.40	A
	3	49	M	neck	38.74	A, B
	4	67	F	abdomen	43.52	A
	5	32	M	chest	24.77	A
	6	34	M	abdomen	40.64	A
	7	67	M	abdomen	32.80	A
	8	16	F	chest	41.90	A, C
	9	64	M	shoulder	25.82	A
	10	28	M	chest	24.85	A
	11	24	F	chest	28.36	A, C
	12	20	F	shoulder	35.43	A
	13	62	F	chest	12.68	A
	14	30	M	shoulder	35.34	A
	15	20	F	chest	48.37	A
	16	65	M	back	35.55	A
	17	38	F	chest	39.68	A
	18	39	F	abdomen		B
	19	75	M	chest		B
	20	21	M	back		B
	21	20	M	back		B
	22	31	M	shoulder		D
	23	20	F	shoulder		D
	24	20	F	chest		C, D
	25	58	F	shoulder		D
	26	24	M	chest		C
	27	41	F	abdomen		B
	28	27	M	chest		C
	29	17	F	chest		C
	30	24	F	chest		C
Normal skin	1	51	F	back	3.76	A, D
	2	45	F	abdomen	2.83	A
	3	47	F	shoulder	2.60	A
	4	51	F	thigh	2.13	A
	5	88	M	back		D
	6	49	F	abdomen		D
	7	51	F	abdomen		D
	8	52	F	chest		C
	9	53	F	chest		C
	10	40	F	abdomen		C
	11	31	F	abdomen		C
	12	52	F	chest		C

[1] The percentage of HtrA1-positive cells was determined using immunohistochemical staining. [2] A, immunohistochemical staining; B, in situ hybridization; C, cell proliferation assay with silencing HtrA1 gene expression or with additional rHtrA1; D, western blotting.

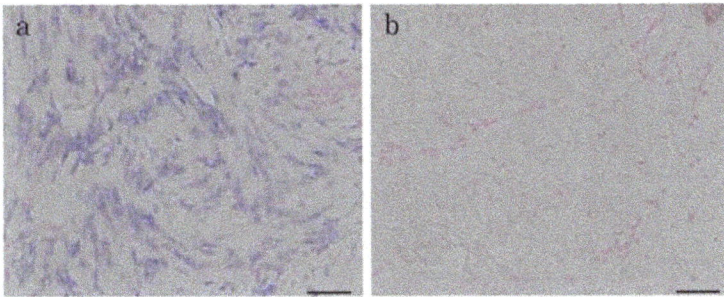

Figure 1. In situ hybridisation for *HtrA1* mRNA in keloid and normal skin. Sections from active keloid lesions (**a**) or unaffected region (**b**) (patient No. 18 in Table 1) were hybridised with a probe specific to *HtrA1* mRNA. Positive signals are visualised in blue. Scale bar = 50 μm.

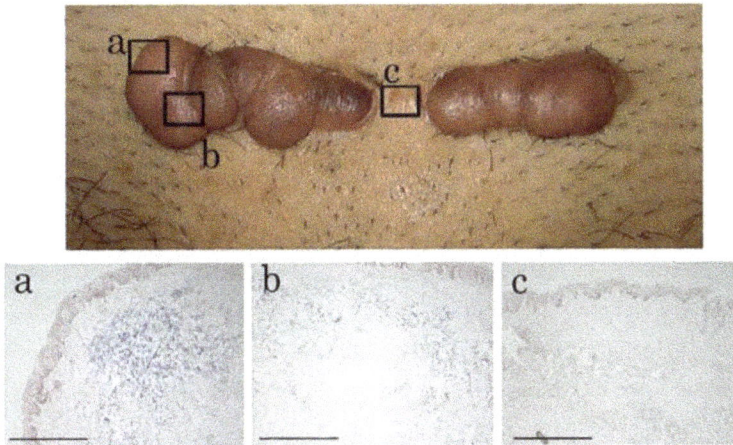

Figure 2. An abdominal keloid after laparoscopic surgery. The activity of keloid was in order a, b and c. Higher activity in regions of the lesion were associated with increased cell proliferation and greater up-regulation of HtrA1. Scale bar = 500 μm.

2.2. Immunohistochemical Staining and Western Blot Analysis of HtrA1

To examine whether the up-regulation of HtrA1 at the mRNA level leads to increases at the protein level, we performed immunohistochemical analysis to detect HtrA1 (Figure 3a,b, Supplementary Figure S2a–f). HtrA1 was clearly detected by immunostaining in keloids (Figure 3a, Supplementary Figure S2a, c, e, while no signals were observed in normal skin (Figure 3b). These data were consistent with the in situ hybridization findings. No positive signal was found in controls not treated with the primary antibody (Supplementary Figure S2b, d, f). Therefore, HtrA1 was strongly up-regulated at the protein level in active areas of the keloid lesions. To confirm the up-regulation of HtrA1 protein, western blot analysis was performed. In all keloid tissue samples from four patients, HtrA1 protein was up-regulated, relative to four normal skin samples (Figure 4). Enumeration of HtrA1-positive cells after immunohistochemical staining indicated that the proportion of cells expressing detectable levels of HtrA1 in keloid tissue ranged from 12.4% to 48.4%, with an average of $31.9 \pm 10.5\%$ (Figure 5). In contrast, the proportion of HtrA1-positive cells in normal skin ranged from 2.1% to 3.8%, with an average of $2.8 \pm 0.6\%$. The proportion of HtrA1-positive cells was significantly higher in keloids than in normal skin ($p < 0.001$). The total number of fibroblasts was much less in normal skin relative to keloid tissue (Figure 3), as previously reported [9]. These results indicate that keloid tissue exhibits an increase in the number of fibroblasts producing HtrA1, as well as an increase in the total number of fibroblasts.

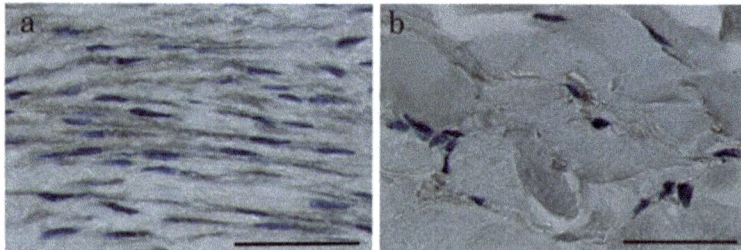

Figure 3. Immunohistochemical staining of HtrA1 protein in keloid (**a**) and normal skin tissue (**b**). Sections from active keloid lesions (**a**) or normal skin (**b**). (**a**) displays the results from patient No. keloid-3 in Table 1, and (**b**) displays the results from patient No. normal skin-1 in Table 1. Positive signals are visualised in brown. Scale bar = 50 μm.

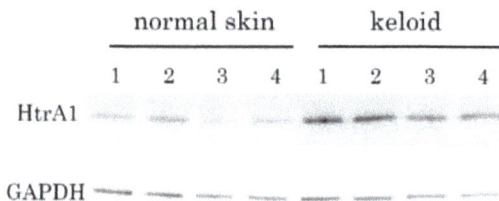

Figure 4. Western blot analysis of HtrA1 in keloid lesions and normal skin tissues. Soluble protein extract (8 μg/lane) was analysed using specific antibodies against HtrA1 or glyceraldehyde-3-phosphate dehydrogenase (GAPDH). Keloid and normal skin samples from four different patients were analysed.

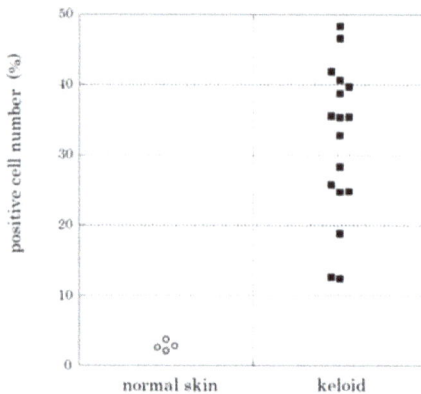

Figure 5. Proportion of fibroblasts expressing HtrA1 protein in keloid lesions and normal skin. The number of fibroblasts with positive signals was counted after immunohistochemical staining of HtrA1 using samples from 17 keloidand 4 unrelated patients. Ten high-power (×400) fields were selected at random from a section and numbers of total and stained fibroblasts were counted. Patient information is described with proportion of HtrA1-positive cells in Table 1.

2.3. HtrA1 Knockdown Inhibits Keloid Cell Proliferation

To investigate role of HtrA1 in keloid pathogenesis, we examined whether HtrA1 affects cell proliferation by silencing *HtrA1* gene expression using specific small interfering RNA (siRNA). Keloid fibroblasts treated with *HtrA1* siRNA exhibited a proliferation rate significantly slower relative to those treated with control siRNA (Figure 6, Supplementary Figure S3). This effect with silencing HtrA1 was also observed in normal fibroblasts, but the inhibition effect was not as pronounced.

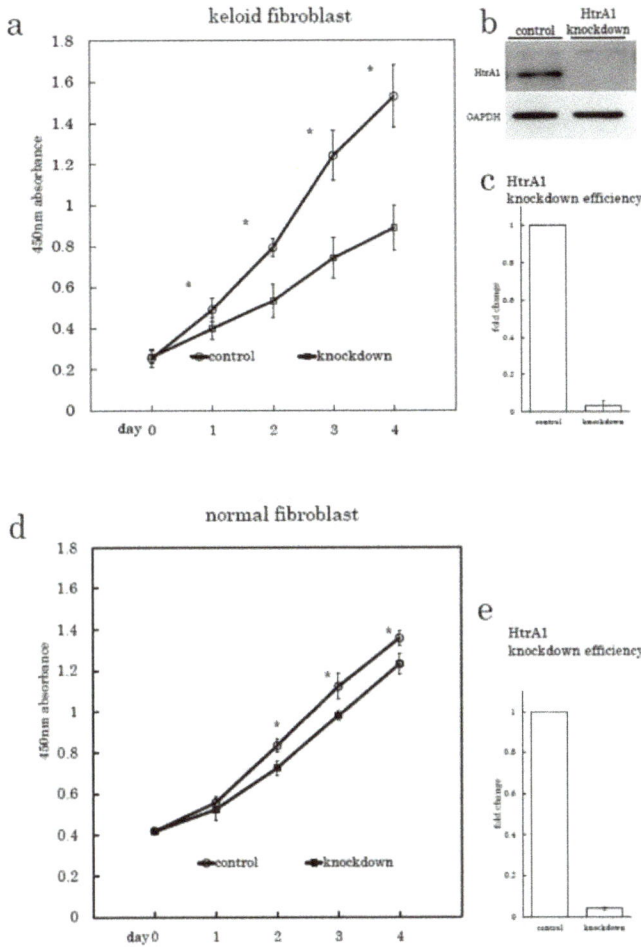

Figure 6. Proliferation rates of keloid fibroblasts and normal fibroblasts transfected with HtrA1 siRNA or control siRNA. Proliferation curves of keloid fibroblasts obtained from keloid sample No. 26 as shown in Table 1 (**a**), (*n* = 3) and normal fibroblasts from sample No. 8 (**d**) transfected with HtrA1 siRNA (knockdown) or control siRNA (control). The efficiency of HtrA1 knockdown in keloid fibroblasts was determined using western blot analysis (**b**) and quantitative PCR (**c**), (*n* = 3). The efficiency of HtrA1 knockdown in normal fibroblasts was similarly determined using quantitative PCR (**e**), *n* = 3. Cell proliferation was analysed using a colorimetric assay with a water-soluble tetrazolium salt as the substrate. Error bars represent standard deviations (*n* = 3). * *p* < 0.001.

2.4. Additional HtrA1 in Culture Medium Stimulates Keloid Cell Proliferation

To confirm the effect of HtrA1 on cell proliferation, we performed a proliferation assay on keloid fibroblasts with the addition of recombinant human HtrA1 in culture medium (Figure 7, Supplementary Figure S4). The addition of HtrA1 stimulated the proliferation of keloid fibroblasts, but not normal fibroblasts. These results suggest that HtrA1 plays an important role in keloid cell proliferation.

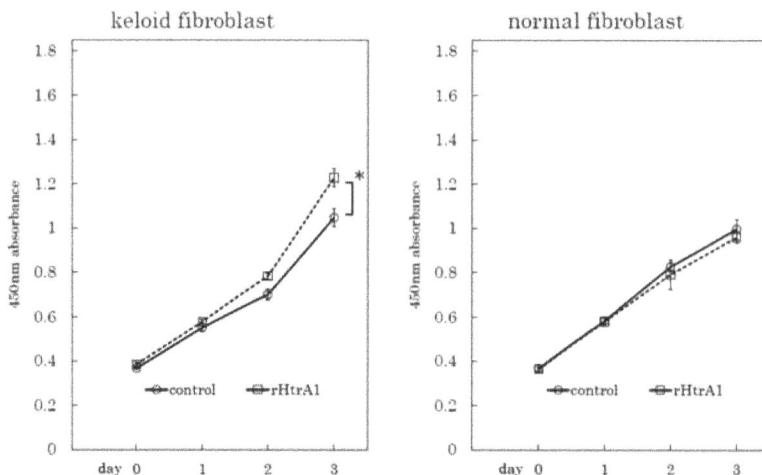

Figure 7. Proliferation rates of keloid fibroblasts and normal fibroblasts incubated with or without recombinant HtrA1. Proliferation curves of keloid fibroblasts obtained from sample No. 29 and normal fibroblasts from sample No. 8 as shown in Table 1, incubated with (rHtrA1) or without (control) recombinant HtrA1. $n = 3$, * $p < 0.01$.

3. Discussion

In the present study, the expression of HtrA1 was strongly up-regulated in active keloid legions as analysed by in situ hybridization and immunohistochemical staining. Previous studies suggested that HtrA1 stimulates arthritis by digesting the ECM [14]. In arthritis, synovial fibroblasts produce abundant HtrA1, and HtrA1 digests cartilage ECM, including fibronectin, collagens, and proteoglycans. ECM fragments produced by HtrA1 digestion reportedly activate synovial fibroblasts and induce the remodelling of cartilage ECM. We propose that HtrA1 functions as a matrix protease that stimulates keloid development because the keloid matrix consists mainly of collagens, fibronectin, and proteoglycans, which are substrates for HtrA1. HtrA1 may degrade keloid matrix and accelerate ECM remodelling in keloid lesions. Matrix protein fragments produced by HtrA1 may activate keloid cells, leading to further progression of the disease. Consistent with this notion, we found HtrA1-knockdown inhibited the proliferation of keloid fibroblasts, and that recombinant HtrA1 added to the culture medium stimulated the proliferation of keloid fibroblasts. Interestingly, the inhibition or stimulation of proliferation with silencing or additional HtrA1 was clearly demonstrated in keloid fibroblasts, but not in normal fibroblasts. These results suggest that HtrA1 is a key molecule of keloid pathogenesis. The more keloid fibroblasts proliferate, the more matrix produced by keloid fibroblasts accumulates in keloid lesions.

Recently, Beaufort et al. reported that HtrA1 facilitates the transforming growth factor-beta (TGF-β) signalling through the processing of latent TGF-β binding protein (LTBP) [13]. Reduced TGF-β activity was observed in embryonic fibroblasts from HtrA1 knockout mice and skin fibroblasts from CARASIL patients caused by *HtrA1* mutations. These observations suggest a role of

HtrA1 in facilitating TGF-β signalling. LTBP functions as a part of the large latency complex (LLC) that anchors TGF-β to the ECM [13,19–22]. Proteolysis of LTBP-1 results in its detachment from ECM, leading to TGF-β release and activation [20,23–25]. HtrA1 cleaves LTBP-1 in the fibronectin binding domain, and this processing occurs in a site-specific manner, distinct from other proteases previously reported [20,23–25]. TGF-β1 is overexpressed and activated in keloid lesions and plays a key role in keloid pathogenesis [26,27]. TGF-β1 stimulates production of abundant ECM including fibronectin and collagens. HtrA1 may facilitate keloid pathogenesis through the activation of TGF-β1 mediated by LTBP-1 cleavage.

HtrA1 has been reported to be a crucial molecule in AMD, a leading cause of irreversible blindness in the elderly [15,16]. AMD is accompanied with choroidal neovascularization and polypoidal choroidal vasculopathy. Analysis of HtrA1 transgenic mice indicated that increased HtrA1 is sufficient to cause hyper-vascularisation and degeneration of elastic laminae in choroidal vessels [15]. Zhang et al. demonstrated that HtrA1 promotes angiogenesis by regulating GDF6, a TGF-β family-protein, using HtrA1 knock-out mice [12]. As in AMD, abundant microvessels are observed in keloid lesions [9]. Thus, HtrA1 may play a role in keloid hypervascularity by modulating TGF-β family signalling.

Taken together, these observations suggest that HtrA1 contributes to the development of keloid lesions as matrix protease by remodelling keloid-specific ECM or cell surface molecules. HtrA1 may be useful as a target of keloid treatment, although further study is required.

4. Materials and Methods

4.1. Tissue Specimens

Between September 2007 and September 2013, 30 keloid patients (aged 16–75 years) and 12 unrelated patients (aged 31–88 years) undergoing surgical treatments were enrolled in this study. With approval from the Institutional Reviewing Board in the Kyoto University Faculty of Medicine (G61, the 14 December 2006), which adheres to the ethical standards as formulated in the Helsinki Declaration, written informed consent was obtained from all the patients. Keloid diagnosis was based on the clinical findings and definitive diagnosis was based on histopathologic data from the operative specimens [3,4]. The skin tissue samples were obtained as the surplus skin at the plastic surgery. Sample information is shown in Table 1.

4.2. Antibodies

Monoclonal anti-human HtrA1 antibody (MAB2916, R&D Systems, Minneapolis, MN, USA) was used for western blotting. The antibody used in immunohistochemical staining was developed in rabbits using a synthetic peptide corresponding to the C-terminal region of human HtrA1 as the immunogen.

4.3. In Situ Hybridization

For in situ hybridization, keloid and surrounding unaffected skin tissue specimens were obtained from the keloid patients at the time of surgical treatment. The specimens were fixed in 4% paraformaldehyde at 4 °C, embedded in paraffin, and Sections 6 µm in thickness were prepared. Deparaffinised sections were fixed in 4% paraformaldehyde in phosphate-buffered saline (PBS) for 15 min and washed with PBS. Sections were treated with 3 µg/mL proteinase K in PBS for 30 min at 37 °C, washed with PBS, refixed with 4% paraformaldehyde in PBS, washed again with PBS, and placed in 0.2 N HCl for 10 min. After washing with PBS, sections were acetylated by incubation in 0.1 M tri-ethanolamine-HCl (pH 8.0)/0.25% acetic anhydride for 10 min. After washing with PBS, sections were dehydrated through a series of ethanol solutions. Hybridization was performed with 1558-2066 of human *HtrA1* gene (Accession # NM_002775) at concentrations of 300 ng/mL in Probe Diluent-1 (Genostaff, Tokyo, Japan) at 60 °C for 16 h. After hybridization, sections were washed in 5× HybriWash (Genostaff) at 60 °C for 20 min, and in 50% formamide with 2× HybriWash at 60 °C for 20 min, followed by RNase treatment with 50 μg/mL RNase A in 10 mM Tris-HCl (pH 8.0)/1 M NaCl/1 mM EDTA for 30 min at 37 °C. Sections were then washed twice with 2× HybriWash at 60 °C for 20 min and twice with 0.2× HybriWash at 60 °C for 20 min. After treatment with 0.5% blocking reagent (Roche Diagnostics, Tokyo, Japan) in TBST (0.05 M Tris-HCl/0.15 M NaCl/0.05% Tween 20) for 30 min, sections were incubated for 2 h at room temperature with anti-DIG alkaline phosphatase conjugate (Roche Diagnostics) diluted 1:1000 with TBST. Sections were washed twice with TBST and then incubated in 100 mM NaCl/50 mM MgCl$_2$/0.1% Tween20/100 mM Tris-HCl (pH 9.5). Colouring reactions were performed with NBT/BCIP solution (Sigma-Aldrich, Saint Louis, MO, USA) overnight, followed by washing with PBS. Sections were counterstained with Kernechtrot stain solution (Muto Pure Chemicals, Tokyo, Japan), dehydrated, and mounted with Malinol (Muto Pure Chemicals).

4.4. Immunohistochemical Analysis

All keloid and normal skin tissue specimens were obtained from the surgical treatment and fixed in 4% paraformaldehyde at 4 °C, and paraffin sections (3 μm) were prepared. Deparaffinised sections were incubated at 90 °C for 10 min in target retrieval solution (pH 9, 1:10, Dako, Glostrup, Denmark). After blocking endogenous peroxidase and non-specific protein binding activities, the sections were incubated with antibody against human HtrA1 (1:400) using LSAB^{TM}2kit/HRP (Dako). After incubation with a peroxidase-conjugated anti-rabbit IgG antibody, sections were stained using a LSAB/HRP kit (Dako) and counterstained with haematoxylin. Microscopic images of sections were obtained by a Biorevo BZ-9000 microscope (Keyence, Osaka, Japan) and counting of total and stained fibroblasts was performed using ten microscopic fields at high-power (×400). The number of cells in the ten fields was determined. Stained fibroblasts per total fibroblasts were assumed as the proportion of HtrA1-positive cells.

4.5. Statistical Analysis

Significance of difference was analysed by the Student's *t*-test. A *p*-value < 0.05 was taken as an indication of statistical significance.

4.6. Western Blot Analysis

Tissue samples were homogenized in RIPA buffer (Takara Bio, Otsu, Japan) containing protease inhibitors at 4 °C using a Polytron homogenizer (Kinematica, Luzern, Switzerland). Following centrifugation (12,000 rpm, 4 °C, 20 min), soluble proteins in the supernatant were separated by SDS-PAGE (gradient gels) and then blotted onto PVDF membranes. The membranes were blocked with 5% Block Ace (DS Pharma Biomedical, Osaka, Japan) in PBS containing 0.05% Tween 20 prior to incubation with anti HtrA1 antibody (1:500, R&D Systems). Specific antibody binding was detected by LAS-3000 (Fuji Photo Film, Tokyo, Japan).

4.7. Knockdown of HtrA1 Gene Expression and Cell Proliferation Assay

Keloid fibroblasts and normal fibroblasts were extracted by the explant method from surgical specimens. Briefly, tissues were cut into 1~2 mm^3 pieces, placed into plastic tissue culture dishes, and cultured in Dulbecco's modified Eagle's medium (DMEM; Sigma-Aldrich, St. Louis, MO, USA) supplemented with 10% fetal calf serum, 10,000 U/mL penicillin G, and 10 mg/mL streptomycin sulphate. Cells were propagated at 37 °C, and semiconfluent cultures of fibroblasts were passaged by trypsinization up to twice prior to analysis. One day before transfection, keloid and normal fibroblasts were plated at 40% confluence at the 3rd passage in DMEM without antibiotics on 10-cm dishes, followed by transfection with *HtrA1* siRNA using Lipofectamine RNAiMAX Reagent, (Life Technologies, Carlsbad, CA, USA). After 48 h, the cell proliferation assay was performed using WST assay reagent (Nacalai Tesque, Kyoto, Japan). The expression levels of target gene and protein were analysed by real-time polymerase chain reaction (PCR) and western blot analysis, respectively.

Int. J. Mol. Sci. **2018**, *19*, 1275

A proliferation assay of keloid and normal fibroblasts was also performed with or without the addition of recombinant human HtrA1 (R&D Systems, Minneapolis, MN, USA) to the culture medium.

4.8. Real-Time PCR Analysis

Total RNA was extracted from cells after the transfection using RNeasy Mini Kit (Qiagen, Venlo, The Netherlands). First-strand cDNA was synthesised using Prime Script RT Reagent Kit with gDNA Eraser (Takara Bio). RT-PCR was performed with cDNA using TaqMan Probe Assay (Applied Biosystems, Foster City, CA, USA). Glyceraldehyde-3-phosphate dehydrogenase was used as a housekeeping control gene. Relative expression was calculated by calibration curve method.

5. Conclusions

In summary, the expression of HtrA1 was revealed, especially in keloid active lesions, and the silencing of HtrA1 suppressed the proliferation of keloid fibroblasts. This effect of silencing HtrA1 was also observed in normal fibroblasts, but the inhibition effect was not so as pronounced. Moreover, the addition of recombinant HtrA1 in culture medium stimulated the proliferation of keloid fibroblasts but not normal fibroblasts. These results suggest that HtrA1 plays an important role in keloid cell proliferation and is a key molecule in keloid pathogenesis.

Supplementary Materials: Supplementary materials can be found at www.mdpi.com/s1.

Author Contributions: Satoko Yamawaki, Motoko Naitoh and Toshihiro Ishiko performed histopathological experiments; Satoko Yamawaki, Motoko Naitoh, Yasuhiro Katayama and Taku Tamura performed molecular and cellular biological experiments; Satoko Yamawaki and Motoko Naitoh analysed the data and performed statistical analysis; Motoko Naitoh and Hiroshi Kubota conceived and designed the experiments; Satoko Yamawaki, Motoko Naitoh, Rino Aya, Yasuhiro Katayama, Toshihiro Ishiko, Katsuhiro Yoshikawa, Tatsuki Enoshiri, Mika Ikeda and Shigehiko Suzuki recruited the patients; Satoko Yamawaki wrote the paper. Motoko Naitoh developed the experimental design and writing.

Acknowledgments: This work was supported by JSPS KAKENHI Grant Number 23592646 and 15H05003.

Conflicts of Interest: The authors declare no conflict of interest.

Abbreviations

HtrA1   High Temperature Requirement Factor A1

References

1. Abergel, R.P.; Pizzurro, D.; Meeker, C.A.; Lask, G.; Matsuoka, L.Y.; Minor, R.R.; Chu, M.L.; Uitto, J. Biochemical composition of the connective tissue in keloids and analysis of collagen metabolism in keloid fibroblast cultures. *J.Investig. Dermatol.* **1985**, *84*, 384–390. [CrossRef] [PubMed]
2. Sidgwick, G.P.; Bayat, A. Extracellular matrix molecules implicated in hypertrophic and keloid scarring. *J. Eur. Acad. Dermatol. Venereol.* **2012**, *26*, 141–152. [CrossRef] [PubMed]
3. Mustoe, T.A.; Cooter, R.D.; Gold, M.H.; Hobbs, F.D.; Ramelet, A.A.; Shakespeare, P.G.; Stella, M.; Teot, L.; Wood, F.M.; Ziegler, U.E.; et al. International clinical recommendations on scar management. *Plast. Reconstruct. Surg.* **2002**, *110*, 560–571. [CrossRef]
4. Alster, T.S.; Tanzi, E.L. Hypertrophic scars and keloids: Etiology and management. *Am. J. Clin. Dermatol.* **2003**, *4*, 235–243. [CrossRef] [PubMed]
5. Van de Kar, A.L.; Kreulen, M.; van Zuijlen, P.P.; Oldenburger, F. The results of surgical excision and adjuvant irradiation for therapy-resistant keloids: A prospective clinical outcome study. *Plast. Reconstruct. Surg.* **2007**, *119*, 2248–2254. [CrossRef] [PubMed]
6. Ogawa, R.; Miyashita, T.; Hyakusoku, H.; Akaishi, S.; Kuribayashi, S.; Tateno, A. Postoperative radiation protocol for keloids and hypertrophic scars: Statistical analysis of 370 sites followed for over 18 months. *Ann. Plast. Surg.* **2007**, *59*, 688–691. [CrossRef] [PubMed]
7. Yamawaki, S.; Naitoh, M.; Ishiko, T.; Muneuchi, G.; Suzuki, S. Keloids can be forced into remission with surgical excision and radiation, followed by adjuvant therapy. *Ann. Plast. Surg.* **2011**, *67*, 402–406. [CrossRef] [PubMed]

8.   Van Leeuwen, M.C.; Stokmans, S.C.; Bulstra, A.E.; Meijer, O.W.; Heymans, M.W.; Ket, J.C.; Ritt, M.J.; van Leeuwen, P.A.; Niessen, F.B. Surgical excision with adjuvant irradiation for treatment of keloid scars: A systematic review. *Plast. Reconstruct. Surg.* **2015**, *3*, e440. [CrossRef] [PubMed]

9.   Naitoh, M.; Kubota, H.; Ikeda, M.; Tanaka, T.; Shirane, H.; Suzuki, S.; Nagata, K. Gene expression in human keloids is altered from dermal to chondrocytic and osteogenic lineage. *Genes Cells* **2005**, *10*, 1081–1091. [CrossRef] [PubMed]

10.   Chien, J.; Campioni, M.; Shridhar, V.; Baldi, A. HtrA serine proteases as potential therapeutic targets in cancer. *Curr. Cancer Drug Targets* **2009**, *9*, 451–468. [CrossRef] [PubMed]

11.   Jiang, J.; Huang, L.; Yu, W.; Wu, X.; Zhou, P.; Li, X. Overexpression of HTRA1 leads to down-regulation of fibronectin and functional changes in RF/6A cells and HUVECs. *PLoS ONE* **2012**, *7*, e46115. [CrossRef] [PubMed]

12.   Zhang, L.; Lim, S.L.; Du, H.; Zhang, M.; Kozak, I.; Hannum, G.; Wang, X.; Ouyang, H.; Hughes, G.; Zhao, L.; et al. High temperature requirement factor A1 (HTRA1) gene regulates angiogenesis through transforming growth factor-β family member growth differentiation factor 6. *J. Biol. Chem.* **2012**, *287*, 1520–1526. [CrossRef] [PubMed]

13.   Beaufort, N.; Scharrer, E.; Kremmer, E.; Lux, V.; Ehrmann, M.; Huber, R.; Houlden, H.; Werring, D.; Haffner, C.; Dichgans, M. Cerebral small vessel disease-related protease HtrA1 processes latent TGF-β binding protein 1 and facilitates TGF-β signaling. *Proc. Natl. Acad. Sci. USA* **2014**, *111*, 16496–16501. [CrossRef] [PubMed]

14.   Grau, S.; Richards, P.J.; Kerr, B.; Hughes, C.; Caterson, B.; Williams, A.S.; Junker, U.; Jones, S.A.; Clausen, T.; Ehrmann, M. The role of human HtrA1 in arthritic disease. *J. Biol. Chem.* **2006**, *281*, 6124–6129. [CrossRef] [PubMed]

15.   Jones, A.; Kumar, S.; Zhang, N.; Tong, Z.; Yang, J.H.; Watt, C.; Anderson, J.; Amrita, J.; Fillerup, H.; McCloskey, M.; et al. Increased expression of multifunctional serine protease, HTRA1, in retinal pigment epithelium induces polypoidal choroidal vasculopathy in mice. *Proc. Natl. Acad. Sci. USA* **2011**, *108*, 14578–14583. [CrossRef] [PubMed]

16.   Vierkotten, S.; Muether, P.S.; Fauser, S. Overexpression of HTRA1 leads to ultrastructural changes in the elastic layer of Bruch's membrane via cleavage of extracellular matrix components. *PLoS ONE* **2011**, *6*, e22959. [CrossRef] [PubMed]

17.   Filliat, G.; Mirsaidi, A.; Tiaden, A.N.; Kuhn, G.A.; Weber, F.E.; Oka, C.; Richards, P.J. Role of HTRA1 in bone formation and regeneration: In vitro and in vivo evaluation. *PLoS ONE* **2017**, *12*, e0181600. [CrossRef] [PubMed]

18.   Tsuchiya, A.; Yano, M.; Tocharus, J.; Kojima, H.; Fukumoto, M.; Kawaichi, M.; Oka, C. Expression of mouse HtrA1 serine protease in normal bone and cartilage and its upregulation in joint cartilage damaged by experimental arthritis. *Bone* **2005**, *37*, 323–336. [CrossRef] [PubMed]

19.   Ruiz-Ortega, M.; Rodriguez-Vita, J.; Sanchez-Lopez, E.; Carvajal, G.; Egido, J. TGF-β signaling in vascular fibrosis. *Cardiovasc. Res.* **2007**, *74*, 196–206. [CrossRef] [PubMed]

20.   Taipale, J.; Miyazono, K.; Heldin, C.-H.; Keski-Oja, J. Latent transforming growth factor-β 1 associates to fibroblast extracellular matrix via latent TGF-β binding protein. *J. Cell Biol.* **1994**, *124*, 171–181. [CrossRef] [PubMed]

21.   Tesseur, I.; Zou, K.; Berber, E.; Zhang, H.; Wyss-Coray, T. Highly sensitive and specific bioassay for measuring bioactive TGF-β. *BMC Cell Biol.* **2006**, *7*, 15. [CrossRef] [PubMed]

22.   Todorovic, V.; Rifkin, D.B. LTBPs, more than just an escort service. *J. Cell. Biochem.* **2012**, *113*, 410–418. [CrossRef] [PubMed]

23.   Dallas, S.L.; Rosser, J.L.; Mundy, G.R.; Bonewald, L.F. Proteolysis of latent transforming growth factor-β (TGF-β)-binding protein-1 by osteoclasts. A cellular mechanism for release of TGF-β from bone matrix. *J. Cell. Biochem.* **2002**, *277*, 21352–21360. [CrossRef]

24.   Ge, G.; Greenspan, D.S. BMP1 controls TGF-β1 activation via cleavage of latent TGF-β-binding protein. *J. Cell Biol.* **2006**, *175*, 111–120. [CrossRef] [PubMed]

25.   Tatti, O.; Vehvilainen, P.; Lehti, K.; Keski-Oja, J. MT1-MMP releases latent TGF-β1 from endothelial cell extracellular matrix via proteolytic processing of LTBP-1. *Exp. Cell Res.* **2008**, *314*, 2501–2514. [CrossRef] [PubMed]

26. Border, W.A.; Noble, N.A. Transforming growth factor β in tissue fibrosis. *N. Engl. J. Med.* **1994**, *331*, 1286–1292. [CrossRef] [PubMed]
27. Peltonen, J.; Hsiao, L.L.; Jaakkola, S.; Sollberg, S.; Aumailley, M.; Timpl, R.; Chu, M.L.; Uitto, J. Activation of collagen gene expression in keloids: Co-localization of type I and VI collagen and transforming growth factor-β 1 mRNA. *J. Investig. Dermatol.* **1991**, *97*, 240–248. [CrossRef] [PubMed]

MDPI

St. Alban-Anlage 66

4052 Basel

Switzerland

Tel. +41 61 683 77 34

Fax +41 61 302 89 18

www.mdpi.com

International Journal of Molecular Sciences Editorial Office

E-mail: ijms@mdpi.com

www.mdpi.com/journal/ijms

www.ingramcontent.com/pod-product-compliance
Lightning Source LLC
Chambersburg PA
CBHW051850210326
41597CB00033B/5849